NATIONAL PARKS AND PROTECTED AREAS

APPROACHES FOR BALANCING SOCIAL, ECONOMIC, AND ECOLOGICAL VALUES

NATIONAL PARKS AND PROTECTED AREAS

APPROACHES FOR BALANCING SOCIAL, ECONOMIC, AND ECOLOGICAL VALUES

Tony Prato
Dan Fagre

Blackwell Publishing

Tony Prato is professor of ecological economics and codirector, Center for Agricultural, Resource and Environmental Systems, University of Missouri, Columbia, Missouri.

Dan Fagre is research ecologist and global change research coordinator, USGS Northern Rocky Mountain Science Center, Glacier National Park, Montana.

Blackwell Publishing Professional
2121 State Avenue, Ames, Iowa 50014, USA

Orders: 1-800-862-6657
Office: 1-515-292-0140
Fax: 1-515-292-3348
Web site: www.blackwellprofessional.com

Blackwell Publishing Ltd
9600 Garsington Road, Oxford OX4 2DQ, UK
Tel.: +44 (0)1865 776868

Blackwell Publishing Asia
550 Swanston Street, Carlton, Victoria 3053, Australia
Tel.: +61 (0)3 8359 1011

First edition, 2005

Cover photograph by Jeff Scholl.

Library of Congress Cataloging-in-Publication Data

Prato, Tony.
National parks and protected areas : approaches for balancing social, economic, and ecological values / Tony Prato and Dan Fagre.
p. cm.
Includes index.
ISBN 0-8138-1248-8 (alk. paper)
1. National parks and reserves—United States—Management. 2. Protected areas—United States—Management. 3. National parks and reserves—Management. 4. Protected areas—Management.
I. Fagre, Dan. II. Title.

SB486.M35P73 2005
333.78'0973—dc22

2004010815

The last digit is the print number: 9 8 7 6 5 4 3 2 1

Contents

Preface

What a beautiful and thrilling specimen for America to preserve and hold up to the view of her refined citizens and the world, in future ages! A nation's park containing man and beast, in all the wild and freshness of their nature's beauty!

—George Catlin, 1831

National parks and reserves are an integral aspect of intelligent use of natural resources. It is the course of wisdom to set aside an ample portion of our natural resources as national parks and reserves, thus ensuring that future generations may know the majesty of the earth as we know it today.

—John F. Kennedy, 1962

Parks are now viewed as agents of conservation rather than of recreation, particularly in that the preservation of natural areas from industrial development is increasingly acknowledged as a critical step in conserving the world's biological diversity.

—Kevin McNamee, 2002

George Catlin in the nineteenth century, John F. Kennedy in the twentieth century, Kevin McNamee at the dawn of the twenty-first century, and many others, have an underlying, consistent theme. National parks and protected areas provide a range of social, economic, and ecological values that are worth preserving in perpetuity.

From their auspicious beginning in 1872 with the creation of Yellowstone National Park to the current day, national parks and protected areas have captured people's admiration, imagination, and support. In the early nineteenth century, several large national parks were established in western North America. While establishment of these parks was motivated by the profits to be earned in railroad-based tourism, the idea of having national parks "embraced the concept of nurturing and protecting Nature—a remarkable reversal from the treatment of natural resources typical of the times" (Sellars 2000). Theodore Roosevelt came out in support of national parks in 1905 when he exclaimed

> Every man who appreciates the majesty and beauty of the wilderness and of wild life, should strike hands with the farsighted men who

> wish to preserve our material resources, in the effort to keep our forests and our game beasts, game-birds, and game-fish—indeed, all the living creatures of prairie and woodland and seashore-from wanton destruction. Above all, we should realize that the effort toward this end is essentially a democratic movement.

From inside and outside their boundaries, national parks and protected areas face threats that impair their capacity to sustain multiple and competing social, economic, and ecological values. Internal threats include steady growth in the number of visitors; greater use of recreational vehicles, especially ATVs, snowmobiles, and personal watercraft; insufficient budgets and personnel; deterioration of transportation, accommodation, and maintenance facilities; overfishing; propagation of nonnative species; and conflicts surrounding the management of threatened and endangered species. External threats include global climate change, conflicts between wildlife conservation and ranching, air pollution, water diversions, and habitat fragmentation caused by economic growth and development in gateway communities. At the time they were established, large western national parks were bits of civilization carved out of the wilderness. They have now become bits of wilderness surrounded by civilization.

A shift in management philosophy from species to ecosystem management; conflicts between social, economic, and ecological values of protected areas; and uncertainty regarding the outcomes of management actions requires an integrated, science-based approach to protected area management. While science provides a powerful and defensible foundation for managing protected areas, it has often been ignored or applied in a strict disciplinary way in which social scientists address social, behavioral, and financial issues, and natural scientists address biological and physical issues facing protected areas.

This book attempts to bridge this disciplinary chasm by examining concepts, principles, and approaches that allow protected areas to be managed in a socially acceptable, economically efficient, and ecologically sensitive manner. Specific concepts, principles, and approaches considered include ecosystem management, ecological economics, adaptive management, natural regulation, carrying capacity, regional scale conservation, benefit-cost analysis, cost effectiveness analysis, economic impact analysis, landscape change detection and analysis, multiple attribute decision making, Bayesian decision making under uncertainty, optimization and simulation, integrated ecological economic modeling, geographic information systems, remote sensing, global positioning systems, interactive mapping, and digital imaging simulation.

Three primary goals of the book are to (1) enhance appreciation and understanding of the complex social, economic, and ecological functions and values of national parks and protected areas; (2) articulate the internal and external forces threatening those functions and values; and (3) demonstrate how various concepts, principles, and approaches can be used to integrate and balance protected area values. Unique features of the book are that it provides a unified treatment of decision-making approaches for national parks and protected areas through the integration of economics and ecology, analytical approaches, and information technologies, and uses case studies to demonstrate the application of those concepts, principles, and approaches to challenges facing protected areas. National parks and protected areas in western North America are the primary geographic focus of the book, although examples are taken from other parts of the world.

This book is designed as a primary or supplemental text for upper division undergraduate courses and introductory graduate courses in conservation and management of national parks and protected areas, natural resource planning and management, natural resource economics, conservation biology, public land management, and related subjects. Planning and management staff of state and national parks and protected areas, federal and state natural resource management agencies, natural resource policy actors, scientists, educators, and environmental organizations should find this book useful in their work.

Purpose of Book

Motivation for this book stems from two premises. First, successful planning and management for protected areas requires managers to balance competing social, economic, and ecological values. Second, protected area managers can do a better job of balancing competing values by integrating scientific knowledge, data, decision-making techniques, and information technologies. The goal of this book is to enhance the science and management of national parks and protected areas by improving the understanding of (1) the causes and consequences of the underlying social, economic, and ecological values of protected areas; (2) how interaction of human and ecological processes influences functions and values of protected areas; and (3) ways to improve the design, evaluation, and implementation of management actions and policies for balancing social, economic, and ecological values of protected areas.

Establishment and design of protected areas involve similar considerations and alternative management actions, and policies for protected areas can be evaluated using common principles. This book elucidates those common principles. Spatial variability in social, cultural, economic, and ecological conditions and values requires that management actions and policies for protected areas be tailored to particular sites and types of protected areas. The spatial context for the book is North America and the protected area categories of primary interest are strict nature reserves (Category Ia), wilderness areas (Category Ib), national parks (Category II), and national monuments (Category III). Considerable attention is paid to national parks for several reasons. First, national parks in North America contain relatively large, complex natural ecosystems that have high cultural, ecological, social, and economic values. Concurrently, national parks contain built-up areas for roads, visitor centers, park headquarters, lodging, restaurants, and maintenance facilities and equipment that have the potential to degrade cultural and ecological values.

Second, growth in the number of visitors to national parks will increase the likelihood of human-ecological conflicts and hence the need for planning and management strategies to alleviate those conflicts. Third, national parks have the largest geographic extent of any category of protected areas. Forty percent of North America's and 30 percent of the world's protected areas are in national parks. Fourth, national parks contain the crown jewels of protected areas as evidenced by places like Banff, Glacier, Grand Canyon, Jasper, Yellowstone, and Yosemite National Parks. Fifth, there is a body of literature on national parks that is more extensive than any other category of protected areas. Yellowstone National Park is used as an example more frequently than other parks because it has been extensively written about, epitomizes many park issues, and is universally recognized as the world's first national park.

Organization of Book

The book has 12 chapters. Chapter 1 (Contemporary Issues in Protected Area Management) addresses four key issues in protected area management (what are protected areas, why do we have protected areas, who manages protected areas, and why is managing protected areas challenging?), and uses three case studies to exemplify value conflicts, management challenges, and opportunities facing protected area managers. Chapter 2 (Role and Status of Protected Areas) provides background information on national parks and protected areas, including origin and history, function and purpose, classification, management objectives, and global designations. Chapter 3 (Cultural Values) discusses how spatial and temporal differences in culture influence conservation attitudes and behavior, role of protected areas in supporting subsistence lifestyles, preserving cultural heritage, conservation challenges in developing countries, sustainable development, collaborative decision making, and conservation and information technology. Chapter 4 (Social, Economic, and Ecological Values) focuses on the sources of social, economic, and ecological values, the public goods nature of national parks and protected areas, the importance of tastes and preferences in determining the value of national parks and protected areas, total economic value, and other concepts.

Chapter 5 (Institutional Environment) examines how the institutional environment (legal, political, and market-based incentives) influences the designation and management of national parks and protected areas, the theory of institutions and new resource economics, statutes and regulations influencing protected area management, the roles of governmental and nongovernmental agencies in management and planning, and other issues. Chapter 6 (Selection and Delineation) addresses selection and delineation decisions for national parks and protected areas, the importance of protected area categories, and national and global system planning.

Chapter 7 (Concepts for Ecosystem Management) addresses ecological and economic concepts needed to understand and apply ecosystem management to national parks and protected areas. Chapter 8 (Approaches to Ecosystem Management) discusses the application of ecosystem management to national parks and protected areas and covers biodiversity conservation, ecological functioning and ecosystem resilience, ecosystem dynamics and complexity, ecological economics, adaptive management, natural regulation, carrying capacity, and regional and continental scale conservation.

Chapter 9 (Case Studies in Protected Area Management—I) and chapter 10 (Case Studies in Protected Area Management—II) contain case studies that examine natural resource issues facing national parks and protected areas, and management approaches for resolving those issues. Chapter 11 (Future Planning and Management) describes a general framework for future planning and management of protected area ecosystems to achieve long-term social, economic, and ecological goals. Chapter 12 (Synthesis) synthesizes the principal thesis and major themes of the book.

It is our hope that the material presented here advances the understanding and science-based management of national parks and protected areas. We concur with Margaret "Mardy" Murie (1902–2003) who said, "I hope the United States of America is not so rich that she can afford to let these wildernesses pass by, or so poor that she cannot afford to keep them."

Reference

Sellars, R. W. 2000. Creating tradition: The roots of national park management. *Wild Earth* 10: 27–32.

NATIONAL PARKS AND PROTECTED AREAS

APPROACHES FOR BALANCING SOCIAL, ECONOMIC, AND ECOLOGICAL VALUES

CHAPTER

1

Contemporary Issues in Protected Area Management

National parks are the best idea we ever had. Absolutely American, absolutely democratic, they reflect us at our best rather than our worst.

—Wallace Stegner

Parks and protected areas are highly prized by the societies that elevated them to special status, but there are differing, and often strongly held, beliefs about their uses, management, and values. Although sacred areas or hunting preserves were established and protected by royalty thousands of years ago, legislatively protected areas for the general populace are recent phenomena. Today, as human populations and leisure time increase, protected areas have come under greater pressure to serve more people and more diverse purposes that were often not envisioned when the protected areas were first created. This has changed the role of park and protected-area managers from being caretakers to being assertive advocates as their areas face greater threats to ecological integrity and play a greater role in local and national economies.

Against this backdrop, the topic of protected area management is introduced by considering four key issues: what are protected areas, why do we have protected areas, who manages protected areas, and why is managing protected areas challenging? While these issues will be addressed in greater depth in subsequent chapters, it is nonetheless helpful to begin our study of the management of national parks and protected areas with a basic understanding of these four issues. Value conflicts, management challenges, and opportunities facing protected area managers are illustrated using three case studies. First, fire management in Yellowstone National Park expounds on the conflicts that arose between economic, aesthetic, and ecological values during the 1988 Yellowstone fires and how that conflict influenced fire-management policy for national parks. In this case, there is a well-established protected area, but public perception about its management forced changes. Second, the debate over oil drilling in the Arctic National Wildlife Refuge demonstrates how the desire to reduce national dependence on oil imports potentially conflicts with preservation of a world-class wildlife refuge. Here, a possible trade-off and a new status for the protected area have to be resolved by society before management direction is changed. Third, designation of the Upper Missouri River Breaks National Monument in Montana exemplifies the land-use conflicts that frequently arise among diverse stakeholders. In this case, resistance to establishing a new protected area had to be reduced before effective management guidelines were implemented. Finally, the purpose and organization of the book are discussed.

What Are Protected Areas?

The World Conservation Union, or IUCN, defines a protected area as "an area of land and/or sea especially dedicated to the protection and maintenance of biological diversity, and of natural and associated cultural resources, and managed through legal or other effective means" (IUCN 1994). Simply stated, a protected area is a location or landscape that is preserved according to legal guidelines and established management objectives. The degree to which protection actually occurs depends on adequate funding, diligent enforcement, and the strength of citizen support. Protected areas encompass a broad range of ecosystems. They include wetland, tropical or deciduous forest, alpine, savanna, marine, and others. Locations of protected areas cover the gamut from densely populated urban areas to unpopulated wilderness areas. They range from a few hundred hectares (acres) to thousands of square kilometers (miles). Most protected areas are publicly owned and managed.

There are many kinds of protected areas, from national parks to marine sanctuaries, around the world. Examples include:

- Yellowstone and Great Smoky Mountains National Parks in the United States and Banff and Wood Buffalo National Parks in Canada
- Bob Marshall Wilderness in Montana and Cranberry Wilderness in West Virginia
- Wrangell-St. Elias National Preserve in Alaska and Mojave National Preserve in California
- Gettysburg National Battlefield in Pennsylvania and Manassas National Battlefield in Virginia
- Golden Gate Park in San Francisco and Gateway Park in New York City
- Cape Cod National Seashore in Massachusetts and Indiana Dunes National Lakeshore in Indiana
- North Fork of the Flathead River in Montana and the Obed River in Tennessee, which are wild and scenic rivers

- Great Barrier Reef in Australia and the Florida Keys National Marine Sanctuary in Florida, which are marine protected areas
- Big Muddy National Fish and Wildlife Refuge in Missouri and the Arctic National Wildlife Refuge in Alaska
- Blue Ridge National Parkway in North Carolina and Virginia and the George Washington Memorial Parkway in Virginia and Maryland
- Statue of Liberty in New York and Devils Tower in Wyoming, which are national monuments

There are 30,350 protected areas worldwide covering more than 13 million km^2 (5 million square miles) (WCPA 2001). The World Conservation Union has established seven categories of protected areas (IUCN 1994):

1. Strict nature reserve (Category Ia)
2. Wilderness area (Category Ib)
3. National park (Category II)
4. National monument (Category III)
5. Habitat/species management area (Category IV)
6. Protected landscape/seascape (Category V)
7. Managed resource protected area (Category VI)

There are four other (non-IUCN) international designations of protected areas, namely Biosphere Reserve, World Heritage Site, Ramsar wetland, and marine. Biosphere Reserves serve three objectives: (1) conservation of landscapes, ecosystems, species and genetic variation; (2) sustainable economic and human development; and (3) support for research, monitoring, education, and information exchange (UNESCO 2001). World Heritage Sites have outstanding values and are categorized as cultural, natural, or mixed (combination of cultural and natural elements). They are protected for the benefit of all humanity through cooperative international arrangements (UNESCO 2000). Ramsar wetlands contribute to waterfowl production and promote the wise use of wetland ecosystems (Ramsar Convention on Wetlands 2001). The UN Law of the Sea provides a legal framework for allocating rights to territorial seas and protecting marine-protected areas.

Each country has its own criteria for protected areas that have evolved through time and are not necessarily consistent. Protected areas, such as national parks, are often established through political processes that include purposes other than preserving significant resources. To maintain appropriate standards for inclusion, the U.S. National Park Service has published criteria and procedures for considering new parklands, which must be ultimately designated by Congress. The United States has over 30 distinct types of federally protected areas, including (1) national park, national monument, national preserve, national historic site, national historic park, national memorial, national battlefield, national parkway, national cemetery, national recreation area, national seashore, national lakeshore, national river, and national trail in the National Park System; (2) forest, rangeland, game refuge, game preserve, wilderness area, national recreation area, research natural area, and national monument in the National Forest System; (3) wild, scenic, and recreation rivers in the Wild and Scenic Rivers System; and (4) national wildlife refuge, waterfowl production area, coordination area, marine protected area, wetland of international importance, wilderness area, research natural area, wild and scenic river, and national natural landmark in the National Wildlife Refuge System. In addition, there are de facto protected areas on various military reserves and Indian reservations (managed by the Bureau of Indian Affairs). Most states also have park systems that confer a degree of protected status to natural areas.

The same type of protected area can be in more than one protected area system. For example, wilderness areas and wild and scenic rivers appear in more than one protected area system, such as the National Park System. A protected area, such as a wilderness, can be embedded in a protected area such as a national park or designated on adjoining federal land, such as a national forest, to functionally create a larger protected area. Protected areas can have more than one global designation or distinction. For instance, Waterton Lakes National Park in Alberta and Glacier National Park in Montana were designated as Waterton-Glacier International Peace Park in 1932 (the first in the world), a Biosphere Reserve in 1974, and a World Heritage Site in 1995 (Gildart et al. 1990).

The geographic distribution of protected areas is skewed. The western United States contains the bulk of the acreage in U.S. national parks and preserves. Alaska alone encloses 58 percent of the acreage in the National Park System (O'Brien 1999). This is due partly to the history of park establishment. Western and Alaskan parks were designated from existing federal lands prior to extensive human settlement and population growth in those areas. It has been much more difficult to establish parks in more developed areas in the central and eastern United States and to purchase private land to augment or create them. The extent (total acreage) and mean size of protected areas is also skewed. U.S. national parks (Category II) have an extent that is 21 times greater and a mean size that is 6.2 times greater than national monuments (Category III) (Green and Paine 1997). Chapter 2 contains more complete details on the type, purpose, extent, and size of protected areas.

Why Do We Have Protected Areas?

Famous western author Wallace Stegner said that creation of national parks was the best thing that ever happened to America. President John F. Kennedy remarked that setting aside an ample portion of our natural resources as national parks allows future generations to know the majesty of the Earth as we know it today. Former National Park Service director Bill Mott said that the future of national parks depends on the public's understanding of their value (Albright et al. 1987). These individuals understood that protected areas provide a wide range of social, economic, and ecological values, such as education, maintenance of cultural/traditional attributes and ecological services, preservation of species and genetic diversity, protection of specific natural and cultural features, scientific research, sustainable use of resources, tourism and recreation, wilderness protection, and others (IUCN 1994). Each protected-area category is managed according to a particular set of objectives. For example, national parks are designated and managed to "(a) protect the ecological integrity of one or more ecosystems for present and future generations, (b) exclude exploitation or occupation inimical to the purposes of designation of the area, and (c) provide a foundation for spiritual, scientific, educational, recreational and visitor opportunities, all of which must be environmentally and culturally compatible" (Davey 1998).

The reasons for establishing and supporting protected areas have evolved as humans have altered the natural landscapes to an ever greater extent. Initially, protected areas focused on setting aside novel or awe-inspiring resources for public enjoyment. Biodiversity or ecosystem integrity was not a relevant concern, and the size of many protected areas was the smallest necessary to protect unique features. Later these areas became increasingly valuable for protecting species and accomplishing other societal goals that required relatively unaltered landscapes. Today parks are being considered in national strategic plans, such as carbon sequestration for mitigating global warming or the creation of continental conservation networks, purposes for which they were not created. Outside the United States, the concept of parks has expanded to include traditional use and impacts on landscapes deemed worthy of preservation. Thus, historic agricultural practices are maintained to continue the interaction of people with the landscape. Some parks have been established in countries with the explicit hope that they will attract tourists and contribute to the national economy. This recognizes the important role parks and protected areas now play economically and has changed the motivation for establishing protected areas.

Who Manages Protected Areas?

Publicly owned protected areas in the United States are managed by federal, state, and local governmental agencies. At the federal level, the National Park Service manages the National Park System, the Forest Service manages the National Forest System, several federal agencies manage units in the National Wild and Scenic Rivers System and National Wilderness Preservation System, the National Oceanic and Atmospheric Administration manages Marine Protected Areas, and the Fish and Wildlife Service manages the National Wildlife Refuge System. Each of these agencies operates under the authority of an organic act that spells out how protected areas are to be managed (see chapter 5 for details). States also designate and manage protected areas. For example, the Missouri Natural Areas System is an extensive statewide network of over 170 natural areas whose management is coordinated by the Missouri Natural Areas Committee. Missouri's natural areas encompass many terrestrial and aquatic communities, including forests, glades, springs, rivers and streams, savannas, wetlands, prairies, and caves (Missouri Department of Conservation and Missouri Department of Natural Resources, undated). In addition, the Missouri Department of Natural Resources manages a state park system that contains more than 80 units (Missouri Department of Natural Resources 2002). Local park commissions manage urban parks, which are typically oriented toward outdoor sports, picnicking, hiking, swimming, and nature viewing. Several nongovernmental organizations purchase land and conservation easements to protect natural values and/or fish and wildlife habitat, including Defenders of Wildlife, The Nature Conservancy, Trout Unlimited, Ducks Unlimited, Rocky Mountain Elk Foundation, and many others. These areas are managed through contracts with leaseholders and organization members or, in some cases, are managed by agencies under an agreement. For example, Defenders of Wildlife and the Wyoming Wildlife Society's Memorial Bear Fund paid a sheep rancher $30,000 to relinquish a grazing allotment in a portion of Idaho's Targhee National Forest that is prime habitat for grizzly bear (*High Country News* 2002). The Nature Conservancy has created the world's largest system of private nature reserves by purchasing critical land parcels since 1951. It has approximately 1,400 reserves that have protected over 6,070,280 ha (15 million acres) in the United States and 40,873,200 ha (101 million acres) outside the United States. The Nature Conservancy often partners with local communities or a public agency for management of these reserves and has an active research program.

Why Is Managing Protected Areas Challenging?

The social, economic, and ecological values provided by protected areas can be diminished by rapid growth in visitation; changes in the mix of recreational and visitor uses (hunting, trapping, motorized/mechanized sports and transportation, hiking, camping, skiing, and air tours); development of roads and visitor facilities; declines in fish and wildlife populations; spread of invasive species; residential and commercial development; resource extraction (logging, grazing, farming, and energy production); air, water, and soil pollution; and others. Although he was speaking about Dinosaur National Monument in Utah and Colorado, Wallace Stegner pointed out that protected areas are valuable because they provide a "sanctuary from a world paved with concrete, jet-propelled, smog-blanketed, sterilized, over-insured, aseptic; a world mass-produced with interchangeable parts, and with every natural beautiful thing endangered by the raw engineering power of the twentieth century" (Stegner 1989).

Some threats to protected areas are internal and others are external. Internal threats result from activities, events, and management actions within protected areas. For example, some scientists contend that the high elk and bison populations in Yellowstone National Park's northern range have resulted in overgrazing of the range, widespread starvation of ungulates, and destruction of plant communities (Kay 1997). Visitor use and competition among recreational uses (e.g., wildlife protection vs. snowmobile use) can adversely affect socioeconomic and environmental values of protected areas.

External threats to protected areas are more difficult to resolve because protected-area managers have less legal authority to control them. A prime example of an external threat to protected areas is residential, commercial, and energy developments on lands adjacent to protected areas. Such developments have fragmented rural landscapes and contributed to loss and impairment of wildlife species and habitats, spread of invasive species, and other impacts. Landscape fragmentation is especially threatening to large carnivores and megaherbivores due to their large ranges (Clark and Minta 1994; Keiter 1996). For instance, human-wildlife encounters, which are more likely in fragmented landscapes, account for more than 70 percent of deaths of postweaning grizzly bears in the U.S. northern Rocky Mountains (Mattson and Clark 2001).

One of the strongest arguments for maintaining the ecological integrity of protected areas is to preserve threatened and endangered species. This argument is particularly important for national parks because they harbor over 60 percent of the endangered species in the United States (NPCA 2002). External threats, such as landscape fragmentation, reduce the capacity of protected areas to serve as refugia for threatened and endangered species. In a 1927 letter to the superintendent of Glacier National Park, Aldo Leopold expressed the need to manage protected areas as follows: "The balance of nature in any strict sense has been upset long ago....The only option we have is to create a new balance objectively determined for each area in accordance with the intended use of that area" (Buchholtz 1976). Protected areas face diverse and sometimes contentious management issues. For example, the National Park Service's management of the Statue of Liberty in New York City involves a very different set of issues than its management of Yellowstone National Park in Wyoming, Idaho, and Montana. Visitors are the common element in both places, but that is where the similarity ends. Visitors to the Statue of Liberty have very different needs, expectations, and behaviors than visitors to Yellowstone National Park. People visit the Statue of Liberty because it is a worldwide symbol of freedom and liberty. In contrast, visitors to Yellowstone National Park want to experience the park's spectacular geologic formations, geothermal features, and wildlife.

Natural processes play a much greater role in Yellowstone National Park than at the Statue of Liberty. Successful management of Yellowstone and other natural-resource-based protected areas requires an understanding of how complex natural processes influence the park's natural resources and how socioeconomic and natural processes influence visitor satisfaction. Natural processes include weather patterns, fire, erosion, spread of nonnative species, variability in wildlife populations, and climate change. Interaction of natural and socioeconomic processes both within and outside protected areas results in specific patterns of timber harvesting, oil and gas exploration/development, livestock grazing, and residential/commercial development. Management actions in the form of predator control, fire management, opening and closing of roads and facilities, access fees, and land-use restrictions interact with natural and socioeconomic processes to determine range conditions for ungulates, the status of threatened and endangered species, visitor satisfaction, and many other ecological economic goods and services provided by protected areas. It is fair to say that managing Yellowstone National Park is more complex, difficult, and controversial than managing the Statue of Liberty.

Legislation for protected areas guides how protected areas are managed. For example, national parks in the United States are managed by the National Park Service under the authority of the National Park Service Organic Act of 1916 (16 U.S.C. §§ 431–33). The Organic Act requires park units to be managed so as "to conserve the scenery and the natural and historic objects and the wildlife therein and to provide for the enjoyment of the same in such manner and by such means as will leave them unimpaired for the enjoyment of future generations" (Rockwood 1988). Everhart (1972) recognized the challenge posed by this mandate in noting that "no facility for public use, no policy of resource management, no proposal of any kind has been advanced but that park people have pondered and examined—and argued—just how best to respect the demanding requirements of preservation and use." Many believe the twin goals of public use/enjoyment and resource conservation mandated by the organic act are incompatible and will continue to conflict over time due to increases in visitation and external threats to protected areas. Although other protected areas have more restrictive or clearer management guidelines, many have similarly broad mandates. Thus, protected area management requires sensitivity and careful balancing of many goals, needs, and demands that are in constant flux.

The challenge of managing national parks and other protected areas for use/enjoyment and conservation is further magnified by uncertainty regarding whether the current state of an ecosystem is sustainable and the impacts of alternative management actions on natural and socioeconomic processes. Consider the General Management Plan for Glacier National Park in northwestern Montana. It calls for reconstruction of Going-to-the-Sun Road (the major east-west route through the park) in a manner that: (1) preserves its historic character and significance, (2) minimizes impacts on visitors, (3) minimizes impacts on the local economy, (4) performs critical repairs before the road fails catastrophically, (5) minimizes the cost of the reconstruction, and (6) minimizes impacts on natural resources (U.S. Department of the Interior, undated).

Achievement of these multiple objectives is difficult for several reasons. First, they are incompatible. Preserving historic character and cultural significance and minimizing impacts on natural resources, visitors, and the local economy are likely to increase construction cost. Second, there is considerable uncertainty regarding how different reconstruction alternatives are likely to influence the objectives, particularly historic character, cultural significance, road failure, and impacts on natural resources, visitors, and the local economy. There is less uncertainty about the costs of reconstruction alternatives. Third, the objectives of minimizing impacts on visitors and the local economy are supported by different stakeholders than the objectives of preserving historic character and significance and minimizing impacts on natural resources. For these reasons, there is not a clear-cut best alternative for reconstruction of Going-to-the-Sun Road. Later chapters discuss several methods for handling uncertainty in natural resource decision making. The three case studies of protected areas discussed below illustrate the nature of the management challenges and opportunities facing protected area managers.

Fire Management in Yellowstone National Park

In the first case, a park has been established followed by a change in the public's view of what is natural and how the park should be managed. Specifically, this case illustrates the conflict that arose between social, economic, and ecological values or objectives during the Yellowstone fires of 1988. Fire is a dynamic, landscape-scale ecological process. Romme and Despain (1989) studied the incidence and extent of burned area in Yellowstone National Park over the past 250–300 years in an area that makes up about 15 percent of the park. Large portions of the study area were burned by major fires that occurred over a period of two decades in the late 1600s and early 1700s. None of the fires that occurred in the study area during this period burned more area in one year than the 1988 Yellowstone fires. From its establishment in 1872 until 1972, Yellowstone National Park operated under a policy of complete fire suppression (Runte 1987). During this 100-year period, there were no extensive burns in the park. In this respect the fire management policy was successful. In 1963, the Leopold Report was issued to the National Park Service (Leopold et al. 1963). It recommended that fire and predator management policies be altered so that ecological processes could operate more freely on the landscape. The Leopold Report redefined fire from a "bad" to a "good" (see "Are Wildfires Good or Bad?" sidebar). Similarly, the Wilderness Act of 1964 (16 U.S.C. §§ 1131–36) prompted the U.S. Forest Service to manage wilderness areas in national forests based on a philosophy of "letting nature take its course" (Keiter and Boyce 1991). A comparable shift in management policy occurred for wolves. In 1926, the last wolf was trapped in the park, and by the 1940s, wolves were extirpated from the park (Mech 1997). In January 1995, 20 years after John Weaver (1978) recommended that wolves be introduced to Yellowstone National Park from Canada or Minnesota, the National Park Service released 14 Canadian wolves in the park (Mech 1997).

Are Wildfires Good or Bad?

The public outcry about how the 1988 Yellowstone fires were managed raises the question "Are wildfires good or bad?" Fire management policy in the United States and particularly human intervention in preventing wildfires are pertinent to this question. Decadal statistics on wildfires (see figure below) indicate that the average number of hectares of wildland area burned was much greater during the decades 1919–1929, 1930–1939, and 1940–1949 than in the decades 1950–1959 through 1990–1999. Why? A contributing factor to the decline in wildland area burned from earlier to more recent decades is the Cooperative Forest Fire Prevention Campaign initiated 60 years ago. It is considered the longest running public service campaign in U.S. history.

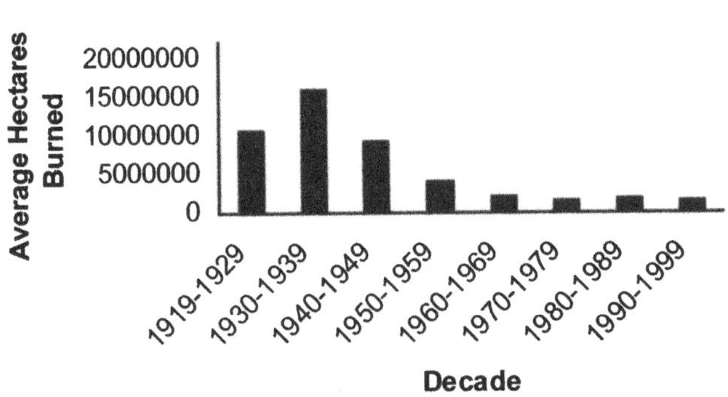

The icon of the campaign was Smokey Bear, created in 1944. For 57 years, Smokey's message was "Only You Can Prevent Forest Fires." In April 2001, the Advertising Council updated the message to "Only You Can Prevent Wildfires" in order to address the increasing number of wildfires in the nation's wildlands. The premise underlying the campaign is that wildfires started by people are bad because they destroy wilderness, property, and lives. Wildfires have many careless, human-related causes, including dis-

carding smoking products before they are extinguished, sparks from operating equipment, campfires, arson, and debris burning. Fire threats to property and lives have grown in recent years with increased building of homes in and around forested areas. The premise of the Smokey campaign was that unplanned and uncontrolled wildland fires were unnecessary and should be prevented.

Another factor contributing to the decadal decline in wildfires is the fire-suppression policy that began in the early twentieth century. This policy was motivated in part by human fear regarding the power and destructiveness of forest fires. By the middle of the twentieth century, fire-suppression efforts had significantly reduced the total area burned. Total fire suppression has caused fuel accumulation in the form of fallen leaves, branches, and excessive plant overgrowth in forest and wildland areas. It has also decreased forest health by stagnating tree growth and short-circuiting the role of fire in shaping and rejuvenating ecosystems. Accumulation of fuels and reduced forest health have contributed to an upward trend since the mid-1970s (indicated by the upward sloping line in the following figure) of more extreme fires that are dangerous to health and property and costly and sometimes impossible to control. For example, the fires of 2000 set records by burning 3.4 million hectares and costing federal agencies $1.6 billion. Much of this money was spent fighting fires that threatened residences in the wildland-urban interface. Another reason for this uptrend is the increased fuels and frequency of dry, hot weather, possibly due to global climate change, which makes fires more extensive and intense.

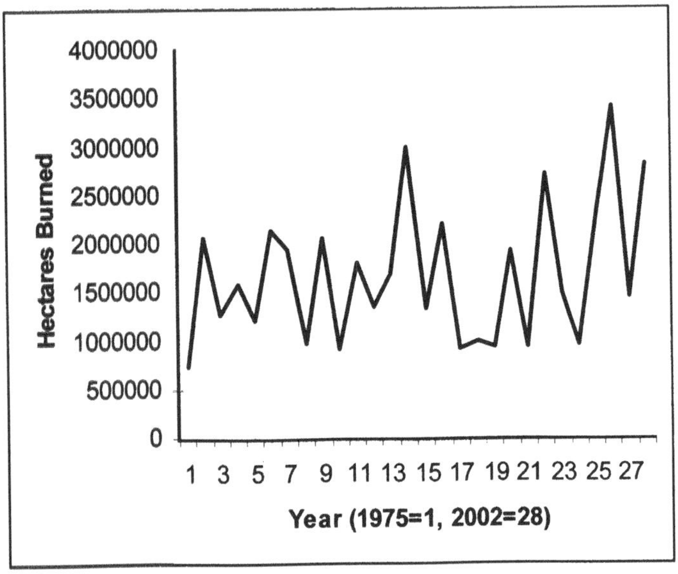

The answer to the question "Are wildfires good or bad?" is not a simple yes or no. Fires are needed to rejuvenate forests and maintain forest health. They are particularly appropriate in national parks and protected areas committed to maintaining the natural functioning of ecosystems with minimum human disturbance. Ecological benefits of wildland fires motivated the natural regulation policy adopted by the National Park Service in 1967. It was this policy that caused Yellowstone Park managers to let the 1988 fires burn until public opinion forced them to fight the fires. From an ecological perspective, wildland fire is good. On the other hand, intense fires can damage soils, streams, and aquatic habitats, even though regeneration of burned areas is often impressive and relatively quick as evidenced by the aftermath of the 1988 Yellowstone fires. Clearly, fire is bad in the sense that it has the potential to take life and damage or destroy property, which occurred, for example, in Montana's Bitterroot Valley during the 2000 fires and in Canberra, Australia, during the 2003 fires. The life and property threats of large, intense fires are likely to be significant in the foreseeable future.

Can the benefits of wildland fire be retained while reducing the costs? Such balance might be possible by continuing something like the natural regulation policy in wildland areas and using a combination of thinning and prescribed burning to reduce forest fuel loads and the risk to life and property in the wildland-urban interface. Two goals of President George W.

Bush's Healthy Forests Initiative is to support legislation that expedites procedures for forest thinning and reduces damages caused by catastrophic wildfires by lowering unnecessary regulatory obstacles that hinder active forest management. Opponents of the Healthy Forests Initiative claim it is a ruse to open up national forests to more logging. Helping communities in the wildland-urban interface to reduce fire hazards could be more vigorously pursued using programs such as Firewise Communities/USA. The latter "is a project of the National Wildfire Coordinating Group's Wildland/Urban Interface Working Team [that] provides citizens with the knowledge necessary to maintain an acceptable level of fire readiness, while ensuring firefighters that they can use equipment more efficiently during a wildland fire emergency" (Firewise Communities/USA 2004). Balancing the ecological benefits against the hazards to life and property of wildland fires is a challenging task, but one that needs to be addressed through the enactment of balanced and effective fire management policies.

Sources: Ferry et al. (1995); Williams (1995); Anderson et al. (1998); Mahoney and Gresswell (1998); General Accounting Office (1999); Turner et al. (1999); Perkins (2001); Arno and Allison-Bunnel (2002); Firewise Communities/USA (2004); National Interagency Fire Center (2003); SmokeyBear.com (2003); The White House (2003).

Not until 1972 did the National Park Service implement the recommendations of the Leopold Report. In that year, a new natural-fire management policy was implemented that permitted lightning-caused fires

> to burn in designated areas unless there is a high potential for wildfire [deaths] and the destruction of human life, human property, or special features of an the area, suppressed all non-prescribed human-caused fires (and any natural fires whose suppression is deemed necessary) in as safe, cost-effective, and environmentally sensitive way as possible, and resort to prescribed burning when and where necessary and practical to reduce hazardous fuels or restore certain types of vegetation. (Knight 1991)

The new policy was grounded on scientific evidence that fire rejuvenates landscapes by maintaining shifting mosaics of vegetation (Knight 1991), preserving biotic diversity and productivity in aquatic ecosystems (Minshall and Brock 1991), and enhancing forage quality and quantity for native ungulates, notably deer, bison, and elk (Van Dyke et al. 1991). The new fire management policy supported the ecosystem management approach whose goal is to maintain, protect, and restore ecological functions and processes that by their very nature are stochastic and dynamic and operate over long time periods (Keiter and Boyce 1991). The Greater Yellowstone Coordinating Committee espoused a similar management philosophy that envisions "a landscape where natural processes are operating with little hindrance on a grand scale...a combination of ecological processes operating with little restraint and humans moderating their activities so they become a reasonable part of, rather than encumbrances upon, those processes" (Greater Yellowstone Coordinating Committee 1990).

The Park Service's natural fire management policy was successful for the first 16 years of its implementation (1972 to 1988). No human lives were lost, no significant injuries occurred, no private property or significant cultural/historical resources were destroyed, and no threatened and endangered species were impaired. The policy faced its first major hurdle in 1988 when dry, windy conditions following several years of drought on the Yellowstone Plateau contributed to the most extensive fires in 300 years (Romme and Despain 1989). Thirty-six percent of the acreage in the park, or 32,200 ha (795,694 acres), experienced some burning (Despain et al. 1989) and 170,000 ha (420,087 acres) outside the park burned (Bunting 1996), thereby threatening life and property. Local residents criticized the natural fire management policy because they feared it would destroy the recreational and aesthetic values of the park, reduce visitation, and negatively impact the regional economy. The latter did not occur as evidenced by a record number of park visitors in 1989 and 1990. Ecologists, on the other hand, held strong to the belief that fire is essential for the proper functioning of natural ecosystems. It was acknowledged that nearly 100 years of fire suppression probably created the potential for more intense and widespread fires than occurred historically. For instance, one drainage sustained such a hot fire that the soil was consumed, increasing postfire erosion and stream sediment loads. Reestablishment of trees did not occur quickly in this drainage as it did elsewhere in the park.

National media coverage of the fires, exaggerated by using maximum estimates for the burned area, and inadequate public understanding of the role of fire in natural ecosystems led to public outrage over the fact that nothing was being done to control the Yellowstone fires and save park wildlife. After public outrage reached the limits of political acceptability, the National Park Service decided to fight the fires. At its peak, interagency fire-fighting efforts involved 9,500 civilian and military fire fighters, dozens of water-dropping helicopters and retardant bombers, and in excess of 100 fire engines at an estimated cost of $120 million. Ironically, while fire-fighting efforts avoided losses of cultural resources and buildings, the burned area would not have been much larger in the absence of fire fighting (Brown 1989) because of limits imposed by weather, fire behavior, and distribution of fuels. Based on recommendations made by an interagency review panel (Fire Management Policy Review Team 1988), the Park Service and Forest Service reached the decision to suppress all fires in the Greater Yellowstone Ecosystem until the existing policy was reviewed. Scientific studies done after the fires supported the position of park ecologists that the fires did not destroy Yellowstone National Park. This result is not surprising. A fundamental ecological process like fire cannot destroy the landscape that it shaped and sustained for thousands of years.

The Yellowstone fires of 1988 caused the National Park Service to modify its natural fire management policy. Current policy allows natural fires to burn in backcountry areas; requires monitoring and evaluation of drought indices, weather forecasts, and potential threats to people and animals before the decision is made to let a fire burn; and permits prescribed fires for resource management (Keiter and Boyce 1991). Use of prescribed fires is not without controversy as evidenced by the 2000 Cerro Grande fire that started in Bandelier National Monument and spread to Los Alamos, New Mexico. That fire caused the destruction of over 200 homes and structures in Los Alamos (National Park Service 2002) and prompted the Park Service to review and revise its policy for prescribed burns.

Developing and implementing a comprehensive fire management policy for national parks and protected areas is challenging because of uncertainty regarding the frequency, location, and extent of fires; greater risk that fires originating in a protected area will spread to nearby residential areas causing harm to humans and damage to property; poor understanding of the benefits and costs of wildfire and other ecological processes; the difficulty of communicating risks and benefits of wildfire to the media and general public; and the difficulty of gauging public reaction to particular wildfires. New tools such as remote sensing greatly improved weather monitoring and forecasting, and computer-aided models of fuel loads and forest responses to fire and climate have enhanced fire management but have not eliminated uncertainty. One thing is certain. The Yellowstone fires of 1988 provided a wealth of knowledge about the socioeconomic and ecological consequences of large wildland fires that is useful in crafting more effective fire management and natural resource management policies for national parks and protected areas. Changes in fire policy in Yellowstone influenced not only other protected areas in the United States, but also units managed by the Canadian Parks Service (Lopoukhine 1991).

Oil Drilling in the Arctic National Wildlife Refuge

The second case exemplifies an established protected area for which the public changed its view of the value of the park when faced with a trade-off. Tucked away in the northeast corner of Alaska is the pristine 7.7 million ha (19.5 million acre) Arctic National Wildlife Refuge (see "Arctic National Wildlife Refuge" sidebar). Defenders of Wildlife describe the Arctic Refuge as "a sweeping expanse of tundra studded with marshes and lagoons and laced with rivers dramatically situated between the rugged foothills of the Brooks Range and the wide, icy waters of the Beaufort Sea" (Defenders of Wildlife 2002a). The Arctic Refuge is called "America's Serengeti" due to its rich biodiversity that includes arctic fox, arctic grayling, Dolly Varden trout, grizzly bear, migratory birds, musk oxen, polar bear, porcupine caribou, snow geese, wolverine, wolf, and many other species. It contains the most important on-shore denning area for polar bears in the United States and is the primary calving grounds for the Porcupine caribou herd whose population of 130,000 is the second largest in the United States.

Arctic National Wildlife Refuge

This is the place for man turned scientist and explorer; poet and artist. Here he can experience a new reverence for life that is outside his own and yet a vital and joyous part of it.

—William O. Douglas, U.S. Supreme Court Justice, 1939–1975

It is a whole place, as true a wilderness as there is anywhere on this continent and unlike any other that I know of.

—Morris Udall, Former U.S. Congressman

Superlatives abound in descriptions of the Arctic National Wildlife Refuge (ANWR). ANWR was established by the president of the United States on May 26, 1952, for

> the purpose of preserving unique wildlife, wilderness and recreational values [in an]...area in northeastern Alaska containing approximately 8,900,000 acres [3,603,239 ha]....The Secretary of the Interior is authorized to permit the hunting and the taking of game animals, birds, and fish in the wildlife range, or parts thereof, as well as the trapping of fur animals. However, no person may hunt, trap, capture, kill, or willfully disturb any wild mammal, wild bird, or fish, or take or destroy the eggs or nests of any such bird or fish within the wildlife range, except as may be prescribed by the Secretary.

The Alaska National Interest Lands Conservation Act (ANILCA) passed in 1980 expanded the refuge to 7.7 million ha (19.5 million acres, which is about the size of South Carolina), designated 3.6 million ha (8.9 million acres) as wilderness under the 1964 Wilderness Act (ANWR is often called the "Last Great Wilderness"), renamed the whole area ANWR, and added four purposes:

1. To conserve fish and wildlife populations and habitats in their natural diversity including, but not limited to, the 120,000 strong porcupine caribou herd, polar bears, grizzly bears, muskox, Dall sheep, wolves, wolverines, snow geese, peregrine falcons, and other migratory birds and Arctic char and grayling
2. To fulfill the international fish and wildlife treaty obligations of the United States
3. To provide the opportunity for continued subsistence uses by local residents
4. To ensure water quality and necessary water quantity within the refuge

ANWR contains habitats in five different regions: lagoons and beaches of the coast; the tundra of the refuge coastal plain; mountains of the Brooks Range; meadows and dwarf spruce south of the mountains; and tall spruce, birch, and aspen of the boreal forest. These diverse habitats make ANWR the most biologically diverse unit of the Wildlife Refuge System in the circumpolar north supporting 36 species of land mammals and 9 species of marine mammals, 36 species of fish found mostly in the biologically rich coastal lagoons, and 180 species of birds from four continents. The refuge is home to muskoxen and Dall sheep, contains dens for all three species of North American bear (black, grizzly, and polar), and is one of the primary calving grounds for the 120,000-strong Porcupine caribou herd that migrates more than 1,194 km (800 miles) to the refuge. ANWR contains over 20 rivers, 3 of which (Sheenjek, Ivishak, and Wind) are designated as wild rivers.

In addition to providing recreational opportunities, ANWR has been a homeland to the Inupiat Eskimos of the north coast and the Athabascan Indians of interior Alaska for thousands of years. ANWR is of value to nonvisitors because it symbolizes "America's vast and remote wilderness–a place of inspiration and beauty–a promise for the future for all Americans." ANWR has scientific value in terms of providing opportunities for studying the natural dynamics of an unaltered landscape.

Source: Matthiessen et al. (2003); U.S. Fish and Wildlife Service (2003).

The Alaska National Interest Lands Conservation Act of 1980 banned oil and gas development in the Arctic Refuge. It states, "production of oil and gas from the Arctic National Wildlife Refuge is prohibited and no leasing or other development leading to production of oil and gas from the [refuge] shall be undertaken until authorized by an act of Congress" (Defenders of Wildlife 2002b). The act required the U.S. Department of the Interior (DOI) to assess the plant and animal life and oil and gas potential of the 1.5 million-acre coastal plain region of the Arctic Refuge named the "1002 area" for the section of the act requiring the assessments. The 1002 area occupies 8 percent of the area of the Arctic Refuge. Based on the Final Legislative Environmental Impact Statement submitted to Congress in April 1987, the U.S. Department of the Interior recommended that the 1002 area be leased for oil and gas exploration. The 1002 area is not designated as wilderness (Matthiessen et al. 2003). Congress did not act on the recommendation due to the devastating economic and environmental impacts of the *Exxon Valdez* oil spill that occurred March 24, 1989, in Prince William Sound near the southern terminus of the Trans-Alaska Pipeline (Congressional Research Service 2002).

Pressure to drill for oil and gas in the Arctic Refuge is strong for several reasons. First, it contains the largest geologic structures in the United States likely to contain petroleum (oil and gas). In 1996, the U.S. Geological Survey estimated that the Arctic Refuge contained between 148 million and 5,150 million barrels of recoverable oil reserves. Second, there is a growing imbalance between domestic oil production and consumption. Domestic oil production peaked in 1973, whereas oil consumption has continued to increase. Current U.S. production of crude oil is 6.5 million barrels per day, which is 28 percent less than peak domestic production. In the last seven years, oil production in Alaska's North Slope declined 25 percent (Congressional Research Service 2002). The shortfall between domestic oil production and consumption has been made up by oil imports. The latter now account for nearly half of U.S. oil consumption.

Third, the George W. Bush administration's energy policy relies heavily on increasing domestic oil production for reasons of national security (reducing dependence on oil imports) and job creation, and favors overturning the 1980 ban on drilling in the Arctic Refuge. President Bush believes that modern technologies would allow drilling to be done without harming the environment (*Columbia Daily Tribune* 2002a). Fourth, producing the expected volume of recoverable oil in the Arctic Refuge would generate billions of dollars of revenue in bonuses, rents, royalties, and sales of gravel, water, and other products. Royalties alone could reach $200 million to $2.5 billion per year during peak production (Congressional Research Service 2002).

Opposition to drilling in the Arctic Refuge occurs for several reasons. First, the law banning oil and gas development in the Arctic Refuge indicates it is to be managed so as to "conserve fish and wildlife populations and habitats in their natural diversity." Other socioeconomic, cultural, and environmental values sustained by the Arctic Refuge include satisfying fish and wildlife treaty obligations, protecting subsistence use of natural resources, maintaining water quantity and quality, supporting sport fishing and hunting, and preserving intrinsic and aesthetic values. For example, the Gwich'in Indians of Alaska depend on the region for clothing, food, and medicine. Oil and gas exploration in the Arctic Refuge would have to be compatible with these values, unless the law was amended.

Second, while the 1002 area covers only 8 percent of the Arctic Refuge, it is the most biologically productive part of the refuge and contains the least disturbed Arctic coastal areas and wildest habitats of its type in the United States (Congressional Research Service 2002). The wells, feeder pipelines, roads, helicopter pads, refineries, living quarters, landfills, water reservoirs, docks and gravel causeways, gas processing facilities and seawater treatment plants that would be required for energy exploration and development throughout the 1002 area would disturb the delicate environment even if modern technologies were employed (Defenders of Wildlife 2002b). To support this position, drilling opponents point to the record for the Prudhoe Bay oilfield, just 91 km (60 miles) west of the Arctic Refuge. That oilfield routinely has spills of oil products and hazardous substances, and noise and air pollution from heavy machinery and equipment. Oil spills have also occurred along the Trans-Alaska Pipeline, which carries crude oil from Prudhoe Bay in the Beaufort Sea to a shipping port in Valdez in Prince William Sound. One of the most devastating oil spills in the world occurred in 1989 when the *Exxon Valdez* tanker leaked more than 10 million gallons of crude oil into Prince William Sound, killing more than 250,000 seabirds, 2,800 sea otters, 300 harbor seals, 250 bald eagles, and nearly two dozen whales (Defenders of Wildlife 2002a). Regarding the *Exxon Valdez* oil spill, the State of Alaska concluded that "the human and natural losses were immense to fisheries, subsistence livelihoods, tourism, [and] wildlife"; the *Exxon Valdez* Oil Spill Trustee Council recognized that 30 species and resources were injured by the spill (State of Alaska 1990). Opponents of oil drilling in the Arctic Refuge argue that the risk to natural resources and the environment of energy exploration and development in the Arctic Refuge is not worth the equivalent of only 6 months supply of oil that would take 10 or more years to bring to market (U.S. Department of Energy 2000). Actually, estimated recoverable oil reserves in the Arctic Refuge (148 million to 5,150 million barrels) are equivalent to between 1 month and 2.2 years worth of current domestic oil production.

Third, opponents believe there are more cost-effective and less environmentally risky ways of reducing U.S. dependence on imported oil than drilling in the Arctic Refuge. They favor development of renewable energy sources and energy conservation. Greater reliance on solar, wind, and hydrogen energy could totally displace the oil that might be discovered in the Arctic Refuge. Furthermore, the United States could save as much oil as it expects to find in the Arctic Refuge by improving average fuel efficiency for motor vehicles by a mere 0.12 km per liter (0.7 miles per gallon) (authors' estimate).

How does the American public view drilling in the Arctic Refuge? A poll taken by the League of Conservation Voters showed that the public supported designating the Arctic Refuge as a national monument and prohibiting drilling by a two-to-one margin (Defenders of Wildlife 2002b). An energy poll sponsored by the Associated Press showed that over one-third favored the Bush administration's energy plan and more than half opposed it (*Columbia Daily Tribune* 2002b). Nonsupport for the energy plan can occur for reasons other than opposition to drilling in the Arctic Refuge.

At the time of this writing, the House of Representatives approved an energy bill that allows drilling in the Arctic Refuge. The Senate has not passed an energy bill. The president will scrutinize whatever compromise energy bill Congress passes. Since the president favors drilling in the Arctic Refuge, he is expected to sign an energy bill that permits drilling, unless other provisions of the bill are unacceptable.

Designation of Upper Missouri River Breaks National Monument

In the third case, the protected area was not established and the resistance, advocacy, and changing views of its value by different parties eventually led to its establishment.

American explorers Lewis and Clark gave the first recorded description of the Upper Missouri River Breaks during their famous Corps of Discovery trek that traveled up the Missouri River, down the Clearwater-Snake-Columbia Rivers to the Pacific Ocean, and back. Clark classified the Upper Missouri River Breaks as one of the Deserts of America and declared it unfit for human settlement. Lewis considered it to be barren country whose only redeeming value was pure air (Cunningham 2001). Two centuries later, a guidebook for the Upper Missouri River Breaks National Monument describes it as "a landscape of living history, a place where a profound appreciation of Montana's past is made possible by a largely undisturbed natural environment" (Wilderness Society 2002). Cunningham (2001) described the Breaks Monument as awe-inspiring with its "meandering river lined with verdant cottonwood bottoms, overseen by soaring cliffs studded with multi-colored rock formations...[containing] hidden canyons...guarded by sandstone spires."

The Breaks Monument is a remnant of the Great Plains that includes a 149-km (98-mile) reach of the Missouri River between Ft. Benton and Kipp, Montana. It contains 91,000 ha (375,000 acres) of federal land managed by the Bureau of Land Management, 16,000 ha (39,000 acres) of state land, and 32,000 ha (80,000 acres) of private land (Missouri River Action 2001). The Breaks Monument is an example of overlapping protected area designations and encompasses segments of the Lewis and Clark National Historic Trail, the Chief Joseph (Nez Perce) National Historic Trail, the Missouri Breaks National Backcountry Byway, Upper Missouri National Wild and Scenic River, the Cow Creek Area of Critical Environmental Concern, and six Wilderness Study Areas (Friends of the Missouri River Breaks Monument 2002b).

The massive white cliffs in the Breaks Monument are made of sedimentary rock (limestone, shale, and sandstone) deposited in a great shallow sea during the Cretaceous period. Upland areas have dense coniferous forests of ponderosa pine, Douglas fir, and juniper. One thousand feet below the cliffs is the Missouri River corridor with its cottonwoods, green ash, and willow. Charles M. Russell National Wildlife Refuge is just east of the monument's eastern border. The Breaks Monument contains 60 mammals, 233 bird species, 20 kinds of amphibians and reptiles, and 48 fish species, including mule deer, sharp-tailed grouse, antelope, grouse, elk, bighorn sheep, eagles, prairie falcons, and hawks. It also provides outstanding opportunities for outdoor recreation in a culturally significant area once inhabited or used by Blackfeet hunting parties, fur trappers, artists, steamboats, woodhawks (people who sold wood to steamboats), horse thieves, outlaws, farmers, ranchers, and merchants (Cunningham 2001). Ranching still occurs in the area.

In August 1999, Secretary of the Interior Bruce Babbitt, Senator Max Baucus (D-MT), and authors Stephen Ambrose and Dayton Duncan floated the wild and scenic river segment of the Missouri River in the Breaks. Secretary Babbitt was so impressed with what he saw that he proposed designating the area a National Conservation Area (Friends of the Missouri River Breaks Monument 2002a). Subsequently, the Central Montana Resource Advisory Council (RAC) held public hearings and, in December 1999, submitted management recommendations to Secretary Babbitt. In spring 2000, Secretary Babbitt organized another meeting that involved the RAC. Fifty-six percent of the comments received from the public favored designating the Breaks area as a national monument (Friends of the Missouri River Breaks Monument 2002b). A subsequent poll of Montana voters taken by the *Great Falls Tribune* indicated that, provided livestock grazing and hunting were allowed to continue in the monument, 64 percent of the respondents favored creation of the Breaks Monument and 26 percent opposed it (Friends of the Missouri River Breaks 2002b). On December 22, 2000, Secretary Babbitt recommended to President Clinton that the Upper Missouri River Breaks National Monument be created (Cunningham 2001).

Not everyone supported the Breaks Monument. Montana's governor, the state's congressional delegation, a local ranching group called the Missouri River Stewards, energy developers, and off-road vehicle users opposed the designation on grounds that it unduly restricts land use and management, reduces private access, and diminishes the ability to earn a living from the land. Opposition of this kind is common in the western United States where most of the public land is located, and there is a history of land-use conflicts. Some environmental groups have expressed concern that creating the monument so close to the 2004-2006 Lewis and Clark Bicentennial Celebration would attract more visitors to the region and damage the environment.

On January 17, 2001, days before leaving office, President Clinton designated 19 national monuments, including the Breaks Monument, and expanded 3 others under the authority of the 1906 Antiquities Act. This action added 2.4 million ha (5.9 million acres) to the U.S. protected-area estate (Montana Wildlife Federation 2002). In an effort to balance competing values, certain traditional land uses, notably livestock grazing, oil and gas production from existing leases, hunting, fishing, hiking, and floating, are permitted in the Breaks Monument. New hard-rock exploration and mining and new oil and gas leases are disallowed. Off-road vehicle use is limited to designated routes on the federally owned portion of the monument. Creation of the Breaks Monument in no way restricts the status of and access to state or private lands in the monument (Montana River Action 2001).

Opponents of the Upper Missouri River Breaks National Monument found a sympathetic ear in the incoming Bush administration. The new administration shared the energy industry's concerns about restricting new mineral (hard rock, oil, and gas) development on lands in the monument because it was out of sync with Bush's energy plan. The latter supported increases in domestic energy production and called for examining the energy production potential of all public lands, including national monuments. Ranchers in the area opposed designation of the Breaks Monument because it includes private land and limits new mineral development. Local and state legislators also opposed the designation (Friends of the Missouri River Breaks Monument 2002b).

Secretary of the Interior Gale Norton "requested the input of the Governor of the State of Montana regarding recent monument designations within the State of Montana" (Executive Order 2002). In 2001, the governor of Montana established the Upper Missouri River Breaks National Monument Task Force to "facilitate public meetings and take public input in deference to the specific points outlined in the letter received from Secretary of Interior Gale Norton." The task force recommended that the boundaries of the Breaks Monument be reduced to include only 53,352 ha (131,838 acres) in the Upper Missouri Wild and Scenic River (Friends of the Missouri River Breaks Monument 2002b). This recommendation amounts to a 73 percent reduction in the size of the Breaks Monument. Environmental groups criticized this strategy because the monument boundaries were carefully drawn to exclude a large area that is promising for oil and gas exploration. Furthermore, existing oil and gas leases were allowed to remain on the 24,281 ha (60,000 acres) of the Breaks Monument under valid lease at the time the monument was created (Wilderness Society 2002). Secretary Norton informed the task force that she did not have the authority to reduce the size of the monument. Events of September 11, 2001, caused the monument designations to take a back seat to America's new war on terrorism (Friends of the Missouri River Breaks Monument 2002b).

Balancing Values

Each of the case studies discussed above involves conflict between two or more social, economic, and/or ecological values of protected areas. Conflicts of this type are frequently resolved by implementing a management action or policy that complies with the law and achieves a reasonable balance among the values. Balancing values does not always mean selecting a management action or policy that achieves a compromise among competing values, although generally it does.

Prior to 1972, the Park Service's fire management policy required that all fires be suppressed. This policy made the forest more vulnerable to larger and more destructive fires and resulted in a more spatially homogenous landscape (Holling 1995). Local interests urged the Park Service to extinguish the 1988 fires for fear they would destroy the tourist-dependent local economies (economic opposition). Society at large believed the fires would destroy Yellowstone National Park (social opposition). An eleventh-hour decision by the Park Service to stay its new policy and fight the fires indicated willingness to trade off the ecological value of fire for greater social and economic values. The 1988 Yellowstone fires prompted the Park Service to reevaluate its fire management policy in hopes of achieving a better balance among social, economic, and ecological values.

Oil drilling in the Arctic National Wildlife Refuge pits socioeconomic values against ecological values. Proponents believe that drilling in the Arctic Refuge will increase domestic oil production, jobs, and income, and reduce U.S. dependence on foreign oil as well as vulnerability to terrorist activities (socioeconomic values). They contend that energy development in the Arctic Refuge will significantly increase socioeconomic values without compromising ecological values. Opponents argue that drilling poses a significant risk to ecological values while achieving only minor gains in socioeconomic values.

Designation of the Upper Missouri River Breaks National Monument involves a conflict between new mineral leasing, ranching, and off-road recreation (socioeconomic values) on the one hand and preservation of the area's unique natural resources (ecological values) on the other hand. Additional opposition arises from those who believe the president should not have the authority, allowed by the Antiquities Act of 1906, to designate an area as a national monument, especially when there is strong local and political opposition.

The institutional framework within which a protected area operates influences how socioeconomic and environmental values are balanced. There is a fundamental difference between the institutional forces shaping fire policy for national parks, such as Yellowstone, and energy production in national refugees like the Arctic Refuge. The National Park Service, which is a bureau of the U.S. Department of the Interior (executive branch of government), has the authority to determine fire policy for national parks. Congress can pass laws to establish broad guidelines for fire policy in national parks, but this did not happen even after the 1988 Yellowstone fires. The Alaska National Interest Lands Conservation Act of 1980 bans energy development in the Arctic Refuge until authorized by Congress. Therefore, the executive branch in the case of fire policy for national parks and Congress in the case of drilling in the Arctic Refuge determine the balance among social, economic, and environmental values.

Energy development in the Arctic Refuge and the Breaks Monument is subject to similar institutional forces. Oil and gas leases cannot be issued in the Arctic Refuge and new mineral leases cannot be let in the Breaks Monument unless authorized by Congress. Finally, the institutional forces influencing fire policy for national parks and energy development in national monuments are fundamentally different. Once again, the Department of the Interior determines fire policy for national parks, whereas only Congress has the authority to modify the boundaries of national monuments to accommodate energy development and other uses.

While the three case studies discussed above involve conflicts or trade-offs between social, economic, and ecological values, not all management actions and policies entail such conflicts. For example, the fire management policy developed after the Yellowstone fires of 1988 allowed natural fires to burn in backcountry areas, required monitoring and evaluation of potential impacts of fire on people and animals in deciding whether or not to let a fire burn, and permitted prescribed fires for resource management (Keiter and Boyce 1991). Since most of the acreage in large national parks is in the backcountry, the post-1988 fire policy should achieve most of the ecological benefits of fire while minimizing human and animal losses and property damage. In other words, the policy appears to advance social, economic, and ecological values.

In contrast, social, economic, and ecological values appear to be in direct conflict for both the drilling controversy in the Arctic Refuge, and the boundary contraction to accommodate new mineral development in the Breaks Monument. Allowing drilling in the Arctic Refuge and decreasing the boundaries of the Breaks Monument appear to advance socioeconomic values at the expense of ecological values. Even in these cases, however, it might be possible to design management actions and policies that minimize conflicts among values.

The three case studies discussed above give the impression that value conflicts are less contentious for national parks than for national refuges and monuments. As a general rule, the nature and extent of the conflicts between social, economic, and ecological values in protected areas are location and time specific. Hence, it is not possible to generalize regarding the nature and degree of value conflicts in different categories of protected areas.

References

Albright, H. M., R. E. Dickerson, and W-P. Mott, Jr. 1987. *National Park Service: The story behind the scenery.* Las Vegas, NV: KC Publishers, Inc.

Anderson, J. E., M. Ellis, and W. H. Romme. 1998. *Establishment, growth, and development of lodgepole pine* (Pinus contorta) *on diverse sites burned in the Yellowstone fires of 1988.* Yellowstone Science *6: 25.*

Arno, S. F., and S. Allison-Bunnel. 2002. Flames in our forest: Disaster or renewal? *Washington, DC: Island Press.*

Brown, J. K. 1989. Could the 1988 fires in Yellowstone have been avoided with a program of prescribed burning? Paper presented at the 155th annual meeting of the American Association for the Advancement of Science, San Francisco, CA, January 19.

Buchholtz, C. W. 1976. *Man in Glacier.* West Glacier, MT: Glacier Natural History Association.

Bunting, S. C. 1996. The use and role of fires in natural areas. In *National parks and protected areas: Their role in environmental protection,* ed. R. G. Wright, 277–302. Cambridge, MA: Blackwell Science.

Clark, T. W., and S. C. Minta. 1994. *Yellowstone's future: Prospects for ecosystem science, management and policy.* Moose, WY: Homestead Publishing.

Columbia Daily Tribune. 2002a. Public divided on energy issue, survey reveals. March 3.

———. 2002b. Drilling in Alaska vital to security, Bush says. February 24.

Congressional Research Service. 2002. *Arctic National Wildlife Refuge.* http://cnie.org/NLE/CRSreports/Biodiversity/biodv-14.cfm#SUMMARY (accessed February 28, 2002).

Cunningham, B. 2001. Upper Missouri Breaks National Monument. *Montana Magazine,* July/August, 12–21.

Davey, A. G. 1998. Protected areas categories and management objectives (Appendix 2), 49–55. *National System Planning for Protected Areas.* Gland, Switzerland, and Cambridge, UK: IUCN.

Defenders of Wildlife. 2002a. *Don't allow big oil to drill in the Arctic National Wildlife Refuge.* http://www.defenders.org/wildlife/arctic/print/overview.html (accessed February 28, 2002).

———. 2002b. *Issue in detail: Saving the "American Serengeti."* http://www.defenders.org/wildlife/arctic/arissue.html (accessed February 28, 2002).

Despain, D., A. Rodman, P. Schullery, and H. Shovic. 1989. Burned area survey of Yellowstone National Park: The fires of 1988. Yellowstone National Park, Division of Research and Geographic Information Systems Laboratory.

Everhart, W. C. 1972. *The National Park Service.* New York: Prager Publishers.

Executive Order establishing the Upper Missouri River Breaks National Monument Task Force. 2002. http://www.discoveringmontana.com/gov2/textonly/eo/14-01.asp (accessed February 26, 2002).

Ferry, G. W., R. G. Clark, R. E. Montgomery, R. W. Mutch, W. P. Leenhouts, and G. T. Zimmeran. 1995. *Altered fire regimes within fire-adapted ecosystems. Our living resources: A report to the nation on the distribution, abundance, and health of U.S. plants, animals, and ecosystems.* Washington, DC: U.S. Department of the Interior, National Biological Survey.

Fire Management Policy Review Team. 1988. *Report on fire management policy.* Washington, DC: U.S. Department of Agriculture and U.S. Department of the Interior.

Firewise Communities/USA. 2004. http://www.firewise.org/usa/ (accessed April 13, 2004).

Friends of the Missouri River Breaks Monument. 2002a. *How the Upper Missouri River Breaks became a national monument.* http://www.missouribreaks.org/designating.htm (accessed February 26, 2002).

———. 2002b. http://www.missouribreaks.org/designating.htm (accessed February 26, 2002).

General Accounting Office. 1999. Western national fires: A cohesive strategy is needed to address catastrophic wildfire threats. U.S. General Accounting Office, House of Representatives, Committee on Resources, Report to the Subcommittee on Forests and Forest Health, RCED-99-65. Washington, DC: Government Printing Office.

Gildart, R. C., et al. 1990. *Glacier country: Montana's Glacier National Park*. Helena, MT: American Geographic Publishing.

Greater Yellowstone Coordinating Committee. 1990. *Vision for the future: A framework for coordination in the Greater Yellowstone Area.* Billings, MT: U.S. National Park Service and U.S. Forest Service.

Green, M. J. B, and J. R. Paine. 1997. *State of the world's protected areas at the end of the twentieth century.* Paper presented at Protected Areas in the 21st Century: From Islands to Networks. Albany, Australia, November 24–27.

High Country News. 2002. Hotline: Idaho buyout for bears. 34: 7 (February 18).

Holling, C. S. 1995. What barriers? What bridges? In *Barriers and bridges to the renewal of ecosystems and institutions*, ed. L. H. Gunderson, C. S. Holling, and S. S. Light, 3–25. New York: Columbia University Press.

IUCN. 1994. *Guidelines for protected area management categories.* Gland, Switzerland, and Cambridge, UK: IUCN Commission on National Parks and Protected Areas with assistance of the World Conservation Monitoring Centre.

Kay, C. E. 1997. Yellowstone ecological malpractice. *PERC Reports Special Issue* 15: 3–39.

Keiter, R. B. 1996. Ecosystem management: Exploring the legal-political framework. In *National parks and protected areas: their role in environmental protection,* ed. R. G. Wright, 63–88. Cambridge, MA: Blackwell Science.

Keiter, R. B., and M. S. Boyce. 1991. Greater Yellowstone's future: Ecosystem management in a wilderness environment. In *The Greater Yellowstone ecosystem: Redefining America's wilderness heritage,* ed. R. B. Keiter and M. S. Boyce, 379–413. New Haven, CT: Yale University Press.

Knight, D. H. 1991. The Yellowstone fire controversy. In *The Greater Yellowstone ecosystem: Redefining America's wilderness heritage,* ed. R. B. Keiter and M. S. Boyce, 87–103. New Haven, CT: Yale University Press.

Leopold, A. S., S. A. Cain, C. M. Cottam, I. N. Garbielson, and T. L Campbell. 1963. Wildlife management in the national parks. *Transactions of the North American Wildlife Resources Conference* 28: 29–44.

Lopoukhine, N. 1991. A Canadian view of fire management in the Greater Yellowstone Area. In *The Greater Yellowstone ecosystem: Redefining America's wilderness heritage,* ed. R. B. Keiter and M. S. Boyce, 149–162. New Haven, CT: Yale University Press.

Mahoney, D. L., and R. E. Gresswell. 1998. Short-term postfire variation in physical habitat, fish populations, and associated aquatic communities in large streams of Yellowstone National Park. *Yellowstone Science* 6: 39.

Matthiessen, P., F. Mauer, W. H. Meadows, D. S. Miller, G. B. Schaller, and D. A. Sibley. 2003. Arctic National Wildlife Refuge, seasons of life and land, a photographic journey by Subhanker Banerjee. Seattle, WA: The Mountaineers Books.

Mattson, D. J., and T. W. Clark. 2001. Conservation of grizzly bears in the Northern U.S. Rockies. In *Human use management in mountain areas,* Conference sponsored by the Mountain Culture of the Banff Centre, Banff, Alberta, ed. L. Taylor and A. Ryall, 201–204.

Mech, L. D. 1997. Returning the wolf to Yellowstone. In *The Greater Yellowstone ecosystem: Redefining America's wilderness heritage,* ed. R. B. Keiter and M. S. Boyce, 309–322. New Haven, CT: Yale University Press.

Minshall, G. W., and J. T. Brock. 1991. Observed and anticipated effects of forest fire on Yellowstone stream ecosystems. In *The Greater Yellowstone ecosystem: Redefining America's wilderness heritage,* ed. R. B. Keiter and M. S. Boyce, 123–135. New Haven, CT: Yale University Press.

Missouri Department of Conservation and Missouri Department of Natural Resources. Undated. *Missouri natural areas system* (brochure).

Missouri Department of Natural Resources. 2002. *State parks*. http://www.dnr.state.mo.us/dsp/index.html (accessed February 14, 2002).

Montana River Action. 2001. *Save the Missouri River Breaks*. http://www.montanariveraction.org/save.missouri.river.breaks.html (accessed February 28, 2002).

Montana Wildlife Federation. 2002. Monument status attack defeated. *Montana Wildlife* 26(Dec/Jan): 15.

National Interagency Fire Center. 2003. *Wildland fire statistics, Boise, ID*. http://www.nifc.gov/stats/wildlandfirestats.html (accessed April 10, 2003).

National Parks Conservation Association. 2002. http://www.npca.org/flash.html (accessed March 5, 2002).

National Park Service. 2002. *Cerro Grande fire, Bandelier National Monument*. http://www.fs.fed.us/r3/sfe/fire/cerrogrande/ (accessed March 5, 2002).

O'Brien, B. R. 1999. *Our national parks and the search for sustainability*. Austin: University of Texas Press.

Perkins, S. 2001. *A nation aflame: After last years' conflagration, can we learn to live with wildfires*. Science News *159: 120–121.*

Ramsar Convention on Wetlands. 2001. http://www.ramsar.org/ (accessed June 11, 2001).

Rockwood, T. H. 1988. Organizational responses to a conflicting mandate: Legitimacy and structural adaptations in the National Park Service. MA Thesis, Washington State University, Pullman.

Romme, W. H., and D. G. Despain. 1989. Historical perspectives on the Yellowstone fires of 1988. *BioScience* 39: 695–699.

Runte, A. 1987. *National parks: The American experience*. Lincoln: University of Nebraska Press.

SmokeyBear.com. 2003. Smokey's vault. http://www.smokeybear.com/vault/default.asp (accessed April 11, 2003).

State of Alaska. 1990. *SPILL: The wreck of the* Exxon Valdez, *final report, Alaska Oil Spill Commission*. http://www.oilspill.state.ak.us/facts/details.html (accessed August 18, 2003).

Stegner, W. 1989. The marks of human passage. In *Mirror in America: Literary encounters with the National Parks*, ed. D. Harmon, 155–172. Boulder, CO: Robert Rinehart Publishers.

Turner, M. G., W. H. Romme, and R. H. Gardner. 1999. *Prefire heterogeneity, fire severity, and early postfire plant reestablishment in subalpine forests of Yellowstone National Park, Wyoming*. International Journal of Wildland Fire *9: 21–36.*

U.S. Department of Energy. 2000. *Potential oil production from the coastal plain of the Arctic National Wildlife Refuge: Updated assessment*. Energy Information Administration Report # SR/O&G/2000–02. http://www.eia.doe.gov/pub/oil_gas/petroleum/analysis_publications/arctic_national_wildlife_refuge/html/anwr101.html.

U.S. Department of the Interior, National Park Service. Undated. *Final general management plan and environmental impact statement—Volume 1*, Glacier National Park, West Glacier, MT.

U.S. Fish and Wildlife Service. 2003. *Arctic National Wildlife Refuge*. http://www.r7.fws.gov/nwr/arctic/arctic.html (accessed April 10, 2003).

UNESCO. 2001. *Man and the Biosphere Program*. MAB Biosphere Reserve Directory-United States of America. http://www.unesco.org/mab/br/brdir/europe-n/USAmap.htm (accessed June 15, 2001).

———. 2000. *World heritage list*. http://www.unesco.org/whc/nwhc/pages/sites/main.htm (accessed June 10, 2001).

Van Dyke, F. G., J. P. Dibenedetto, and S. C. Thomas. 1991. Vegetation and elk response to prescribed burning in south-central Montana. In *The Greater Yellowstone ecosystem: Redefining America's wilderness heritage*. ed. R. B. Keiter and M. S. Boyce, 163–179. New Haven, CT: Yale University Press.

Weaver, J. 1978. The wolves of Yellowstone. Natural Resources Report 14. Washington, DC: National Park Service.

White House. 2003. Healthy Forests Initiative. http://www.whitehouse.gov/infocus/healthyforests/ (accessed April 11, 2003).
Wilderness Society. 2002. *Upper Missouri River Breaks National Monument, Montana.* http://wilderness.org/newsroom (accessed February 26, 2002).
Williams, J. 1995. *Firefighter safety in changing forest ecosystems.* Fire Management Notes *55: 6–8.*
World Commission on Protected Areas (WCPA). 2001. *About protected areas.* http://wcpa.iucn.org/wcpainfo/protectedareas.html (accessed June 11, 2001).

Chapter 2

Role and Status of Protected Areas

Protected areas are a civilized society's strongest signs of prosperity, foresightedness, and greatness of spirit.

—Carl Safina

Origin and History

The primary motivation for establishing natural or protected areas was to limit adverse impacts of resource and economic development on natural resources for the common good. Initially, the common good referred to royalty. In 700 B.C., Assyrian noblemen created reserves for hunting and horseback riding. Rome and medieval Europe likewise reserved areas for exclusive use of the ruling classes (Runte 1979). Plato expressed concern about the environmental impacts of economic development. He noted that removal of forest cover from the hills of Attica in Greece had increased soil erosion and degraded watercourses, causing these areas to appear "like the skeleton of a body wasted by disease" (Talbot 1984).

Other early examples of protected areas include a Lithuanian reserve for European bison established in 1551 and a Swiss reserve for chamois formed in 1569. In the eighteenth and nineteenth centuries, England's ruling class set up game preserves for hunting, the most notable being Sherwood Forest. The concept of game reserves was brought to North America (Canada, United States, and Mexico) with European settlement. By 1900, five hundred relatively small (less than 1,500 ha, or 3,705 acres) wildlife preserves had been established by private sport-hunting clubs and other groups. Preserves were generally closed to the public and managed for game species using invasive management practices, such as predator control, habitat alteration, and introduction of exotic species (Wright and Mattson 1996).

In 1832, George Catlin, famed artist of the American Indian, suggested that the wilderness expanse surrounding the confluence of the Yellowstone and Missouri Rivers in Montana be designated as a nation's park. It never happened. In that same year, Hot Springs Reservation was established in Arkansas. In 1921, the reservation was designated as a national park.

India created the first protected forest areas and instituted the first recorded game laws to protect mammals, birds, and fish. In 1935, India established Corbett National Park, which is Asia's first national park. Concern and awe for protected areas continued in India. During the First World Conference on National Parks in Seattle, Washington, in 1962, M. A. Badshah, wildlife officer for the State of Madras, referred to national parks as "the nearest approach to a paradise on earth... which contain a combination of forested hills and valleys, sparkling multicolored lakes, crystal clear streams, rippling brooks, placid rivers, silvery beaches, scented flowers, luscious fruits, sweet berries, snow-clad mountains, and wildlife" (Wright and Mattson 1996).

Wilderness Proponents

Ralph Waldo Emerson, Henry David Thoreau, and John Muir were notable American proponents of natural landscape conservation. Others include George Perkins Marsh, Aldo Leopold, Bob Marshall, and George Wright. Ralph Waldo Emerson (1803–1882) was a transcendentalist who argued against the absolute separation between human consciousness and nature. Transcendentalism is the philosophy that God's existence can be deduced from the beauty and design of nature. He viewed nature as the primal influence on the human mind that confirms the existence of God. As a theist, Emerson believed in the eternal, transcendent creator and the Judeo-Christian belief that humans are God's ultimate creation (anthropocentrism). A wilderness odyssey to Emerson was not simply a sensual experience (sights, sounds, and smells). It afforded an opportunity for the human mind to contemplate the significance of nature and confirm the existence of God. Emerson believed that a person could rediscover the truth of the Bible (e.g., creation story of Genesis) through an encounter with nature, and that God's existence can be rationalized based on transcendental reason rather than faith (Oelschlaeger 1991).

Henry David Thoreau (1817–1862) was a natural classicist and one of America's greatest nature writers. He is best known for essays on life with nature, especially *Walden*, which he was inspired to write during his 26-month residency at Walden Pond west of Concord, Massachusetts. In *Walden*, Thoreau rejected the conventional separation between wilderness and civilization and pleaded for reintegration of human consciousness and nature. He espoused a unique wilderness philosophy that integrates matter, life, and consciousness. Thoreau is known for two statements he made. In *Economy*, he criticized the fact that economic forces dominated the lives of people by noting, "the mass of men lead lives of quiet desperation." In *Walking*, Thoreau espoused the philosophy that freedom increases as people live closer to nature. His wilderness philosophy is encapsulated in a few words that capture the essence of the wilderness preservation movement, "In wildness is the preservation of the world" (Howarth 1982; Oelschlaeger 1991).

John Muir (1838–1914) is considered the father of the American preservation movement. He is best known for his love of mountains and wilderness preservation ethic. Muir was one of the founders of the Sierra Club and its first president (Ehrlich 2000). He lobbied for the creation of six of America's premier national parks including Yosemite and Grand Canyon. Muir believed that humanity needed wild places, as reflected in his statement, "everyone needs beauty as well as bread, places to play in and pray in, where Nature may heal and cheer and give strength to body and soul." In Muir's pantheistic view, God and creation are synonymous because God suffuses the natural world. No aspect of creation is above any other (biocentrism), which implies that all creation is sacred (Oelschlaeger 1991).

Other individuals who contributed ideas and actions that were critical to the conservation and management of natural resources and protected areas include Aldo Leopold, Bob Marshall, Rachael Carson, Edward Abbey, Paul Ehrlich, E. F. Schumacher, David Brower, and Arne Naess.

Birth of American National Parks

The birth of national parks and protected areas in America occurred during a period of time (starting in the 1860s) when rapid exploitation and development of natural resources were the order of the day (Limerick 1988). The Homestead Act encouraged agricultural development. Vast tracks of public land, which exceeded the combined area of France and the United Kingdom (O'Brien 1999), were transferred to railroads to stimulate their westward expansion.

Yosemite Valley and Yellowstone National Park are the first two rural natural parks. The Yosemite Park Act of 1864 transferred Yosemite Valley and the Mariposa Grove (giant sequoias) to the State of California. Two conditions facilitated this federal-to-state land transfer. First, there was little private interest in exploiting the valley's agricultural, timber, and mineral resources. Second, a Supreme Court ruling prohibited homesteading in the valley. Due to negligent management of its natural resources, Yosemite Valley was transferred back to federal ownership and was made part of Yosemite National Park in 1906. The park itself was established in 1890 (O'Brien 1999; Ehrlich 2000).

After his participation in the Lewis and Clark Expedition, John Colter explored the Yellowstone region in 1807. His accounts of Yellowstone's geologic wonders seemed unbelievable to most people. The first expedition in the Yellowstone region was scheduled for 1859, but was cancelled due to the Civil War. The Washburn-Doane Expedition of 1870 was charged with confirming Yellowstone's scenic wonders. Unlike previous expeditions, it established credibility with the public and Congress by including Nathaniel P. Langford, a politician and entrepreneur who later became the first superintendent of Yellowstone National Park.

The birth of the national park idea is credited to an event that supposedly occurred near the end of the Washburn-Doane Expedition. While disputed by some historians (Schullery 1997), the story goes that on the evening of September 19, 1870, while gathered around a campfire near the Madison Junction, the expedition party recounted the wonders of Yellowstone. While the party entertained thoughts of staking private land claims and developing the region as a tourist attraction, they had a change of heart. As Sellars notes, "in a moment of high altruism, the explorers agreed that Yellowstone's awe-inspiring geysers, waterfalls, and canyons should be preserved as a public park" (Sellars 2000).

The Northern Pacific Railroad Company was instrumental in the creation of Yellowstone National Park. In addition to supporting Langford's participation in the Washburn-Doane Expedition, the Northern Pacific maneuvered to block private land claims and limit competition for tourism in Yellowstone. Their goal was to gain a monopoly on rail transportation in the southern Montana territory. In 1871, the Northern Pacific financed artist Thomas Moran's participation in the Hayden Survey led by geologist Ferdinand V. Hayden. The government-sponsored Hayden Survey recorded the region's geology, flora, and fauna (Langford 1972). Moran's sketches of the Yellowstone region (his paintings were yet to be completed) were displayed in the nation's capital and were instrumental in convincing Congress to reserve lands in the Yellowstone region.

After strong lobbying by the Northern Pacific and minimum opposition from economic interests, Congress passed the Yellowstone Park Act on March 1, 1872, which established Yellowstone National Park, the world's first national park. Yellowstone was set aside as "a public park or pleasuring ground for the benefit and enjoyment of the people." The vast wilderness and abundant wildlife contained in the park were not the primary motivation for its establishment. Boundaries enclosing the park's 0.89 million ha (2.2 million acres) were determined based on land Ferdinand Hayden identified as containing spectacular geothermal features. While the act protected vast wilderness tracts and outlawed "wanton destruction of the fish and game," it provided little guidance on how to manage the area's biological resources. Following the establishment of Yellowstone National Park, the United States developed an extensive, culturally and ecologically diverse system of national parks and protected areas.

The World Commission on Protected Areas of the International Union for the Conservation of Nature and Natural Resources (IUCN, or World Conservation Union) defines a protected area as "an area of land and/or sea especially dedicated to the protection and maintenance of biological diversity, and of natural and associated cultural resources, and managed through legal or other effective means" (IUCN 1994). Protected areas encompass many ecosystem types, including wetland, tropical or deciduous forest, alpine region, savanna, marine, and others. The IUCN has established seven categories of protected area (see the sidebar and table 2.1). National parks comprise one category. They are defined as a "natural area of land and/or sea, designated to (a) protect the ecological integrity of one or more ecosystems for present and future generations, (b) exclude exploitation or occupation inimical to the purposes of designation of the area and (c) provide a foundation for spiritual, scientific, educational, recreational and visitor opportunities, all of which must be environmentally and culturally compatible" (Davey 1998).

The Multifaceted World Conservation Union (IUCN)

The World Conservation Union (short for International Union for the Conservation of Nature and Natural Resources, or IUCN) was founded in 1948 as a group of states, government agencies, and nongovernmental organizations. Its mission is "to influence, encourage and assist societies throughout the world to conserve the integrity and diversity of nature and to ensure that any use of natural resources is equitable and ecologically sustainable." Its vision is to achieve "a just world that values and conserves nature." IUCN's conservation goals are to:

1. Reduce the loss of diversity within species, between species, and in ecosystems
2. Maintain and if necessary restore ecosystem integrity
3. Promote sustainable and equitable use of natural resources

Membership in IUCN is open to states, governmental agencies, political and/or economic integration organizations, and international and national nongovernmental organizations. IUCN currently has 980 members representing 140 countries.

IUCN programs address several conservation themes including business management, climate change, economics, ecosystem management, education and communication, environmental law, forest conservation, gender, marine areas, monitoring and evaluation, policy/biodiversity/international agreements, protected areas, social policy, species, sustainable resource use, wetlands and water, and others.

The IUCN incorporates six commissions, namely the Commission on Education and Communication; Commission on Environmental, Economic, and Social Policy; Commission on Environmental Law; Commission on Ecosystem Management; Species Survival Commission; and World Commission on Protected Areas. Through its commissions, IUCN has access to over ten thousand expert volunteers organized in project teams and action groups that address a wide range of issues dealing with particular species, biodiversity conservation, and the management of habitats and natural resources. National conservation strategies have been developed for many countries with assistance from IUCN. Most IUCN operations are decentralized in a network of regional programs and country offices, located for the most part in developing countries. Currently there are regional programs in Mesoamerica, South America, West and Central Asia, North Africa, West Africa, Central Africa, Eastern Africa, Southern Africa, Asia, and Europe

In carrying out its mission, IUCN "seeks to influence, encourage and assist societies throughout the world to conserve the integrity and diversity of nature and to ensure that any use of natural resources is equitable and ecologically sustainable." Coordination of the IUCN program is handled by a central secretariat that represents the views of IUCN members throughout the world and provides them with strategies, services, scientific knowledge, and technical support required to achieve their goals. For example, IUCN publishes a news and advocacy journal called *World Conservation* three times a year, a bimonthly newsletter called *IUCN Today*, and technical papers from IUCN commissions. In addition, the IUCN *Publications Catalog* contains numerous titles covering policy and technical conservation issues and developments around the world. IUCN's red list of threatened plants and animals is a definitive guide to endangered species around the world. IUCN does annual assessments of World Heritage Sites under the World Heritage Convention and develops definitions of the UN World Protected Areas Categories as shown in table 2.1. IUCN also conducts independent, scientifically based assessment of projects, plans, and proposals related to protected areas; conducts seminars and workshops and joint projects at the regional, national, and international levels; and collaborates closely with the IUCN worldwide regional and country offices.

Some noteworthy accomplishments of IUCN include the provision of scientific advice, technical support, and/or policy analysis related to international conventions such as the Convention on Trade in Endangered Species (CITES), the Convention on Wetlands of International Importance (RAMSAR), and the Convention on Biological Diversity. IUCN facilitates interaction between protected-area specialists throughout the world via contacts in Ministries of Foreign Affairs, and Ministries of Environment or Natural Resources.

Source: IUCN (2003).

Function and Purpose

National parks and protected areas serve a wide range of functions and purposes. These include providing habitat for threatened and endangered species (conservation of biodiversity); supplying clean water to communities; supporting subsistence lifestyles of indigenous peoples; providing opportunities for recreation, education, and scientific research; and others. Protected areas stimulate tourism, recreation, and second-home development that increase local and regional economic growth.

Conservation of biodiversity has become an important objective of national parks and protected areas. Article 8 of the Convention on Biological Diversity (CBD) encourages signatory nations to establish and manage systems of protected areas in support of natural and cultural resource conservation, sustainable use, and equitable sharing of benefits. The CBD encourages nations to "establish a system of protected areas or areas where special measures need to be taken to conserve biodiversity" (Article 8a) and "develop, where necessary, guidelines for the selection, establishment and management of protected areas" (Article 8b) (Glowka et al. 1994). Over 170 signatory nations, including the United States, have signed the CBD.

Nature, in the form of air, soil, water, forests, fish, wildlife, fossil fuels, and other natural resources, supports production of a wide range of consumer and ecological goods and services that sustain life. Natural resources are used to produce consumer goods and services, such as automobiles, houses, food, clothing, education, and recreation. The value of consumer goods and services (Gross National Product) are a major indicator of economic performance. Ecological goods include timber, forage (grazing), seafood, biomass fuels (coal, oil, and natural gas), natural fiber, pharmaceuticals, industrial products (waxes, rubber and oils), and others. Ecological services include purification of air and water, generation and renewal of soil and soil fertility, conservation of biodiversity, partial stabilization of climate, and many others (Daily 1997). Provision of ecological goods and services is supported by complex natural cycles (nitrogen, phosphorus, carbon, bacterial, hydrologic, etc.), which are driven by solar energy. Disruption of these cycles can have adverse consequences. Emissions of CO_2 and other greenhouse gases are altering the carbon cycle resulting in climate change, including global warming. Ecological goods and services are vital to human welfare and ultimately human existence (Barbier et al. 1995).

National parks and protected areas provide ecological goods and services that generate numerous social, economic, and ecological values. Consider the multiple values of protected areas. First, national parks and protected areas provide benchmarks from which to measure the impacts of resource extraction, environmental degradation, and climate change on ecological functions and processes. Second, they generate unique recreational, aesthetic, and spiritual values not provided by the built-up environments in which most people live. Third, national parks and protected areas located in headwater regions are sources of clean and abundant supplies of water and provide habitat for many threatened and endangered species of fish and wildlife. Fourth, national parks and protected areas provide income, employment, and tax benefits to gateway communities as indicated above.

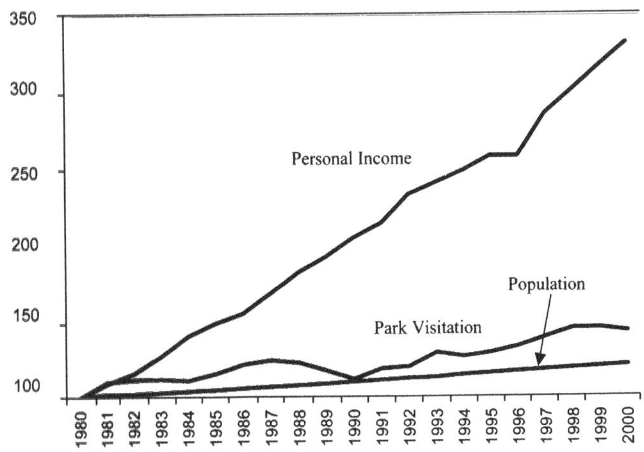

FIG. 2.1. Trends in annual visitation to national parks, disposable personal income (December), and resident population (December 1), United States, 1980–2000 (1980 = 100). Disposable personal income from U.S. Department of Commerce (2002); park visitation from National Park Service (2002); resident population from U.S. Census Bureau (2001)

Growth in consumer and ecological goods and services and the forces driving that growth (population, income, and technology) influence natural parks and protected areas in three major ways. First, increases in population and consumer income increase demand for outdoor recreation and tourism. That is why visitation in national parks and protected areas has increased faster than population, although not as fast as personal income. During the 1980–2000 period, resident population increased 21 percent, visitation to national parks increased 43 percent, and personal income increased 332 percent in the United States (see fig. 2.1). While increased use of national parks and protected areas has generated substantial social benefits, it has reduced the capacity of these areas to sustain ecological values. This problem has been exacerbated by inadequate funding of agencies that manage national parks and protected areas.

Second, increases in national income and wealth combined with the environmental amenities afforded by national parks and protected areas have caused rapid growth in gateway communities (communities surrounding national parks and protected areas). While this growth generates income, employment, and tax benefits (Howe et al. 1997), it also causes landscape fragmentation in buffer areas for national parks and protected areas. Landscape fragmentation degrades fish and wildlife habitat and increases wildlife mortality and fire hazard in the rural-wildlife interface.

Third, natural resource development and use in areas distant from and adjacent to national parks and protected areas compromises their ecological integrity.

Adverse impacts emanating from distant areas include loss of biodiversity, melting of glaciers and increased incidence of forest fires from global warming, air pollution and loss in visibility from power plants and industrial facilities, and damage to forests and aquatic ecosystems due to acid deposition caused by the burning of fossil fuels. Negative impacts from adjacent areas include degradation of habitats for threatened/endangered and other species from timber harvesting, oil and gas development, and livestock grazing.

In summary, growth in visitation and use of national parks, protected areas, and gateway communities provides substantial social and economic benefits. Unfortunately, the natural resource development, use, and degradation that often accompanies this growth have diminished the ecological integrity of national parks and protected areas and their capacity to conserve and protect natural ecosystems. As a result, the ecological value of national parks and protected areas has diminished. This apparent trade-off between increases in social and economic values on the one hand and decreases in ecological value on the other hand calls for development of management strategies that achieve a better balance among the social, economic, and ecological values generated by national parks and protected areas.

Classification

Classification of protected areas has evolved over time. The IUCN defined 5 categories of protected areas in 1978 (IUCN 1978), 10 in 1984 (IUCN 1984), and 7 in 1994 (IUCN 1994) (see table 2.1 for the 1994 categories). A category's title (e.g., Strict Nature Reserve for Ia) indicates the function of the protected area and the primary management objective indicates the primary purpose for which the protected area is to be managed (e.g., Science for Ia). Critical elements of each category are explained in the definition. The order of categories from lowest to highest human exploitation is Ia/Ib, II, III, VI, IV, and V (Green and Paine 1997). Category VI appears out of numerical order because it was added at the time of IUCN's 1994 revision. If a site does not meet the IUCN definition of a protected area, application of a category is not appropriate.

While IUCN categories are recognized around the world, the names given to protected areas do not always conform to the IUCN classification. This is evidenced by the fact that over 1,388 different names have been used to designate protected areas around the world. A national park in North America generally imposes greater limitation on human habitation and resource use than one in Europe. For this reason, national parks in America fall in Category II (National Parks), whereas England's national parks fall in Category V (Protected Landscape/Seascape). Countries use different labels for protected areas that essentially perform the same function. For example, the following names have been given to protected areas used primarily for research (Green and Paine 1997):

- Managed nature reserves in the Bahamas
- Strict nature reserves in Bhutan
- Nature reserves in Ontario, Canada
- National nature reserves in the Czech Republic

Table 2.1. IUCN Protected Area Categories

Category and primary management objective	Definition
CATEGORY Ia. Strict Nature Reserve: *Science*	Area of land and/or sea possessing some outstanding or representative ecosystems and geological or physiological features and/or species, available primarily for scientific research and/or environmental monitoring
CATEGORY Ib. Wilderness Area: *Wilderness protection*	Large area of unmodified or slightly modified land and/or sea, retaining its natural character and influence, without permanent or significant habitation, which is protected and managed so as to preserve its natural condition
CATEGORY II. National Park: *Ecosystem conservation and recreation*	Natural area of land and/or sea, designated to (a) protect the ecological integrity of one or more ecosystems for present and future generations; (b) exclude exploitation or occupation inimical to the purposes of designation of the area; and (c) provide a foundation for spiritual, scientific, educational, recreational, and visitor opportunities, all of which must be environmentally and culturally compatible
CATEGORY III. Natural Monument: *Conservation of specific natural features*	Area containing one or more specific natural or natural/cultural feature, which is of outstanding or unique value because of its inherent rarity, representative or aesthetic qualities, or cultural significance
CATEGORY IV. Habitat/Species Management Area: *Conservation through active management*	Area of land and/or sea subject to active intervention for management purposes so as to ensure the maintenance of habitats and/or to meet the requirements of specific species
CATEGORY V. Protected Landscape/Seascape: *Landscape/seascape conservation and recreation*	Area of land, with coast and sea as appropriate, where the interaction of people and nature over time has produced an area of distinct character with significant aesthetic, ecological, and/or cultural value, and often with high biological diversity. Safeguarding the integrity of this traditional interaction is vital to the protection, maintenance, and evolution of such an area.
CATEGORY VI. Managed Resource Protected Area: *Sustainable use of natural ecosystems*	Area containing predominantly unmodified natural systems, managed to ensure long-term protection and maintenance of biological diversity, while providing at the same time a sustainable flow of natural products and services to meet community needs.

Source: Adapted from IUCN (1994).

- Nature reserves and marine nature reserves in Indonesia
- Nature conservation areas in Japan
- Strict natural reserves in Sri Lanka

While IUCN does not expect each country to develop protected areas in all seven categories, it supports creation of a system of protected areas that reflects a country's natural resource base, needs and desires, and constraints. The distribution of protected areas by category and extent for North America and the world is shown in table 2.2.

On a global scale, there are 30,350 protected areas covering more than 13.2 million km^2 (5 million mi^2) of the world's land area, which is equivalent to the combined areas of India and China (WCPA, undated a). Terrestrial protected areas account for 11.6 million km^2 (4.5 million mi^2) or 7.78 percent of the Earth's total land area, whereas marine protected areas account for only 1.6 million km^2 (0.6 million mi^2), which is less than 1 percent of the Earth's total marine area.

North America has 6,714 protected areas covering 4 million km^2 (1.5 million mi^2). Protected areas in Category Ia (strict nature reserve) account for a substantially smaller percentage of total protected areas in North America than in the world (1.44 percent vs. 7.42 percent). North America has almost twice as much of its land in protected areas than the world (17.44 percent vs. 8.84 percent). National parks (Category II) make up the largest category of protected areas in terms of surface area, namely 40 percent in North America and 30 percent in the world.

There is substantial variation in the global number, size, and extent of protected areas across categories as shown in figures 2.2, 2.3, and 2.4, respectively.

Table 2.2. Number and Extent of Protected Areas in North America and the World, 1997

	Category							
	Ia	Ib	II	III	IV	V	VI	Total
				North America				
Number	661	630	1,286	342	1,249	2,085	461	6,714
Extent (1,000 km^2)	59	392	1,634	58	823	245	877	4,088
Percentage PA[a]	1.44	9.6	39.9	1.43	20.12	6	21.45	17.44[c]
				World[d]				
Number	4,395	806	3,383	2,112	11,169	5,578	2,897	30,350
Extent (1,000 km^2)	982	940	4,001	193	2,460	1,057	3,601	13,236
Percentage PA[b]	7.42	7.1	30.23	1.46	18.58	7.99	27.21	8.84[e]

Source: Green and Paine (1997).

[a] Area in PA category as a percentage of total area in North America in protected areas.

[b] Area in PA category as a percentage of total area in world in protected areas.

[c] Total area in protected areas in North America as a percentage of total area in North America.

[d] The world includes the 15 regions defined by the World Commission on Protected Areas (WCPA).

[e] Total area in protected status in the world as a percentage of total area in the world.

Category V (Protected Landscape/Seascape) accounts for the largest number and Category II (National Park) for the smallest number of protected areas. Protected areas in Category II have the highest extent and mean size of all categories, namely 3.8 times larger extent and 6.2 times greater mean size than protected areas in Category V. The size distribution of protected areas is highly skewed; 17,892 protected areas less than 1,000 ha (2,471 acres) in size account for 0.2 percent of the global protected area, and 1,673 protected areas greater than 1,000 km^2 (386 mi^2) in size account for 87 percent of the global protected area.

The number and extent of protected areas have risen steadily and dramatically over the period 1980–2000 (see fig. 2.5). Part of this growth was stimulated by greater world concern for environmental degradation and human-induced losses in biodiversity.

Management Objectives

Protected areas have numerous management objectives, namely (IUCN 1994):

- Education
- Maintenance of cultural/traditional attributes
- Maintenance of environmental services
- Preservation of species and genetic diversity
- Protection of specific natural and cultural features
- Scientific research

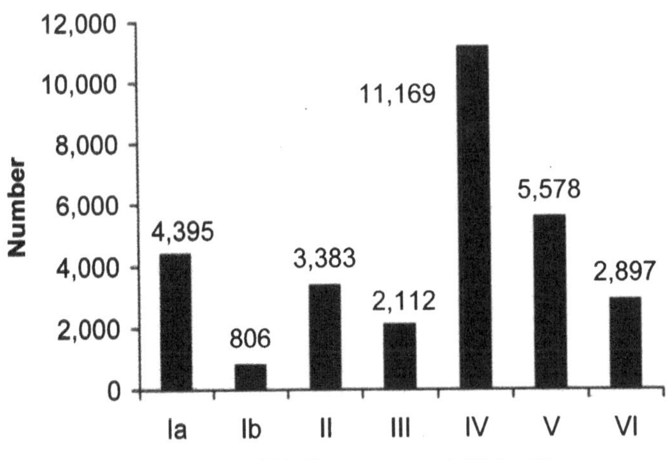

FIG. 2.2. Number of global protected areas, by management objective, 1994. (Data from IUCN 1994)

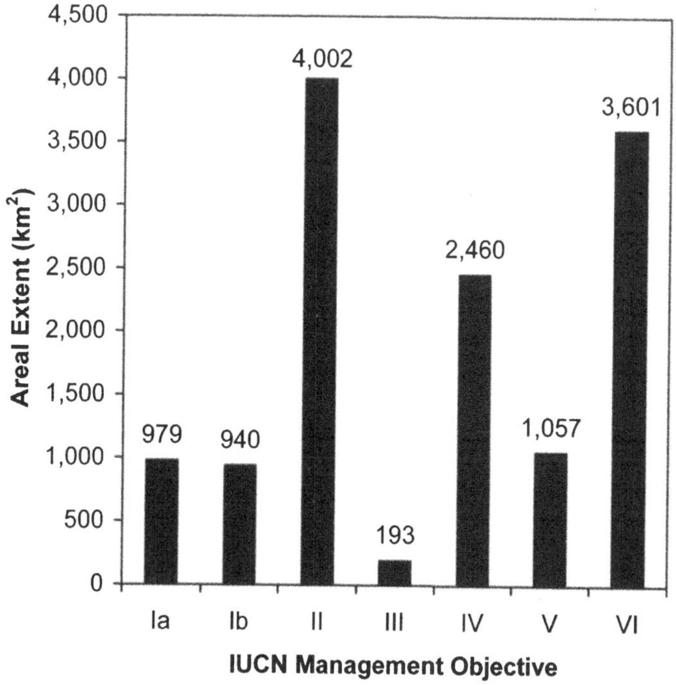

FIG. 2.3. Extent of global protected areas, by management objective, 1994. (Data from IUCN 1994)

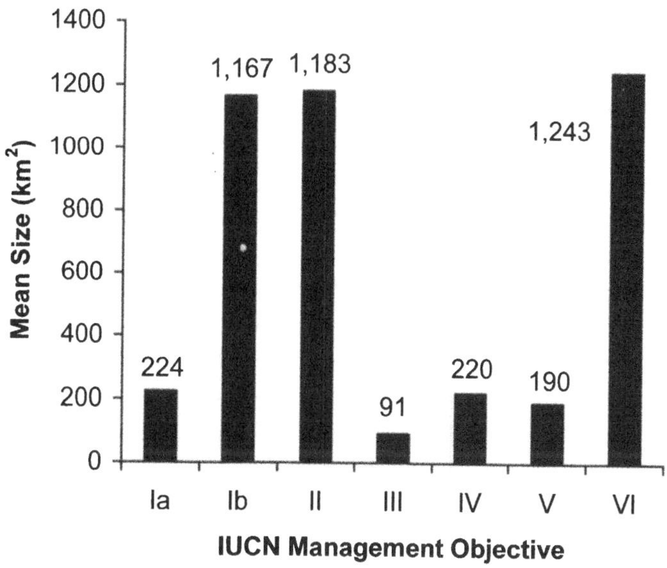

FIG. 2.4. Mean sizes of global protected areas, by management objective, 1994. (Data from IUCN 1994)

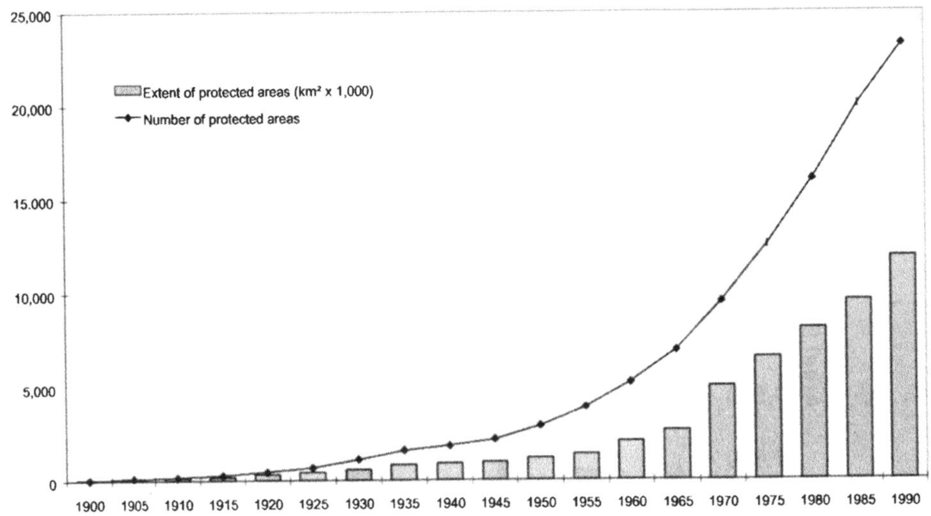

FIG. 2.5. Trend in number and extent of global protected areas, 1900–1990. (From Green and Paine 1997)

- Sustainable use of resources from natural ecosystems
- Tourism and recreation
- Wilderness protection

IUCN has specified appropriate management objectives for each of the seven protected area categories (see table 2.3). No more than three objectives are listed for any single protected area. Deciding which management objectives to pursue in an area is important because it allows managers to determine which uses of a protected area to allow and disallow. It is not always this straightforward because special interest groups often have different opinions about which uses are compatible and incompatible with established objectives. For example, tourism and recreation, species/genetic diversity, and environmental services are three primary objectives of national parks (see table 2.3). Use of snowmobiles and personal watercraft is recreation. Many conservationists believe this form of recreation is incompatible with protecting species/genetic diversity and provision of ecological services. This gives rise to conflicts between use and preservation.

National Park Management Objectives

IUCN management objectives for national parks (Category II) are:

- To protect natural and scenic areas of national and international significance for spiritual, scientific, educational, recreational, or tourist purposes

Table 2.3. Management Objectives for IUCN Protected Area Categories

Management objective	Category[a]						
	Ia	Ib	II	III	IV	V	VI
1. Scientific research	1[b]	3	2	2	—	2	3
2. Wilderness protection	2	1	2	3	3	—	2
3. Preservation of species and genetic diversity	1	2	1	1	1	2	1
4. Maintenance of environmental services	2	1	1	—	1	2	1
5. Protection of specific natural and cultural features	—	—	2	1	3	1	3
6. Tourism and recreation	—	2	1	1	3	1	3
7. Education	—	—	2	2	2	2	3
8. Sustainable use of resources from natural ecosystems		3	3	—	2	2	1
9. Maintenance of cultural and traditional attributes	—	—	—	—	—	1	2

Source: Davey (1998) and IUCN (1994).
[a]Description of categories is given in table 2.1.
[b]1 designates primary objective, 2 designates secondary objective, 3 designates potentially applicable, and a dash (—) designates not applicable.

- To perpetuate, in as natural a state as possible, representative examples of physiographic regions, biotic communities, genetic resources, and species, to provide ecological stability and diversity
- To manage visitor use for inspirational, educational, cultural, and recreational purposes at a level that will maintain the area in a natural or near natural state
- To eliminate and thereafter prevent exploitation or occupation inimical to the purposes of the designation
- To maintain respect for the ecological, geomorphologic, sacred, or aesthetic attributes that warranted designation
- To take into account the needs of indigenous people, including subsistence resource use, in so far as these will not adversely affect the other objectives of management

National Park Selection Criteria

IUCN management selection criteria for national parks (Category II) are as follows:

- The area should contain a representative sample of major natural regions, features, or scenery, where plant and animal species, habitats, and geomorphological sites are of special spiritual, scientific, educational, recreational, and tourist significance

- The area should be large enough to contain one or more entire ecosystem not materially altered by current human occupation or exploitation.

Global and Other Designations

Three primary international agreements govern the designation of protected areas of global significance. The UNESCO Man and the Biosphere (MAB) Program governs Biosphere Reserves, the World Heritage Convention governs World Heritage Sites, and the Ramsar Convention on Wetlands governs international wetlands. A protected area can have more than one of these global designations.

Biosphere Reserves

The concept of Biosphere Reserves was developed in 1974 by a task force of the UNESCO Man and the Biosphere Program. Designation of Biosphere Reserves began in 1976. In order for an area to be designated as a Biosphere Reserve, it must be nominated by a national government, meet minimum criteria, and adhere to a minimal set of conditions. Biosphere Reserves consist of a core area, a buffer zone, and a transition area. Only the core area requires legal protection. Typically, core areas are public land, but they can contain land owned in private or by nongovernmental organizations (NGOs). It is common for the buffer zone and transition area to be in community or private ownership and for a Biosphere Reserve to overlap with areas protected under other systems, such as national parks or nature reserves, World Heritage Sites, or Ramsar wetlands (UNESCO undated).

Biosphere Reserves cover a wide range of biomes including coastal, forest, desert, island, mountain, plains, polar, and tropical. Collectively, Biosphere Reserves form a world network. Individual Biosphere Reserves remain under the sovereign jurisdiction of the countries in which they are located. There were 352 Biosphere Reserves covering more than 2.2 million km^2 (0.85 million mi^2) worldwide in November 1997 (Green and Paine 1997).

Biosphere Reserves fulfill three complementary and mutually reinforcing functions, namely conservation, development, and logistic. Conservation function supports the conservation of landscapes, ecosystems, species, and genetic variation. Development function fosters economic and human development that is socioculturally and ecologically sustainable. Logistic function provides support for research, monitoring, education, and information exchange related to local, national, and global conservation and development issues (UNESCO 2001).

World Heritage Sites

The Convention Concerning the Protection of the World Cultural and Natural Heritage of 1972, or World Heritage Convention (WHC), authorized World

Heritage Sites. Listed sites have outstanding values and are protected for the benefit of all humanity through cooperative international arrangements. There are three types of World Heritage Sites. A cultural site includes a monument, group of buildings, or site of historical, aesthetic, archaeological, scientific, ethnological, or anthropological value. A natural site has outstanding physical, biological, and geological features and includes habitats of threatened plants or animal species and areas having scientific, aesthetic, or conservation value. A mixed site contains elements of the first two. Of the 730 inscriptions on the World Heritage List, 563 are cultural, 144 are natural, and 23 are mixed properties in 125 countries (UNESCO World Heritage Centre 2002). North America contains 53 World Heritage Sites (see table 2.4). World Heritage Sites are eligible for funding unlike Biosphere Reserves.

Ramsar Convention

The Ramsar Convention on Wetlands is an intergovernmental treaty signed in 1971 in Ramsar, Iran. The mission of the Ramsar Convention is the "conservation and wise use of wetlands by national action and international cooperation as a means to achieving sustainable development throughout the world" (Conference of the Contracting Parties 1996). The convention's mission is accomplished by encouraging signatories to mutually agree on designation of wetland-protected areas that are important to waterfowl and that promote the wise use of wetland ecosystems. The treaty involves 124 Contracting Parties and protects 1,069 wetland sites covering a total of 81.2 million ha (201 million acres). Protected sites appear on the Ramsar List of Wetlands of International Importance. There are two kinds of Ramsar wetlands: those containing representative, rare, or unique wetland types, and those of international importance for conserving biological diversity (Ramsar Convention on Wetlands 2001). The Ramsar Convention has the distinction of being the only global environmental treaty that deals with a particular ecosystem (United Nations System-Wide Earthwatch 2001).

Marine and Coastal Areas

The UN Law of the Sea provides a legal framework for establishing marine protected areas by allocating rights to territorial seas. There is a major disparity between the proportions of the Earth's terrestrial and marine surfaces in protected status. Almost 9 percent of the Earth's land surface is in protected areas. In contrast, less than 1 percent of the ocean surface is protected (WCPA undated b). This disparity is significant because only 30 percent of the Earth's surface is land whereas 70 percent is ocean. Most of the area in marine protected areas is in Great Barrier Reef Marine Park in Australia and Greenland National Park.

Table 2.4. World Heritage Sites in North America, November 2000

Canada
- 1978 L'Anse aux Meadows National Historic Site
- 1978 Nahanni National Park
- 1979 Dinosaur Provincial Park
- 1981 SGaang Gwaii (Anthony Island)
- 1981 Head-Smashed-In Buffalo Jump
- 1983 Wood Buffalo National Park
- 1984 Canadian Rocky Mountain Parks
- 1985 Historic District of Québec
- 1987 Gros Morne National Park
- 1995 Old Town Lunenburg
- 1999 Miguasha Park

Canada and the United States
- 1979 Kluane/Wrangell-St. Elias/Glacier Bay/Tatshenshini-Alsek
- 1995 Waterton Glacier International Peace Park

Mexico
- 1987 Sian Ka'an
- 1987 Pre-Hispanic City and National Park of Palenque
- 1987 Historic Centre of Mexico City and Xochimilco
- 1987 Pre-Hispanic City of Teotihuacan
- 1987 Historic Centre of Oaxaca and Archaeological Site of Monte Alban
- 1987 Historic Centre of Puebla
- 1988 Historic Town of Guanajuato and Adjacent Mines
- 1988 Pre-Hispanic City of Chichen-Itza
- 1991 Historic Centre of Morelia
- 1992 El Tajin, Pre-Hispanic City
- 1993 Whale Sanctuary of El Vizcaino
- 1993 Historic Centre of Zacatecas
- 1993 Rock Paintings of the Sierra de San Francisco
- 1994 Earliest 16th-Century Monasteries on the Slopes of Popocatepetl
- 1996 Prehispanic Town of Uxmal
- 1996 Historic Monuments Zone of Querétaro
- 1997 Hospicio Cabañas, Guadalajara
- 1998 Historic Monuments Zone of Tlacotalpan
- 1998 Archeological Zone of Paquimé, Casas Grandes
- 1999 Historic Fortified Town of Campeche
- 1999 Archaeological Monuments Zone of Xochicalco

United States
- 1978 Mesa Verde
- 1978 Yellowstone
- 1979 Grand Canyon National Park
- 1979 Everglades National Park
- 1979 Independence Hall
- 1980 Redwood National Park
- 1981 Mammoth Cave National Park
- 1981 Olympic National Park
- 1982 Cahokia Mounds State Historic Site
- 1983 Great Smoky Mountains National Park
- 1983 La Fortaleza and San Juan Historic Site in Puerto Rico
- 1984 Statue of Liberty
- 1984 Yosemite National Park
- 1987 Monticello and University of Virginia in Charlottesville

1987 Chaco Culture National Historic Park
1987 Hawaii Volcanoes National Park
1992 Pueblo de Taos
1995 Carlsbad Caverns National Park and Glacier National Park

---

Others
1982 National History Park–Citadel, Sans-Souci, Ramiers (Haiti)

Source: UNESCO (2002). Courtesy of UNESCO World Heritage Center, whc.unesco.org.

Summary

National parks and protected areas harbor natural ecosystems and resources that provide multiple social, economic, and ecological values. These values include production of consumer goods and services, such as recreation, tourism, education, and research, and ecological goods and services, such as timber, oil, gas, purification of air and water, conservation of biodiversity, water supply, and others. While use and enjoyment of national parks and protected areas generates significant social and economic values, it has degraded the ecological integrity and ecological value of national parks and protected areas. This situation calls for the development of management strategies that balance the social, economic, and ecological values of national parks and protected areas.

The earliest protected areas were game reserves created for the benefit of ruling classes. India established the first protected forest areas and first recorded game laws. American wilderness preservationists like Ralph Waldo Emerson, Henry David Thoreau, and John Muir promoted conservation of wildlands and formation of protected areas. In 1872, the United States established Yellowstone National Park as "a public park or pleasuring ground for the benefit and enjoyment of the people." Yellowstone is the world's first national park.

National parks and protected areas have a wide range of functions and purposes including providing habitat for threatened and endangered species (biodiversity); purifying air and water; supporting natural resource-based subsistence lifestyles of indigenous peoples; providing opportunities for recreation, education, and scientific research; and many others.

The International Union for the Conservation of Nature and Natural Resources (IUCN-World Conservation Union) established seven categories of protected areas and one to three management objectives for each category. The world has 30,350 protected areas covering more than 13.2 million km^2 (5 million mi^2), and North America has 6,714 protected areas covering 4 million km^2 (1.5 million mi^2). National parks make up a disproportionate share of the total acreage in terrestrial protected areas (30 percent worldwide and 40 percent in North America). Global designations of protected areas include Biosphere Reserves, World Heritage Sites, and Ramsar wetlands. Marine protected areas cover proportionately less area than terrestrial protected areas (1 percent vs. 9 percent).

References

Barbier, E. B., J. C. Burgess, and C. Folke. 1995. *Paradise lost? The ecological economics of biodiversity,* 3–41. London, England: Earthscan Publications, Ltd.

Conference of the Contracting Parties. 1996. Resolutions of the 6th Meeting of the Conference of the Contracting Parties, Ramsar Convention, Brisbane, Australia, March 19–27. http://www.ramsar.org/(accessed April 9, 2004).

Daily, G. C. 1997. Introduction: What are ecosystem services. In *Nature's services: Societal dependence on natural ecosystems,* ed. G. C. Daily, 3–6. Washington, DC: Island Press.

Davey, A. G. 1998. Protected areas categories and management objectives (appendix 2). *National system planning for protected areas,* 49–55. Gland, Switzerland, and Cambridge, UK: IUCN.

Ehrlich, G. 2000. *John Muir: Nature's visionary.* Washington, DC: National Geographic Society.

Glowka, L., F. Burhenne-Guilmin, H. Synge, J. A. McNeely, and L. Gundling. 1994. *A guide to the Convention on Biological Diversity.* Environmental Policy and Law Paper No. 30, IUCN Environmental Law Centre, Gland, Switzerland.

Green, M. J. B., and J. R. Paine. 1997. *State of the world's protected areas at the end of the twentieth century.* Paper presented at Protected Areas in the 21st Century: From Islands to Networks. Albany, Australia, November 24–27.

Howarth, W. (commentor). 1982. *Thoreau in the mountains: Writings by Henry David Thoreau.* Toronto, Canada: McGraw Hill Ryerson Ltd.

Howe, J., E. McMahon, and L. Propst. 1997. *Balancing nature and commerce in gateway communities.* Washington, DC: Island Press.

IUCN. 1978. *Categories, objectives, and criteria for protected areas.* Commission in National Parks and Protected Areas, Committee on Criteria and Nomenclature. Gland, Switzerland.

———. 1984. Categories, objectives, and criteria for protected areas. In *National parks, conservation, and development: The role of protected areas in sustaining society,* ed. J. A. McNeely and K. R. Miller. Washington, DC: Smithsonian Institution Press.

———. 1994. Guidelines for protected area management categories. IUCN Commission on National Parks and Protected Areas with assistance of the World Conservation Monitoring Centre. Gland, Switzerland, and Cambridge, UK.

IUCN-World Conservation Union. 2003. *The program.* http://www.iucn.org/ourwork/programme.htm (accessed April 9, 2003).

Langford, N. P. 1972. *The discovery of Yellowstone National Park.* Lincoln: University of Nebraska Press; reprint of 1905 ed. printed by the Haynes Foundation, St. Paul, MN.

Limerick, P. N. 1988. *The legacy of conquest: The unbroken past of the American West.* New York: W.W. Norton & Company.

National Park Service. 2001. *Public Use Statistics Office.* http://www2.nature.nps.gov/stats/ (accessed April 2, 2001).

O'Brien, B. R. 1999. *Our national parks and the search for sustainability.* Austin: University of Texas Press.

Oelschlaeger, M. 1991. *The idea of wilderness.* New Haven: Yale University Press.

Ramsar Convention on Wetlands. 2001. http://www.ramsar.org/ (accessed June 11, 2001).

Runte, A. 1979. *National parks: The American experience.* Lincoln: University of Nebraska Press.

Schullery, P. 1997. *Searching for Yellowstone: Ecology and wonder in the last wilderness.* New York: Houghton Mifflin Co.

Sellars, R. W. 2000. Creating tradition: The roots of national park management. *Wild Earth* 10: 27–32.

Talbot, L. M. 1984. The role of protected areas in implementation of the world conservation strategy. In *National parks, conservation, and development: The role of protected areas in sustaining society,* ed. J. A. McNeely and K. R. Miller. Washington, DC: Smithsonian Institution Press.

U.S. Census Bureau. 2001. *Population Estimates Program, Population Division.* http://www.census.gov/population/estimates/nation/intfile1-1.txt (accessed April 2, 2001).

U.S. Department of Commerce. 2002. Bureau of Economic Analysis (Economagic.com): http://www.Economagic.com/em-cgi/data.exe/fedstl/dspi+1) (accessed April 2, 2001).

UNESCO. 2001. *Man and the Biosphere Program.* MAB Biosphere Reserve Directory-United States of America. http://www.unesco.org/mab/br/brdir/europe-n/USAmap.htm (accessed June 15, 2001).

———. 2003. MAB Programme, World Network of Biosphere Reserves. http://www.unesco.org/mab/brlist.htm (accessed April 13, 2004).

UNESCO World Heritage Centre. 2002. *World Heritage List.* http://www.unesco.org/whc/heritage.htm (accessed December 8, 2002).

United Nations System-Wide Earthwatch. 2001. Background profile for the second report on International Scientific Advisory Processes on the Environment and Sustainable Development. http://ioc.unesco.org/igospartners/scpRamsar.htm (accessed June 11, 2001).

World Commission on Protected Areas (WCPA). Undated a. *About protected areas.* http://wcpa.iucn.org/wcpainfo/protectedareas.html (accessed June 11, 2001).

———. Undated b. *WCPA marine program.* http://wcpa.iucn.org/theme/marine/marine.html (accessed June 11, 2001).

Wright R. G., and D. J. Mattson. 1996. The origin and purpose of national parks and protected areas. In *National parks and protected areas: Their role in environmental protection,* ed. R. Wright Gerald, 3–14. Cambridge, MA: Blackwell Science.

CHAPTER

3

Cultural Values

Many of the environmental challenges we face today—loss of biodiversity, global climate change, depletion of renewable resources, and others—are rooted and maintained in culture.

—Frances Westley et al.

Culture shapes human attitudes and behaviors toward natural resource conservation and the development and management of protected areas. Similarly, human impacts on natural landscapes influence human behaviors and public policies toward natural resource conservation and protected areas. Relationships between human culture and the natural environment and attitudes and behaviors toward protected areas in particular vary over space and time. If the nature of these relationships is well understood and reflected in the design and management of protected areas, then public support for protected areas is likely to be strong. On the other hand, if the nature of these relationships is poorly understood, then protected areas are likely to be mismanaged and public support weak. This chapter discusses the nature of the relationships between culture and protected area management and how knowledge of those relationships can be used in the management of protected areas.

What Is Culture?

An understanding of culture is important for protected area management because it molds the relationship between humans and the natural environment. Furthermore, individual and social attitudes and behaviors toward the natural world are transmitted through culture (Furze et al. 1996). Culture is the product of complex interactions among ethnicity or race, language/dialect, religion, political ideology, natural resource endowments, social mobility, idioms, and local traditions (Furze et al. 1996; Grotzbach and Stadel 1997). Each element of culture can be further subdivided. For example, the mountainous regions of Asia contain several religious groups including Animist, Buddhist, Christian, Hindu, Muslim, and Sikh. The Alps contain people whose native language is German, French, or

Italian. German-speaking residents are further distinguished as Swiss, Tyroleans, Bavarians, and others (Grotzbach and Stadel 1997).

Human attitudes and behaviors are driven by cultural values. For example, countries that aspired to dominate the world, such as Assyria, the Roman Empire, and the former Soviet Union, were aggressive toward other nations. Losses in innocent lives and environmental quality resulting from such aggression were viewed as unavoidable collateral damage. Such losses are common in time of war. Resource degradation and associated losses in biodiversity are caused by other human activities, such as extensive clear-cutting in the Amazon basin of South America and the Pacific Northwest of the United States.

Cultural diversity refers to variability in the elements of a culture. For example, America is culturally diverse and its cultural diversity varies from one part of the country to another. The dominant ethnic heritage of Americans is European, particularly northern European, British, German, Scandinavian, French, Dutch, and Irish, and the culture contains several minority groups, namely African-Americans (12.1 percent of the total population), Latinos (9 percent), Asian Americans (3 percent), and Native Americans (less than 1 percent) (deBuys et al. 1999). The concentration of African Americans is greater in the southeast than in the northwest United States.

Cultural Diversity and Biodiversity

Preservation of cultural diversity and conservation of biodiversity is closely related (McNeely 1995). The Keystone Dialogue on Biological Diversity and the Global Biodiversity Strategy both subsume human cultural diversity in biodiversity. The Global Biodiversity Strategy recognizes that nomadism and shifting cultivation represent a cultural adaptation to environmental conditions that favors survival (WRI, IUCN, UNEP 1992). Cultural adaptation is an aspect of cultural diversity that refers to the human responses to environmental (relief, exposure, and altitude) and economic (prices of land, labor, and capital) conditions. For example, vertical arrangement of land parcels devoted to crop/livestock production and human settlements is very common in mountainous regions having traditional cultures (Grotzbach and Stadel 1997). Farming in steep mountainous areas of China and other parts of the world is done in terraces in order to reduce soil erosion and maintain the land's long-term capacity to produce food and fiber. Terraced farming is labor-intensive and capital extensive in accordance with the low cost of labor relative to capital. In contrast, terraces are not used in the grain-producing northern Great Plains of the United States due to the relatively flat terrain and dry conditions. Production methods in this region are capital intensive rather than labor intensive due to the high cost of labor relative to capital and economies of size. Part of the variability in farming practices is explained by ethnic/religious background. The religion of the Amish culture in the United States prohibits the use of fuel-powered farm machinery and transportation, and electricity.

Cultural adaptation varies within the same region. In the western United States, which contains the bulk of the nation's national park acreage, valleys are used for housing, commercial and industrial development, and transportation networks (highways, rail lines, airports, and pipelines). Upland areas are devoted primarily to timber harvesting, fish and wildlife habitat, livestock grazing, outdoor recreation, protected areas, and dispersed residential housing.

Incorporating cultural values in protected area management rests on the premise that "cultures do not develop independent of ecosystems, nor do most ecosystems exist independent of human influence," or equivalently that "human, cultural and social systems are parts of the ecosystems in which they occur" (deBuys et al. 1999). This widely held position (Booth and Kessler 1996; Bonnicksen et al. 1999) underlies the "archeology of place," which implies "places are, to a significant degree, products of culture and society" and "human societies and culture are, to a degree, products of the places in which they develop" (deBuys et al. 1999). Interdependence between natural areas and culture suggests that local communities should be involved in management decisions for protected areas, especially those decisions having potentially large socioeconomic and environmental impacts.

One approach for involving local communities is collaborative decision making in which "stakeholders [become involved] in forums designed to expose concerns, build relationships, establish trust, and encourage creative problem solving" (Yaffee and Wondolleck 1997). This point is corroborated by Halvorson (1996), who notes that "management of landscapes that will provide for the long-term well-being of both natural and human populations requires the cooperative efforts of all who live in and manage that landscape," and "failure to develop protected landscapes that safeguard natural biodiversity and provide for conservation of natural resources, while at the same time encouraging harmony between humans and nature, may allow development and resource exploitation to overrun natural areas." The most significant finding of the President's Commission on Sustainable Development was "the potential power of and growing desire for decision processes that promote direct and meaningful interaction involving people in decisions that affect them" (Bruzaaelli and Lash 1996). Therefore, a key element in achieving harmony between humans and nature in protected area management is to ensure that cultural issues are addressed in all phases from site selection and delineation to management.

There are many examples of what happens when the values embodied in local culture are ignored in establishing and managing protected areas. Kluane National Park in northern Canada is a case in point. The park contains Canada's highest mountains, the largest nonpolar ice field in the world, and large and significant populations of big game and other species. Kluane became a park reserve in 1942 and a game sanctuary in 1943 without consulting local people, big-game hunters, and subsistence resource users. The park's first management plan, developed in 1980, disappointed local groups for not delivering on its promise to provide road and tram access. In contrast, a 1990 revision of the management plan involved extensive public consultation and gave adequate consideration to local desires, including access (Parks Canada 2002). Several U.S. national parks have been established in areas that are culturally significant to Native Americans. While park managers have attempted to mitigate negative cultural impacts of protected area designations, tensions still remain between Native Americans and park managers.

National environmental groups have expressed concern that placing too much emphasis on cultural values could lead to management decisions that erode ecological values, particularly when the two are in conflict. A case in point is the National Park Service's initial decision to phase out snow machine use in Yellowstone and Grand Teton National Parks by the winter of 2003–2004. This decision reflects the high priority on preserving natural resources in national park units having outstanding natural values. In this case, the ban is expected to protect wildlife, reduce air and noise pollution, and favor more environmentally benign recreation, such as snowshoeing and cross-country skiing. The "snow machine culture" and local businesses strongly oppose the ban. Snow machine users contend the ban infringes on their right of access to national parks and the potential wildlife benefits of the ban are uncertain. The snowmobile industry has sued the National Park Service over this decision. Businesses oppose the ban for economic reasons. One business claims the ban would reduce demand for its services by 30 percent (Long 2001). This conflict pits the "snow machine culture" against the "preservation culture." A similar conflict exists in Denali National Park and Preserve in Alaska. Whether or not the next generation of snow machines significantly reduces the noise and air pollution has an important bearing on the resolution of this conflict

Cultural Dynamics

Culture and associated resource conservation attitudes and behaviors vary over space and time. An example of spatial variability is the cultural differences between Switzerland and the former Soviet Union. Switzerland is a politically neutral country that has hosted numerous meetings that have resulted in international agreements, such as the Geneva Convention for the protection of victims of war adopted in 1949. In addition, Switzerland places a high value on environmental protection. In contrast, the former Soviet Union was militarily aggressive and demonstrated little regard for environmental quality. Consequently, many former Soviet Bloc countries, particularly those in Eastern Europe, have serious air, water, and soil pollution problems.

Temporal variability in culture impacts natural resource use and conservation. Consider the dynamic interaction between culture and landscape conservation in North America. Prior to human habitation of North America, geologic, climatic, and biological processes were the principal drivers of landscape dynamics. Human habitation of North America commenced about 12,000 years ago. During the prehistoric period (prior to 1492), Native Americans intentionally started fires in order to change the size and spatial distribution of animal and plant species and other attributes of the landscape in ways that favored human survival. In particular, fire was used to control plant diseases and insect infestations, increase the frequency and range of more desirable plant species, enhance desirable characteristics of plants, minimize the severity and number of uncontrollable wildfires, and improve hunting conditions by reducing undergrowth (Lewis 1993). Fires prevented certain plant species from reaching late seral stages, which increased the frequency of the more desirable early seral stages. Other manipulations included sowing and broadcasting of seed, transplanting shrubs and small trees, pruning and coppicing of plant communities, and installing water diversion structures to control soil erosion (Blackburn and Anderson 1993; Periman et al. 1999).

After European settlement of North America, landscape manipulation continued with the adoption of large-volume harvesting and winter storage of resources in the Northwest (Shalk 1986), construction of fields, large-scale wood harvesting for fuel and construction, and development of earthworks and water distribution canals (Periman et al. 1999). Many other technologies were developed that increased the capacity of human cultures to manipulate their environment and dramatically increase the productivity of managed ecosystems. Examples include progressive development of energy sources from wood to coal to petroleum, conversion from draft animals to tractors in agriculture, and technologies like the steam engine, moldboard plow, cotton gin, and internal combustion engine.

Cultural Values and Natural Resource Use

Use and value of natural resources can only be understood within the context of a particular culture because landscapes are expressions of culture and society and cultures and societies are influenced by the landscapes in which they operate (deBuys et al. 1999). Furze et al. (1996) indicate that "our relationship to the natural world is by definition a social one" and that "our relationship to nature is socially constructed and socially patterned." Consider how U.S. public land management and environmental policy coevolved with attitudes, awareness, and understanding regarding the impacts of economic development on natural resources. During the period of public land disposition (1862–1891), the federal government transferred vast tracts of public land to private and commercial interests in order to stimulate economic development. Creation of the world's first national park (Yellowstone) in 1872 was unopposed because it provided a spectacular setting for public recreation and enjoyment without locking up mineral and energy resources needed for economic development. In other words, Yellowstone National Park was expected to stimulate economic development in the region, which is precisely what happened.

Several events intensified public support for natural resource conservation. George Perkins Marsh (1801–1882), John Muir (1838–1914), President Theodore Roosevelt (1858–1919), Aldo Leopold (1886–1948) and others warned about the negative impacts of population growth and human activity on natural resources. Public acceptance of these warnings coupled with cessation of the public land disposition programs ushered in the period of public land reservation (1892–1934) and the subsequent period of public land retention and management (1935 to present) (Keiter et al. 1999). In the 1930s, a combination of farming systems ill suited for dryland farming and a drought resulted in the Dust Bowl, which was memorialized in John Steinbeck's *The Grapes of Wrath.* The Dust Bowl resulted in massive soil erosion on farmlands in the southwestern Great Plains region of the United States and caused tens of thousands of farmers to abandon their land. In an effort to stem recurrence of a dust bowl, the U.S. Department of Agriculture created the Soil Erosion Service and charged it with developing and implementing programs and policies to reduce excessive soil erosion on farmland and other areas. The Soil Erosion Service evolved into the Soil Conservation Service. The Soil Conservation Service was renamed the Natural Resources Conservation Service.

Beginning in the early 1960s, several writers and their works riveted public attention on the negative environmental impacts of economic development. Influential works include Carson's *Silent Spring* (1962), Meadows and others' *Limits to Growth* (1972) and *Beyond the Limits* (1992), Barney's *Global 2000 Report* (1980), Ehrlich and Ehrlich's *Healing the Planet* (1991), and Terbourgh's *Requiem for Nature* (2000). *Limits to Growth* argued that economic growth is constrained by the availability of natural resources and the waste assimilative capacity of the environment. Other works argued that technological advancement would be more than sufficient to increase the productivity of labor and human-made capital and control pollution. These include Kahn's *The Next 200 Years* (1976), Simon and Kahn's *The Resourceful Earth* (1984), and Repetto's *The Global Possible* (1986).

The limits to growth debate continues to this day as evidence mounts that economic growth is taking a toll on natural capital. For example, Wilson argues that the "rate of extinction is now about 400 times that recorded through recent geological time and is accelerating rapidly" (Wilson 1985) and that extant species are vanishing by as much as 10,000 times faster than new ones are born (Wilson 2002). Annual extinction rates in the tropics over the next several decades have been estimated at 10,000 to 150,000 (Wilson 1988; Diamond 1990). Natural communities in the United States are being destroyed at a faster rate than occurs in tropical rain forests. Cases in point include freshwater habitats in California (Moyle and Williams 1990) and old growth forests in the Pacific Northwest (Norse 1990).

Starting in the late 1960s, Americans and Congress became alarmed over the adverse impacts of economic development on natural resources and the environment. Numerous environmental laws were enacted that reinforced sociocultural expectations regarding natural resource conservation and environmental protection. These laws ranged from legal requirements to evaluate the environmental impacts of major federal actions (National Environmental Policy Act) to protection of particular resources (Clean Water Act, Clean Air Act, and Endangered Species Act). For example, the National Environmental Policy Act requires preparation of an environmental impact statement when resource management plans are developed for public lands, such as national forests, national parks, and wilderness areas. Acceleration of species extinction due to human-caused global change and urbanization have propelled protected areas into the forefront as biologically significant refugia for threatened and endangered species, led to the Endangered Species Act, and increased scientific interest in using protected areas as benchmarks for measuring biophysical responses to global climate change. Therefore, in addition to their cultural value in tourism and outdoor recreation, protected areas are very important for preserving biodiversity and understanding the impacts of global change.

Paralleling the development of environmental policy has been the proliferation of environmental organizations that lobby for improved environmental conditions and the emergence of new academic disciplines that expand knowledge and understanding of adverse environmental impacts of human activity. In the 1980s, the discipline of conservation biology merged multiple natural and social science disciplines for the purpose of conserving biological diversity worldwide (Meffe and Carroll 2004). In the 1990s, the transboundary discipline of ecological economics took up the banner for advancing knowledge regarding the complex relationships between natural and economic systems and public policies for preserving natural capital. The 1992 Rio Earth Summit examined the relationships between global economic and ecological systems and championed sustainable development (United Nations Conference on Environment and Development 2004). The most noteworthy product of the Rio Earth Summit is Agenda 21.

Sacred Places and Spiritual Values

Many protected areas contain sacred places that sustain spiritual and religious values. Such values are important due to their deep-seated nature, laws protecting freedom of religion, and strong desire to increase all aspects of cultural diversity. Sigurd Olson (undated) felt very strongly about the spiritual values of national parks. He noted: "The National Parks, I feel, have an overriding purpose, and that can be encompassed in one word—spiritual. Spiritual values are values that affect your emotions, that affect your happiness, that affect your culture. They are hard to define, hard to pinpoint, but they are there....no matter how you look at it, it is the spiritual value of the National Park that the visitor carries with him." Mountain landscapes, perhaps more than any other, epitomize spiritual values by invoking a sense of wonder, awe, and power in native and contemporary cultures. The International NGO Consultation on the Mountain Agenda (1995) observed that "the sacred values of mountains are of utmost sensitivity to people [having certain] cultural and religious traditions." Many outstanding mountain protected areas around the world convey deep spiritual meanings.

Maintaining healthy relationships between people and sacred landscapes contributes to social cohesion and vitality, whereas destroying or compromising those landscapes undermines cultural values (Bernbaum 1997). Additionally, deBuys et al. point out, "American society in general must accept the embeddedness of indigenous religion in culture and its connections to a wide range of landscapes and places" (deBuys et al. 1999). There are many examples of mountains as temples or places of worship. Mount Sinai is a sacred mountain because it is where God gave Moses the Ten Commandments. The Zuni in North America allow only a few worthy individuals to enter their sacred places and have knowledge of sacred beliefs and practices (Bernbaum 1997). John Muir (1838–1914), who believed that God's existence can be deduced from the beauty and design of nature (transcendentalism), held that humans are an empathetic part of nature rather than a scientific observer of it and that all living things are sacred (Oelschlaeger 1991). Naturalist John Muir adamantly supported protecting Yosemite Valley and lobbied vociferously for the creation of Yosemite National Park and five other national parks. Muir strongly opposed the construction of a dam that would flood Hetch Hetchy Valley in order to supply water to San Francisco and was greatly dismayed when the dam was constructed.

Several national park units in the United States contain features that are spiritually significant to Native Americans. In describing the sacredness of the Rocky Mountains to the Stoney Indians, Chief Joseph exclaimed, "these mountains are our temples, our sanctuaries, and our resting places. They are a place of hope, a place of vision, a very special and holy place where the Great Spirit speaks to us. Therefore these mountains are our sacred places" (Snow 1977). Glacier National Park in Montana contains several sacred places, including Chief Mountain and Two Medicine Campground. These places are sacred and spiritually significant to the Blackfeet, Kootenai, and Flathead Indians (Hanna 1987). Blackfeet and Flathead Indians climbed to the 9,000-foot summit of Chief Mountain in vision quests, a right of passage in which young braves enter the wilderness to pray and fast with the hope of receiving power from the Spirit World. Kootenai and Stoney Indians frequented the northwestern sections of what became Glacier National Park to camp, fish, and hunt. Legend has it that one year the Pikuni tribe and Kai'na or Blood tribe of the Blackfeet erected medicine lodges next to one another in an area and lake that became known as Two Medicine Campground and Two Medicine Lake, respectively (Hanna 1987). When George Bird Grinnell visited the Swiftcurrent Valley on the west side of the Glacier National Park in 1887, he discovered evidence that the Blackfeet constructed trails and hunting camps in the valley (Schultz 1962). In contemporary times, the cultures of the park and Blackfeet Nation have come into conflict. For instance, cattle owned by the Blackfeet, whose reservation adjoins the eastern side of the park, cross into the park to graze in the Aspen parklands.

Native Americans inhabited the Yellowstone region of northwestern Wyoming for at least 7,500 years, including the Assiniboine, Bannock, Blackfeet, Crow, Flathead, Kalispel, Nez Perce, Pend d'Oreille, and Shoshone. A Lakota Baha'i by the name of Locke pointed out that "the pre-eminence of the Yellowstone basin as a site of particular spiritual potency invoking awe, wonderment and spiritual upliftment for thousands of years is indisputable" (cited in Weixelman 2001). In other words, Native Americans believed Yellowstone was suffused with the Creator's powers. American Indians used Yellowstone's geyser basins for praying, meditation, and bathing (in hot springs). Mud from mudpots was used to clean and purify skin. The Bannock and Shoshone did vision questing in the Yellowstone region. Indigenous use of the Yellowstone region was systematically terminated after Yellowstone National Park was established in 1872. In 1883, a council ceded the mountainous northern region of the park, which had been Crow territory, and relocated the tribal agency outside the park. In *Ward v. Race Horse* (1896), the Supreme Court terminated Native Americans' hunting rights in the Yellowstone region. During the army's management of the park (1886–1918), certain tribes were ordered to stay out of the park (Weixelman 2001).

Creation of national parks is by no means the only factor that disrupted spiritual and cultural traditions, lifestyles, and the economy of Native Americans. European settlers engaged in behavior that dramatically reduced the capacity of Native Americans to coexist with and derive sustenance from the physical environment. Prominent examples include near extermination of the 60-million strong bison herds that once roamed the Great Plains, trapping fur-bearing animals for the fur trade, adoption of the horse and firearms, farming, removal of Indians from homelands, and forced relocation of Indians to reservations. Introduction of alcoholic beverages and transmission of diseases from settlers to Indians had a debilitating and devastating impact on Indian populations and culture. While most of these impacts were irreversible, considerable effort has been made to provide restitution for some of the losses incurred by Native Americans as a result of European settlement.

The spiritual significance of protected areas is not limited to indigenous peoples. As countries become more urbanized, materialistic, and competitive, people's lives become more tense, hectic, and stressful. Protected areas counterbalance contemporary lifestyles by providing outstanding opportunities for spiritual and physical renewal in the form of solace, tranquility, inspiration, clean air and water, and physically challenging activities. In support of this point, Bernbaum (1997) observes, "the inspirational value of mountains, in particular, has played a vital role in the establishment of national parks and is one of the most effective tools for the conservation of wilderness areas."

Maintaining the spiritual values of protected areas poses a quandary. Increases in tourism and recreation that accompany the creation of a protected area can cause social and ecological carrying capacities of those areas to be exceeded. When this occurs, protected areas are less able to support spiritual values. For example, significant overcrowding occurred in Yosemite Valley and Grand Canyon National Parks before the National Park Service developed a plan to transport the bulk of visitors to the south rim of the Grand Canyon via light rail and to Yosemite Valley in Yosemite National Park via buses (Cahalan 2001; Clarke 2001). Reducing overcrowding is expected to enhance a protected area's capacity to provide spiritual values. Another place where spiritual values are being threatened by recreation is in Devil's Tower National Monument, an 867-foot tall volcanic core in Wyoming. Out of respect for the Native American belief that the tower is sacred and that June is one of the holiest months, the National Park Service instituted a voluntary suspension of climbing in June, which has been in effect since 1996. Commercial climbers and the Bear Lodge Multiple Use Association reacted negatively to the suspension by suing the National Park Service on grounds that the climbing suspension violated the constitutional separation of church and state (deBuys et al. 1999). In 1999, the Tenth Circuit Court of Appeals issued a finding that stated, in part, that "because they have alleged no injury as a result of their claim the Final Climbing Management Plan improperly establishes religion, we hold the Climbers have no standing to sue in this case." The Court of Appeals' decision was appealed to the U.S. Supreme Court, which refused to hear the case. While designating an area as protected is likely to increase visitation and thereby raise the risk of exceeding carrying capacities and reducing spiritual values, not bestowing protected status and allowing economic development and resource extraction to occur in pristine natural areas is likely to permanently reduce spiritual, cultural, and ecological values. In light of this predicament, park managers must be diligent to designate and manage protected areas in a manner that sustains spiritual and other cultural values.

Cultural Landscapes

A cultural landscape is an integrative construct based on the coevolutionary relationship between sociocultural values and the human use of landscapes. A cultural landscape is the result of culture operating on a natural landscape (Sauer 1963) and reflects a place's historical, cultural, and ecological values (Mitchell 1997). In 1972, the World Heritage Commission, which implements the World Heritage Convention, revised the convention's guidelines to affirm that cultural landscapes have "outstanding universal value" that are distinct from the scientific values and scenic qualities of natural landscapes. The commission defined a cultural landscape as "a diversity of manifestations of the interaction between humankind [or culture] and its natural environment." The commission recognized that cultural landscapes reflect the "combined works of nature and of man...[and] are illustrative of the evolution of human society and settlement over time, under the influence of the physical constraints and/or opportunities presented by their natural environment and of successive social, economic and cultural forces, both external and internal." The commission recognized the synergistic relationship between cultural landscapes and natural values in noting, "protection of cultural landscapes can... maintain or enhance natural values in the landscape. The continued existence of traditional forms of land-use supports biological diversity in many regions of the world. The protection of traditional cultural landscapes is therefore helpful in maintaining biological diversity" (Mitchell and Buggy 2000).

The National Park Service employs a broad definition of cultural landscapes, namely "a geographic area including both natural and cultural resources and the wildlife or domestic animals therein, associated with a historic event, activity, or person or exhibiting other cultural or aesthetic values" (National Park Service 1994). Perhaps the best-known class of protected areas in terms of historical and cultural significance is national military battlefields. Examples of famous American military battlefields are Gettysburg in Pennsylvania and Manassas in Virginia and of famous national monuments are the Lincoln and Jefferson Memorials in Washington, DC. Even places valued primarily for their spectacular natural landscapes contain significant cultural resources. Mount Rainier National Park in Washington contains in excess of one hundred structures listed on the National Register of Historic Places of which eight are national historic monuments (Toothman 1996).

A cultural landscape is a useful social construct because (1) it draws attention to historical and cultural elements of landscapes, (2) a landscape's historical significance is valuable in understanding existing conditions and impacts of past land use on natural systems and informing future planning, (3) understanding landscape dynamics improves evaluation and management, and (4) it facilitates a more comprehensive approach to landscape preservation (Mitchell 1997).

Desire to protect cultural landscapes has prompted some public agencies to alter or even reverse their management strategies. A case in point is Boxley Valley, a traditional agricultural valley in Arkansas' Ozark Mountains. Boxley Valley is in the upper portion of the Buffalo National River, which is managed by the National Park Service. The valley contains 3,237 cultivated ha (8,000 cultivated acres) and two hundred structures. In 1982, the National Park Service purchased

75 percent of the private lands in the valley. This action caused many farms to be abandoned, neglected, or dismantled and the valley's population to drop precipitously. After recognizing the action resulted in the decline of a working landscape, the Park Service and public jointly developed a land use plan for Boxley Valley that called for transferring farmland back to private ownership while retaining easements that specified farm management practices, protection of historic structures and water quality, and limitations on visitor access to private property. The goal of the plan is to "provide the opportunity for continued traditional use of the landscape by the people who have distinctive associations with the land, while preserving the historic character of the valley" (Mitchell 1997).

Another shift in management philosophy occurred in a cattle-producing region that straddles the border between Arizona and New Mexico and is part of the Gila River Basin. Prior to 1998, grazing allotments issued by the U.S. Forest Service allowed ranchers in the region to graze their cattle along more than a dozen rivers resulting in severe degradation of riparian areas and water quality. In 1998, a legal settlement of lawsuits filed by the Center for Biological Diversity and Forest Guardians resulted in a legal agreement between the two environmental groups and the U.S. Forest Service. The agreement required ranchers to fence cattle off 152 km (230 miles) of rivers and streams (except for selected water access points for cattle), and in some cases, reduce their herds in an effort to restore fish and bird species. As many as 15,000 cattle were removed from the region causing significant economic losses for ranchers. The settlement constitutes one of the largest removals of cattle from public lands since passage of the Taylor Grazing Act. One of the environmental groups that won the lawsuits claimed that cessation of livestock grazing in riparian areas will eventually boost the region's populations of cuckoos, otters, songbirds, and jaguars (Davis 2001). In the case of Boxley Valley, the Park Service's new policy restored the working landscape that existed before they purchased area farms. In the case of the Gila River Basin, the Forest Service restricted grazing rights to public lands in order to restore fish and wildlife habitat and populations. Both shifts in policy connected people and the natural environment in a new way and redefined the cultural landscape.

It should come as no surprise that traditional and cultural uses of landscapes often conflict with the desire to preserve natural landscapes. When such conflicts involve public land, which is often the case for protected and multiple-use areas, the solution is often determined by the relative economic and political power of parties to the conflict. In the case of Boxley Valley, the purchase of private land by the National Park Service had negative consequences, namely community resentment, declining resident population, and reduced economic activity. The economic stress and political agitation caused by this action impelled the Park Service to work collaboratively with the local community in developing a plan to transfer public land back to private ownership, albeit with caveats. In the case of livestock grazing in the Gila River Basin, the historically strong economic and political power of livestock interests (the region was a hotbed for the wise use movement) yielded to the successful efforts of environmental groups to reverse the decline in fish and wildlife habitat and populations.

The National Park Service has created several programs and policies to foster the preservation of cultural landscapes. In 1981, the Park Service established the Applied Ethnology Program to develop data that would facilitate consultation with traditional communities and identify policies for improving relationships between people and parks (deBuys et al. 1999). In developing management policies, the Park Service is committed to acting in ways that "safeguard cultural and natural resources while reflecting informed concern for the contemporary peoples and cultures traditionally associated with them" (National Park Service 1988). The Park Service's Ecosystem Management Working Group adopted an integrative collaborative approach to natural and cultural resource management, which rests on the premise that "preservation of cultural traditions and historic, archaeological, and other cultural places is a goal valued by all Americans" (National Park Service 1994).

Legislation and Court Decrees

Several laws require the National Park Service and other federal agencies to consider how their management actions and policies influence cultural resources and values, especially for traditional or subsistence cultures (see the sidebar). The most general statue in this regard is the National Environmental Policy Act of 1969 (42 U.S.C. §§ 4321 et seq.). This act requires federal agencies to consider the extent to which their actions significantly affect the quality of the human environment, which includes cultural resources. The National Historic Preservation Act of 1966 as amended (16 U.S.C. §§ 470 et seq.) requires the Park Service to "inventory, evaluate, and protect the abundance of prehistoric and historic cultural properties within their boundaries" (Toothman 1996). Park Service programs that facilitate identification and protection of cultural resources include the Archaeological Inventory Program, which implements the requirements of the Archaeological Resources Protection Act of 1979 (16 U.S.C. §§ 470aa–470mm), List of Classified Structures, and Cultural Landscapes Inventory (Toothman 1996).

Traditional or subsistence cultures are highly dependent on natural resources "essential to the survival of family, community, or regional cultural systems" (National Park Service 1988). The most noteworthy law in this regard is the Alaska National Interest Lands Conservation Act of 1980 (16 U.S.C. §§ 410hh 1 to 410hh 5, et seq.), which contains model provisions for preserving subsistence resource uses in national parks and preserves (Wright 1996). Certain statutes are designed to reduce the impacts of protected areas on Native Americans including

Subsistence Cultures and National Parks

Most national parks in North America are embedded in a formal economy in which people use earned wages to purchase goods and services in the marketplace. National parks located in formal economies typically prohibit resource extraction in the form of hunting, trapping, and harvesting of trees and plants, although fishing is typically allowed. In contrast, many national parks and protected areas located in the circumpolar biome (characterized by northern coniferous forest or taiga) are located in mixed economies. In a mixed economy, a formal economy and informal or subsistence economy "co-exist and function in mutually beneficial ways." Subsistence economies often include First Nation or indigenous people who live off the land and directly exchange or barter food, clothing, shelter, and other necessities with one another. People in subsistence cultures typically have part-time jobs that generate cash to buy products that are not provided by subsistence economies.

While subsistence cultures in the taiga are often considered impoverished, many people are attracted to the taiga because of its cooperative social system. Berger describes the allure and deep satisfaction of subsistence economies as follows:

> This intricate mixture of cultural activities, co-operation, and economic enterprise provides a profound sense of belonging to the land. It offers the ultimate connection with nature, a deep sense of membership in the surrounding ecological community—as a participating citizen. Cultural fulfillment comes in ways that no other lifestyle can produce. Life within a subsistence economy allows a practice of values, a deep sense of community, a spirit of receiving and sharing, a quality of health and vigor, and a constantly varying mixture of activity and challenges.

The informal economy is economically significant. Foods harvested from nature (country foods) contributed more than $10 million to the Yukon economy in 2000. Sweden's informal economy involves the exchange of wild meats, garden produce, homes, heating systems, heating fuels, crafts, and technical skills.

Since access to natural resources is vital to subsistence cultures, it is important that protected area managers take account of the unique resource needs of subsistence cultures in establishing protected areas and approving management plans. As explained in this chapter, the first management plan for Canada's Kluane National Park had to be revised because it did not consider the needs of local people including subsistence cultures. The subsistence needs of First

Nations were given explicit consideration in establishing Denali National Park and Preserve in Alaska. In particular, the Alaska National Interest Lands Conservation Act that established the park allows local rural residents to utilize park resources for customary and traditional subsistence purposes as long as it does not degrade the health of fish and wildlife resources.

In summary, accounting for the unique needs of nearby subsistence cultures in managing national parks not only reduces friction between the park and those cultures, but also provides an opportunity for national parks to support more diverse cultural values.

Sources: Berger (1985); Henry (2002); U.S. Department of the Interior (undated).

the American Indian Religious Freedom Act (42 U.S.C. §§ 1996 et seq.) and the Native American Graves Protection and Repatriation Act (25 U.S.C. §§ 3001 et seq.).

Numerous court cases have favored Native American rights to utilize natural resources on and off the reservations. For example, a turning point in recovering salmon in the Columbia River System of the Pacific Northwest Region of the United States came in the late 1970s when Indian tribes won a lawsuit that allowed them to exercise fishing rights stipulated in prior treaties. One consequence of this legal victory is the establishment in 1982 of the Columbia River Basin Fish and Wildlife Program. Prior to the program, generation of hydroelectric power was deemed a more important value of the river system than fish and wildlife. As a result of the program, fish and wildlife were given "equitable treatment" relative to hydropower (Dietrich 1995). This and other conflicts in the Columbia River System are essentially a search for a "culturally, economically and ecologically viable relationship between people and the environment they inhabit" (Lee 1995). In other words, sustainable management of cultural landscapes requires balancing competing interests, a task complicated by the dynamic nature of the interactions among sociocultural, economic, political, and ecological conditions.

Protected Areas and Cultural Heritage

The preservation of cultural landscapes has motivated the establishment of protected areas and organizations dedicated to conserving cultural heritage on a worldwide basis. Under the authority of the Convention Concerning the Protection of World Cultural and Natural Heritage of 1972, the World Heritage Commission authorizes World Heritage Sites. Cultural sites include a monument;

group of buildings; or site of historical, aesthetic, archaeological, scientific, ethnological, or anthropological value. Natural sites have outstanding physical, biological, and geological features and encompass habitats for threatened plants or animal species and areas exhibiting scientific, aesthetic, or conservation value. Mixed sites have a combination of significant cultural and natural features. Of the 690 World Heritage Sites that existed in November 2000, 529 were cultural, 138 were natural, and 23 were mixed (UNESCO 2000).

Another way protected area managers preserve cultural heritage is to pay attention to cultural resources in delineating protected areas and developing management plans for these areas. A prime U.S. example of this approach is the preservation of subsistence lifestyles in Denali National Park and Preserve in Alaska. For the past 11,000 years, five Athabaskan cultural groups and other Alaska natives depended on the area in and near the park and preserve for traditional subsistence uses, including hunting or trapping caribou, sheep, moose, bear, and beaver for income, meat, and clothing; gathering berries, edible plants, and firewood; using sod and moss for home insulation; cutting logs for homes; collecting bark for baskets, dyes, and handicrafts; gardening; fishing; and others. The Alaska National Interest Lands Conservation Act (ANILCA) of 1980 afforded protection for about 66 million ha (100 million acres) of land and resources in Alaska and added 1.6 million ha (4 million acres) to Mt. McKinley National Park, tripling its size. The expanded area was renamed Denali National Park and Preserve. ANILCA allows local rural residents to use the 66 million ha (100 million acres) for customary and traditional subsistence purposes provided fish and wildlife resources remain in a natural and healthy state (United States Department of the Interior, undated).

Based on Section 203(3) of ANILCA, the National Park Service's Organic Act, and the organic act for Denali National Park and Preserve, the Park Service has adopted the following mission statement related to subsistence uses in the park and preserve:

- Protect the opportunity for qualified local rural residents to continue traditional subsistence activities;
- Recognize that subsistence ways of life differ from region to region and are continuing to evolve, and where appropriate, park management practices may reflect regional diversity and evolution;
- Promote local involvement and participation in processes associated with subsistence management;
- Ensure that management practices involving the utilization of public lands adequately consider the potential for restriction of subsistence uses and impacts upon subsistence resources;
- Ensure that management of park resources is consistent with the conservation of unimpaired ecosystems and natural and healthy populations of fish and

wildlife, incorporating scientific data and principles with traditional knowledge and cultural values; and
- Promote effective communication and mutual understanding of subsistence uses and related cultural and social values, and park purposes and protection, between the NPS, subsistence users, the State of Alaska and the public (Denali National Park and Preserve 1999)

While this mission statement is unique to Denali National Park and Preserve, it can be adapted to other protected areas.

Raish and others (1999) identified a shorter and more generic set of key tasks for incorporating diverse sociocultural values and viewpoints into natural resource management, namely:

1. Determine which groups are in the ecosystem.
2. Communicate with them to gather information on their values, traditions, and customary resource-use practices.
3. Incorporate their information into land management plans and policies.

Determining the inhabitants of an ecosystem might appear to be a simple task, but not necessarily so when the ecosystem is large and diverse. Furthermore, it is not only the inhabitants of a protected area ecosystem that need to be consulted, but also major stakeholders regardless of their residence. This is especially important for protected areas and/or natural resource issues having a national or international constituency, which is typically the case for national parks. Communication between managers and stakeholders is difficult when there are conflicting opinions regarding the uses and values of natural and cultural resources, differences in economic and political power, insufficient trans-cultural knowledge, and poor communication (Raish et al. 1999). Incorporation of information might seem like a minor task given rapid advancements in information technologies. However, the task of acquiring the requisite information from disparate stakeholders, integrating that information in a decision-making framework, and communicating the expected impacts of alternative management actions to stakeholders in an understandable and meaningful way is a major challenge.

Federal land management agencies have programs to both educate visitors about how various cultures use protected areas and increase visitation rates by culturally diverse groups. In particular, there is interest in overcoming the stigma that parks and other natural areas are playgrounds for white people. This concern is likely to increase in importance with further changes in the ethnic and cultural makeup of the population. As the U.S. Census of 2000 shows, groups formerly considered "minority" are now in the "majority" in certain regions of the country.

An important issue is the extent to which programs in units of the National Park System encourage participation and visibility of ethnic minorities. Based on surveys of nine park units, Roberts and Rodriguez (2001) determined that parks must "cast their nets more broadly to interest diverse communities," and improve their understanding of why parks are less appealing to ethnic minorities than whites, and to domestic than international visitors. Units in their study attempted to improve their responsiveness to ethnic minorities by hiring community liaison staff, employing advisory boards to actively seek input from ethnic minority communities, developing community partnerships, instituting interpretive and educational programs to raise awareness of the unit's national significance, and expanding opportunities for participation by all people. For example, many of the units had interpretive programs that explained the role of African Americans in the westward expansion of the United States, a topic often overlooked in historical accounts. Rocky Mountain National Park in Colorado has two successful programs that reach out to ethnically diverse students in Denver schools, namely Wonders in Nature and Wonders in Parks. The study concluded that the capacity of park units to expand outreach to underrepresented communities could be enhanced by increasing funding and staff levels, diversifying staff, educating staff about the needs of underrepresented communities, and providing ways for diverse ethnic groups to travel to the parks for pleasure or work.

Summary

Culture is the nexus of complex interactions among ethnicity or race, language/dialect, religion, political ideology, natural resource endowments, social mobility, idioms, and local traditions. Protected area managers must pay attention to culture because it influences individual attitudes and behaviors toward the natural world and how humans relate to the natural environment. In particular, protected area managers need to consider the relationships between human culture and natural systems in making resource management decisions. Cultural diversity, which refers to the variability in the attributes of a culture, is an important component of biodiversity. When cultural diversity diminishes, biodiversity decreases. Cultures adapt to environmental conditions in order to survive.

Since human culture and natural ecosystems are interdependent and coevolutionary, protected area managers need to make management decisions that contribute to the long-term well-being of local culture and the natural environment. Using a collaborative approach that involves local communities in planning and management decisions enhances achievement of this objective. Some environmental groups have expressed concern that placing too much emphasis on cultural values might degrade ecological values, particularly when the two sets of values are in conflict.

Uses and values of protected areas and natural resources are influenced by culture. Changes in the culture explain the three dominant periods of land management in the United States: public land disposition (1862–1891), public land reservation (1892–1934), and public land retention and management (1935 to present).

Public support for natural resource conservation in the United States grew in leaps and bounds from the early 1960s onward. Growth was precipitated by publication of Carson's *Silent Spring* (1962) and Meadows and others' *Limits to Growth* (1972) and *Beyond the Limits* (1992). Other works took exception to the idea that natural resources constrained economic growth, notably Kahn's *The Next 200 Years* (1976), Simon and Kahn's *The Resourceful Earth* (1984), and Repetto's *The Global Possible* (1986). Numerous laws, court decrees, environmental activism, new disciplines (conservation biology and ecological economics), and international agreements, such as the Agenda 21, support natural resource and environmental protection.

Protection of sacred places is a high priority for public land management agencies because it emanates from deep-seated spiritual values that contribute to social cohesion and vitality and cultural diversity. In addition, spiritual values are protected by freedom of religion in many parts of the world. Two events disrupted spiritual and cultural traditions and lifestyles of Native Americans, namely, creation of national parks in areas used by those cultures and European settlement. The latter wreaked havoc on the ability of Native Americans to coexist with and derive sustenance from the physical environment. By providing solace, tranquillity, inspiration, clean air and water, and physically challenging activities, protected areas afford outstanding opportunities for individuals in urbanized, materialistic, and competitive societies to be spiritually and physically renewed.

A useful construct for preserving the cultural and social values is the cultural landscape, which reflects a place's historical, cultural, and ecological values. National battlefields and national parks are prime examples of cultural landscapes. The U.S. National Park Service has several programs and policies for preserving cultural landscapes, including the Applied Ethnology Program and the Ecosystem Management Working Group. The Alaska National Interest Lands Conservation Act is a landmark statute that serves as a useful template for protecting natural resource use by traditional subsistence cultures. Finally, the National Park Service and other federal land management agencies have instituted programs to educate visitors about how traditional cultures use protected areas and to attract more culturally diverse visitors.

References

Barney, G. O. 1980. *The global 2000 report to the President of the United States.* Oxford, England: Pergamon Press.

Berger, T. R. 1985. *Village journey: The report of the Alaska Native Review Commission.* New York: Hill and Wang.

Bernbaum, E. 1997. The spiritual and cultural significance of mountains. In *Mountains of the world: A global priority,* ed. B. Masserli and J. D. Ives, 39–60. London, England: The Parthenon Publishing Company.

Blackburn, T., and K. Anderson. 1993. Introduction: Managing the domesticated environment. In *Before the wilderness: Environmental management by native Californians,* ed. T. Blackburn and K. Anderson, 1–17. Menlo Park, CA: Ballena Press.

Bonnicksen, T. M., M. K. Anderson, H. T. Lewis, C. E. Kay, and R. Knudson. 1999. Native American influences on the development of forest ecosystems. In *Mountains of the world: A global priority,* ed. B. Masserli and J. D. Ives, 439–470. London, England: The Parthenon Publishing Company.

Booth, A. L., and W. B. Kessler. 1996. Understanding people, natural resources and ecosystem health. In *Natural resource management: The human dimension,* ed. A. W. Ewert, 231–248. Boulder, CO: Westview Press.

Bruzaaelli, D. T., and J. Lash. 1996. *Sustainable America.* Report of the President's Council on Sustainable Development. Washington, DC: U.S. Government Printing Office.

Cahalan, J. M. 2001. Walking with a ghost. *National Parks* 75(November/December): 40–41.

Carson, R. 1962. *Silent spring.* New York: Houghton Mifflin Co.

Clarke, W. M. 2001. The national park or parking system? *National Parks* 75(July/August): 34–37.

Davis, T. 2001. Healing the Gila. *High Country News* 33(October): 8–10.

deBuys, W., M. Crespi, S. H. Lees, D. Meredith, and T. Strong. 1999. Cultural and social diversity and resource use. In *Mountains of the world: A global priority,* ed. B. Masserli and J. D. Ives, 190–208. London, England: The Parthenon Publishing Company.

Denali National Park and Preserve. 1999. Subsistence users guide: Denali National Park and Preserve, Subsistence Resource Commission, Denali Park, AK.

Diamond, J. M. 1990. Playing dice with megadeath. *Discover* (April): 55–59.

Dietrich, W. 1995. *Northwest passage: The Greater Columbia River.* New York: Simon & Schuster.

Ehrlich, P. R., and A. H. Ehrlich. 1991. *Healing the planet.* Reading, MA: Addison-Wesley.

Furze, B., T. De Lacy, and J. Birkhead. 1996. *Culture, conservation and biodiversity.* West Sussex, England: John Wiley & Sons, Ltd.

Grotzbach, E., and C. Stadel. 1997. Mountain peoples and culture. In *Mountains of the world: A global priority,* ed. B. Masserli and J. D. Ives, 17–38. London, England: The Parthenon Publishing Company.

Halvorson, W. L. 1996. Changes in landscape values and expectations: What do we want and how do we measure it? In *National parks and protected areas: Their role in environmental protection,* ed. R. Gerald Wright, 3–14. Cambridge, MA: Blackwell Science.

Hanna, W. L. 1987. Montana's many-splendored glacierland: All you've ever wanted to know about Glacier National Park. Grand Forks: University of North Dakota Foundation.

Henry, J. D. 2002. *Canada's boreal forest.* Washington, DC: Smithsonian Institution Press.

International NGO Consultation on the Mountain Agenda. 1995. Summary report and recommendations to the United Nations Commission on Sustainable Development (Lima, Peru). Available from The Mountain Institute, Franklin, WV.

Kahn, H. 1976. *The next 200 years: A scenario for America and the World.* New York: William Morrow.

Keiter, R. B., T. Boling, and L. Milkman. 1999. Legal perspectives on ecosystem management: Legitimizing a new federal land management policy. In *Ecological stewardship: A common reference for ecosystem management,* ed. W. T. Sexton, A. J. Malk, R. C. Szaro, and N. C. Johnson, vol. III, 10–41. Oxford, UK: Elsevier Science.

Lee, K. N. 1995. Deliberately seeking sustainability in the Columbia River Basin. In *Barriers and bridges to the renewal of ecosystems and institutions,* ed. L. H. Gunderson, C. S. Holling, and S. S. Light, 214–238. New York: Columbia University Press.

Lewis, H. T. 1993. Patterns of Indian burning in California: Ecology and ethnohistory. In *Before the wilderness: Environmental management by native Californians,* ed. T. Blackburn and K. Anderson, 55–116. Menlo Park, CA: Ballena Press.

Long, B. 2001. Desperately Seeking Silence. *National Parks* 75(November/December): 28–32.
McNeely, J. 1995, Feb. Conserving diversity in mountain environments: Biological and cultural approaches, 11. Presentation to Mountain Agenda Consultation. Lima, Peru.
Meadows, D., L. D. Meadows, and J. Randers. 1992. *Beyond the limits: Confronting global collapse, envisioning a sustainable future.* Post Mills, VT: Chelsia Green Publishing Co.
Meadows, D. H., et al. 1972. *The limits to growth.* New York: Universe Books.
Meffe, G. K., and C. R. Carroll. 2004. *Conservation biology.* Sunderland, MA: Sinauer Associates, Inc.
Mitchell, N., and S. Buggey. 2000. Protected landscapes and cultural landscapes: Taking advantage of diverse approaches. *George Wright Forum*: 17: 35–46.
Mitchell, N. J. 1997. Protecting landscapes: Contributions from landscape preservation to management of parks and reserves. Proceedings of the 9th Conference on Research and Resource Management in Parks and on Public Lands, 285–289. George Wright Society, Hancock, MI.
Moyle, P. B., and J. E. Williams. 1990. Biodiversity loss in the temperate zone: Decline of the native fish fauna of California. *Conservation Biology* 4: 475–484.
National Park Service. 1988. *Management policies.* Washington, DC: U.S. Department of the Interior.
———. 1994, September. Ecosystem Management Working Group, Resource Stewardship Team, Vail Agenda. Ecosystem Management in the National Park Service (draft).
———. 1994. *Cultural resource management guideline, NPS-28.* Washington, DC: U.S. Department of the Interior.
Norse, E. A. 1990. *Ancient forests of the Pacific Northwest.* Washington, DC: The Wilderness Society and Island Press.
Oelschlaeger, M. 1991. *The idea of wilderness.* New Haven: Yale University Press.
Parks Canada. 2002. *Kluane National Park and Reserve of Canada.* http://parkscan.harbour.com/kluane/. (accessed January 22, 2002).
Periman, R. D. 1996. The influence of prehistoric Anazasi cobble-mulch agricultural features on northern Rio Grande landscapes. In *Desired future conditions for Southwestern riparian ecosystems: Bringing interests and concerns together,* ed. D. W. Shaw and D. F. Finch, 181–188. Ft. Collins, CO: U.S. Department of Agriculture, Forest Service, Rocky Mountain Forest and Range Experiment Station.
Periman, R. D., C. Reid, M. K. Zweifel, G. McVicker, and D. Huff. 1999. Human influences on the development of North American landscapes: Applications for ecosystem management. In *Ecological stewardship: A common reference for ecosystem management,* ed. R. C. Szaro, N. C. Johnson, W. T. Sexton, and A. J. Malk, 471–492. Oxford, UK: Elsevier Science, Ltd.
Raish, C., L. Engdahl, W. Anderson, D. Carpenter, M. Crespi, P. Johnson, L. McConnell, and E. Neiler. 1999. Resource management strategies for working with cultural and social diversity. In *Mountains of the world: A global priority,* ed. B. Masserli and J. D. Ives, 209–225. London, England: The Parthenon Publishing Company.
Repetto, R. (ed.). 1986. *The global possible: Resources, development and the mew century.* New Haven, CT: Yale University Press.
Roberts, N. S., and D. A. Rodriguez. 2001. Reaching out. *National Parks* 75(September/October): 38–39.
Sauer, C. O. 1963. The morphology of landscape. In *Land and life: A selection of writings of Care Otwin Sauer,* ed. J. Leighly, 315–350. Berkeley, CA: University of California Press.
Schultz, J. W. 1962. *Blackfeet and buffalo.* Norman, OK: University of Oklahoma Press.
Shalk, R. F. 1986. Estimating salmon and steelhead usage in the Columbia Basin before 1850: The anthropological perspective. *The Northwest Environmental Journal* 2: 1–29.
Simon, J., and H. Kahn. 1984. *The resourceful Earth: A response to Global 2000.* Oxford, England: Basil Blackwell.
Snow, Chief John. 1977. *These mountains are our sacred places: The story of the Stoney people.* Toronto, Canada: Samuel Stevens and Company.
Terbourgh, J. 2000. *Requiem for nature.* Covelo, CA: Island Press.

Toothman, S. S. 1996. The cultural and natural resource management needs of parks and protected areas: Is there an appropriate balance? In *National parks and protected areas: Their role in environmental protection,* ed. R. G. Wright, 355–368. Cambridge, MA: Blackwell Science.

U.S. Department of the Interior, National Park Service. Undated. Subsistence: A vital connection to the land. Denali National Park and Preserve, AK.

UNESCO. 2000. *World heritage list.* http://www.unesco.org/whc/nwhc/pages/sites/main.htm (accessed June 10, 2001).

United Nations Conference on Environment. 1992. Earth Summit, Rio de Janeiro, Brazil. http://www.un.org/geninfo/bp/enviro.html (accessed April 9, 2004).

Weixelman, J. O. 2001. Fear or reverence? Native Americans and the geysers of Yellowstone. *Yellowstone Science* 9: 2–11.

Wilson, E. O. 1985. The biological diversity crisis. *BioScience* 35: 700–706.

———. The current state of biological diversity. In *Biodiversity,* ed. E. O. Wilson, 3–18. Washington, DC: National Academy Press.

———. 2002. Hotspots. *National Geographic* 201(January): 86–89.

World Resources Institute (WRI), World Conservation Union (IUCN), and United Nations Environment Programme (UNEP). 1992. Global biodiversity strategy: Guidelines for actions to save, study, and use Earth's biotic wealth sustainably and equitably. Washington, DC: WRI, IUCN and UNEP.

Wright, R. G. 1996. Expansion of the U.S. National Park System in Alaska. In *National parks and protected areas: Their role in environmental protection,* ed. R. G. Wright, 165–171. Cambridge, MA: Blackwell Science.

Yaffee, S. L., and J. M. Wondolleck. 1997. Building bridges across agency boundaries. In *Creating forestry for the 21st century: The science of ecosystem management,* ed. K. A. Kohm and J. F. Franklin, 381–396. Washington, DC: Island Press.

CHAPTER

4

Social, Economic, and Ecological Values

The potential for conflict between the needs of human systems and ecological systems has become a central defining issue of contemporary conservation.

—R. A. Haeuber and V. H. Dale

This chapter discusses several values and concepts important for the management of protected areas. Of particular interest are economic concepts for characterizing tastes and preferences for ecological goods and services, representing trade-offs between social, economic, and ecological values, and determining preferred management actions for different stakeholders. Specific topics addressed in the chapter include properties of public goods; challenges in managing protected areas; tastes and preferences; use, non-use, intrinsic, spiritual, and ethical values; demand and supply; private vs. external costs; externalities; social value; economic efficiency; distributional equity; and ecological sustainability. This chapter also discusses the efficient combinations of social, economic, and ecological values and associated management actions for a protected area, and the most preferred management action for a stakeholder having particular preferences for these values.

Public Goods

Production and distribution of consumer goods and services, like food, fuel, and automobile repair, are governed by market demand and supply. The interaction of market demand and supply determines the prices of consumer goods and services. Market prices guide the allocation of resources, like land, labor, and capital, in the production of goods and services, and their distribution to consumers. Demand and supply are influenced by population, income, prices, tastes and preferences, and technology. Protected areas provide ecological goods and services. Ecological goods contained in protected areas include fish and wildlife, timber, minerals, and water. These goods can be extracted only for subsistence uses. Ecological services of protected areas include recreation, tourism, water supply, habitat for fish and wildlife, conservation of biodiversity, purification of air and water, and many others. The provision of ecological goods and services by protected areas are not totally governed by demand and supply the way they are for consumer goods and services.

Ecological goods and services of protected areas are public goods. Public goods have two unique properties. First, the quantity consumed by one person does not diminish the amount available to other persons. This is known as the *nonrivalry* property of public goods. For example, a scenic view in a national park is a *nonrival good* as long as the enjoyment of current viewers is not diminished by the addition of new viewers. Scenic viewing becomes a *rival good* when the number of viewers exceeds a threshold level beyond which the satisfaction of current viewers is diminished as new viewers are added. Therefore, scenic viewing is a nonrival good as long as the level of use is below the threshold.

A second property of public goods is that it usually is not cost effective to exclude additional persons from using them. This property of public goods is known as *nonexcludability*. It is possible to ration scenic views by limiting the number of individuals at each view site. Rationing could be accomplished by assigning viewing periods on a first-come-first-served basis. The cost of rationing, however, is likely to exceed the benefits of reduced congestion, particularly in large protected areas that have numerous viewing sites and limited staff. When rationing is not cost effective, the good has the property of nonexcludability. Nonexcludability of many of the ecological services provided by protected areas is one of the reasons that the admission fees to national parks are set low relative to the value of the experience.

An ecological good or service can have one property but not the other. Backcountry camping in national parks in North America is nonrival and *excludable*. It is nonrival as long as camping by one party does not diminish the enjoyment of other parties. The nonrival nature of backcountry camping is achieved by limiting the number of backcountry campsites and requiring backcountry camping permits. The permit system makes backcountry camping excludable. Requiring backcountry permits not only maintains the quality of backcountry camping, but also reduces natural resource impacts of camping. In this case, the benefit of restricting the number of backcountry camping parties (excluding parties) exceeds the cost of issuing permits and monitoring for violations (parties that camp without permits). It is possible for ecological goods to be nonexcludable and rival. Game hunting in a wilderness area is a rival good because an animal bagged by one hunter is not available for other hunters. Yet, hunting can be nonexcludable when it is not cost effective to limit the number of hunters in the area.

Challenges in Managing Protected Areas

The public goods nature of protected areas gives rise to several management challenges. First, it means the ecological goods and services provided by protected areas are not subject to the market forces of demand and supply. Hence, prices are not available for making management decisions in protected areas. Just because ecological goods and services do not have market prices does not mean they are not valuable. As evidenced by many nonmarket valuation studies (e.g., Costanza et al. 1997), the aggregate value of ecological services and natural resources is quite substantial. Nonmarket valuation is discussed later in the chapter.

Second, management of protected areas is challenging because there is almost always disagreement about the most desirable mix of ecological goods and services to produce. Disagreement is almost unavoidable for national parks because of their dual mandate of providing public enjoyment and conserving natural and cultural resources for future generations. In other words, the goals of public enjoyment and resource conservation for national parks can easily conflict.

Third, management of protected areas is challenging because of potential conflicts between economic development and resource protection. On the one hand, economic development in or near protected areas can have negative impacts on resource protection. Large carnivores, some of which are threatened and endangered species, are especially vulnerable to landscape fragmentation caused by economic development, particularly when it occurs in their home ranges. Landscape fragmentation threatens the survival of large carnivore species (Newmark 1995). Banff National Park in Alberta, Canada, is a prime example of competition between economic development and resource protection. The town of Banff is located in the park and the Trans Canada highway runs through the park. Both the town and the highway have seriously degraded wildlife habitat and increased wildlife mortality. On the other hand, economic development is important to gateway communities because it generates income, employment, and tax revenue.

Fourth, the management of protected areas is challenging because it is normally based on long planning horizons, which is not the case for most consumer goods and services. The private decision of whether to travel to work by car or public transportation is a short-term decision. If a person has a car, then the decision can be made on a daily basis. Timeframes for managing the public goods provided by protected areas are multigenerational.

Tastes and Preferences

People's *tastes and preferences* influence their purchases and use of consumer and public goods and services. Consider once again an individual who is deciding whether to travel to work by car or public transportation. Driving is typically more convenient and provides greater flexibility than public transportation. On the other hand, public transportation is usually less stressful (no need to fight the traffic) and more energy efficient. The individual's tastes and preferences for convenience, flexibility, stress reduction, and energy conservation influence the decision of whether to drive or use public transportation. An individual who places a high premium on being able to stop at the athletic club and/or supermarket on the way to or from work is likely to prefer the convenience and flexibility of driving over the stress reduction and energy efficiency afforded by public transportation. Such a person is more likely to drive to work than use public transportation. In contrast, a person who places a high premium on stress reduction and energy conservation is more likely to use public transportation.

Preferences for the ecological goods and services provided by protected areas are driven by tastes and preferences. Consider recreational use of snow machines in national parks. Snow machine advocates evidently prefer riding their snow machines to cross-country skiing and snowshoeing. Cross-country skiers and snowshoe enthusiasts may object to the noise, air pollution, and wildlife disturbance caused by the use of snow machines. In this case, snow machine use in a national park diminishes the recreational experience for cross-country skiers and snowshoe enthusiasts. The situation is further complicated by a third group of stakeholders, namely businesses in gateway communities for national parks. Businesses are likely to oppose the banning of snow machines from the park because it would reduce business sales associated with snowmobiling.

Sources and Nature of Value

Protected areas provide ecological goods and services that have a wide range of social, economic, and ecological values. It is important to understand the sources and nature of these values because they influence the establishment, use, and management of protected areas as well as the satisfaction of users and nonusers. Protected areas generate four primary values—utilitarian, intrinsic, spiritual, and ethical. Each value is discussed below.

Utilitarian Value

Utilitarianism implies that goods and services have value because they satisfy human wants or contribute to human endeavor. The philosophical basis for utilitarianism is *anthropocentrism*, which maintains that humans are the dominant species and defines ethical behavior in terms of human wants and needs (Watson 1983). Anthropocentrism is central to *shallow ecology*, which places humans above (dominant over) nature, treats human satisfaction as the source of all value, and considers nature as having only instrumental or use value as opposed to intrinsic value (Capra 1995). In contrast, *deep ecology* emphasizes the intrinsic value of nature. Norwegian philosopher Arne Naess distinguished between shallow ecology and deep ecology as follows: "The shallow ecology movement tends to talk only about resources for humans, whereas in deep ecology we talk about resources for each species" (Bodian 1995). Satisfaction of human-centered values is also a paramount concern of libertarianism, which is rooted in an ethical system that emphasizes individual rights and private property. In the utilitarian theory of value, an individual's welfare is completely determined by the consumption of goods and services. Utilitarianism is a pivotal construct in neoclassical economics and benefit-cost analysis.

Utilitarian values are classified as follows:

- Use Values
  - Direct
    - Consumptive
    - Nonconsumptive
  - Indirect
- Nonuse Values
  - Option
  - Quasi-option
  - Existence
- Bequest

Use Value

Use value is defined with reference to a person's utility function. A *utility function* represents an individual's tastes and preferences for goods and services. It is used to rank from most preferred to least preferred all combinations of goods and services available to an individual. A hypothetical total utility function for a single good (q_1) is shown in figure 4.1a. It assumes that consumption of all other goods and services is constant. As consumption of q_1 increases, total utility increases, reaches a maximum at q^*_1 and then decreases. Figure 4.1b depicts the marginal utility function for q_1. It shows the relationship between the change in total utility with respect to a change in consumption of q_1 (slope of the total utility curve) on the vertical axis and consumption of q_1 on the horizontal axis. The law of diminishing marginal utility says that marginal utility decreases as consumption increases. This implies that increasing amounts of q_1 provide decreasing increments of satisfaction to the individual. Marginal utility is positive for values of q_1 less than q^*_1 and negative for values of q_1 greater than q^*_1 (negative values of marginal utility are not shown in fig. 4.1b). A rational consumer would not consume more than q^*_1 because it would decrease total utility or satisfaction.

Utilitarianism states that an individual consumes those quantities of goods and services that maximize total utility subject to a budget constraint. The latter requires total expenditures on all goods and services to be less than or equal to disposable income. Maximizing total utility subject to a budget constraint results in individual demand curves for goods and services. Along a *demand curve*, the quantity demanded of a good or service decreases (increases) as price increases (decreases) when all other factors affecting demand are held constant. Hence, demand curves are negatively sloped like the one shown in figure 4.2. Changes in

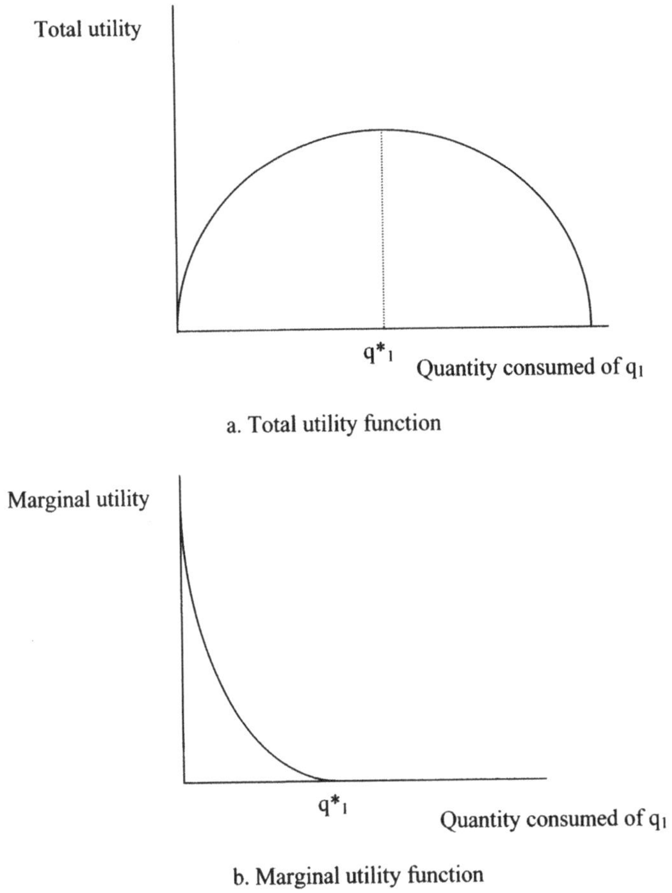

FIG. 4.1. Hypothetical total utility function (panel a) and marginal utility function (panel b) for q_1.

tastes and preferences, market prices of other goods and services, and disposable income shift demand curves to the right or left, which can alter consumption patterns.

Returning to the winter recreation example, the utility function for the snow machine enthusiast would show that personal satisfaction increases with additional days spent on a snow machine up to the satiation level where utility is a maximum. The utility function for the cross-country skier/snowshoe enthusiast would show that personal satisfaction increases with the number of days spent in these recreational activities up to the satiation level and, perhaps, decreases as snow machine use increases.

There are two forms of utilitarian value, direct use and indirect use. *Direct use value* is generated by direct consumption of ecological goods (timber, energy min-

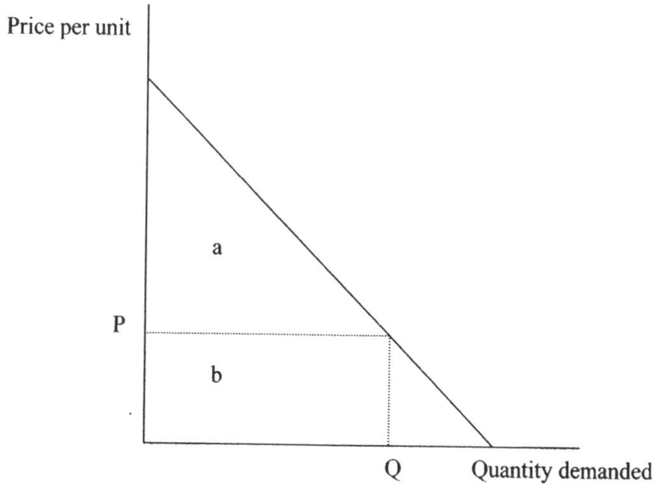

FIG. 4.2. Market demand curve and use value (area a + area b) for tents.

erals, and water), or direct use of ecological services (hiking, camping, nature viewing). Direct use value can be consumptive or nonconsumptive. Cutting timber and extracting oil and gas are consumptive uses. Hiking, camping, and nature viewing are considered nonconsumptive uses. Since most ecological services of protected areas are not directly consumed, they are said to have *indirect use value*. Conservation of biodiversity and purification of air and water are ecological services of a protected area that have indirect use value. Some ecological goods, such as timber, oil, and gas, have market prices. Ecological services of protected areas do not have market prices because they are not traded in the marketplace.

Total use value of a marketed good or service is determined from the *market demand curve* for that good or service. In theory, a market demand curve is the sum of the demand curves for consumers of a good or service. In practice, market demand curves are estimated statistically using data on market prices and quantities and other variables. A hypothetical market demand curve for tents is shown in figure 4.2. The demand curve shows that when the market price is P, total quantity demanded is Q. Use value of tents is area a + area b. Area a is consumer surplus and area b is total consumer expenditure for tents. Consumer surplus is the value of a good or service above and beyond what consumers pay for it. Social value, which is discussed later in this chapter, includes consumer surplus.

Nonuse Values

Nonuse value is the second category of utilitarian value. There are four types of nonuse value: option, quasi-option, existence, and bequest. As the name implies, nonuse values do not involve use of the good, resource, or service.

Option value is the risk premium an individual is willing to pay to preserve the option of having an irreplaceable asset, such as a protected area, available for his or her own future use. Risk averse individuals have a propensity to avoid risk, which results in positive option value. Individuals who are not risk averse do not care about avoiding risk and have zero option value (Goulder and Kennedy 1997). Option value is estimated by asking people how much they are willing to pay for the option of having future access to a resource. It is relevant when there is uncertainty regarding use and/or availability of the resource, use of the resource is infrequent, and/or it is very costly to restore the resource to its original use once it has been damaged. A resource with the latter property is said to be *irreplaceable* (Weisbrod 1964). Irreplaceability is a relevant consideration for protected areas when certain changes in resource use make it very costly or infeasible to return it to its original use. Rejection of a proposal to build a commercial jetport in Everglades National Park was based on its irreplaceability. Opponents of the proposal argued that construction and operation of the proposed jetport would cause severe and irreversible ecological damage to the park.

Option value came into play in preserving Sequoia National Park. It was argued that the social value of the park should be determined by comparing the sum of use value and option value of recreation to the value of harvesting the park's redwood trees (Weisbrod 1964). It was argued that ignoring the option value of park recreation would result in underestimation of recreational value and increase the likelihood of activities that would degrade the park. While the area was not a protected area, the option value for Prince William Sound in Alaska was used in the *Exxon Valdez* court case. Exxon was sued for various criminal charges and civil claims brought by the federal and state governments for recovery of natural resource damages caused by a major oil spill that occurred in March 1989 when the *Exxon Valdez* oil tanker plowed into a reef spilling 10.8 million gallons of crude oil into the sound (*Exxon Valdez* Oil Spill Council 2003).

Uncertainty about resource use arises when an individual is not sure whether she or he will utilize the resource in the future. Uncertainty about resource availability is relevant when the value of a resource or area can be eroded by future conditions such as economic development and global climate change. Uncertainty about resource use is a characteristic of the user of a resource not the resource itself. In contrast, uncertainty about resource availability is a characteristic of a resource. The latter is relevant to protected areas facing internal threats, such as high rates of visitation, and external threats, such as degradation of wildlife habitat due to landscape fragmentation.

Quasi-option value is a variant of option value (Arrow and Fisher 1974; Henry 1974). It represents the amount people are willing to pay to have an irreplaceable resource preserved until such time as sufficient information is available to make a decision about how best to use or manage it. Quasi-option value is relevant to the decision about whether or not to convert a forested area to agricultural or residential uses. Such land-use changes run the risk of destroying certain plant species that could have potentially significant medicinal benefits, for example, protecting areas containing the Pacific yew tree whose bark was used to treat certain types of cancer. The value of delaying an irreversible use of a forest until better information is available regarding whether it contains potentially valuable species is the quasi-option value of the forest. Quasi-option value forms the basis for a *no-regrets policy*. For example quasi-option value of climate resources is the willingness to pay to limit carbon emissions to levels that do not degrade climate resources until such time as better information is available regarding the impacts of those emissions on climate resources and human activity. A no-regrets policy takes action now to avoid adverse impacts on climate resources and is justified when quasi-option value is positive.

Existence value is how much people are willing to pay to ensure that a resource is preserved in a particular state above and beyond its direct use value or option value (Randall and Stoll 1983; Pearce 1993). Positive existence value provides the rationale for not harvesting humpback whales to the point of extinction, preserving Australia's Great Barrier Reef in perpetuity, and prosecuting individuals who poach lion, elephant, rhinoceros, and giraffe in South Africa's Kruger National Park. Organizations like the National Wildlife Federation, Audubon Society, Sierra Club, and Wilderness Society draw their membership from individuals for whom wild animals and wild places have existence value. Existence value can exceed use value for unique natural assets. One study estimated that existence value for improved visibility in the Grand Canyon region was 60+ times greater than its use value (Schultz et al. 1983).

Bequest value is the value of preserving a resource for future generations (Krutilla 1967). It is estimated by how much an individual is willing to pay to ensure that a resource is available for future generations. Bequest value is an important source of value for protecting unique natural areas and is similar to existence value. Both bequest value and existence value are rooted in the satisfaction of knowing a resource will be preserved into the future. It is not easy to disentangle existence and bequest values. Both users and nonusers of a resource can have bequest value. For instance, people may be willing to pay a certain amount of money to ensure that Shenandoah National Park is available to future generations, perhaps their offspring, whether or not they have ever been there or ever intend to go there. Option, existence, and bequest values are relevant only when a resource and the ecological goods and services it provides are threatened or damaged.

Utilitarianism can be either weak or strong (Goulder and Kennedy 1997). *Weak utilitarianism* states that the value of a good or service to an individual depends solely on how much personal satisfaction (utility) it provides. *Strong utilitarianism* states that the value of a resource to society is the sum of the values derived by all individuals that benefit from it. Strong utilitarianism is the basis for *cost-benefit analysis*. Cost-benefit analyzes economic efficiency of alternative actions and policies by comparing their benefits and costs. Chapter 8 provides a more detailed treatment of cost-benefit analysis. Since benefit-cost analysis is an efficiency-based criterion, it does not consider how the benefits and costs of an action or policy are distributed among different social groups and the ethical implications of that distribution. For example, the benefits of banning snow machines in a national park are likely to accrue to cross-country skiers, snowshoe enthusiasts, and wildlife. However, the costs of banning are borne by users of snow machines (loss of recreational value) and gateway communities (loss of income, employment, and tax revenues). Objections have been raised to this limitation of cost-benefit analysis and the strong utilitarian foundation on which it rests.

Another shortcoming of cost-benefit analysis is that it does not account for *externalities*. An externality occurs when one party (individual or group) engages in an activity that affects the welfare of another party in a positive or negative way, and the party causing the externality is not concerned with how its activities affect the welfare of others (Prato 1998). Externalities can occur between firms, between individuals, and between individuals and firms. External costs result from negative externalities generated by the production of goods and services. For example, suppose air pollution from coal-fired electrical generating plants reduces visibility in national parks downwind from those plants. Reduced visibility imposes an external cost or negative externality on park visitors and resources. Conversely, positive externalities generate external benefits. An example of a positive externality is the aesthetic and scenic amenities experienced by persons living near a national park or scenic river.

Ignoring externalities has social consequences. For instance, net benefit (benefits minus costs) of economic development in gateway communities for a protected area is overstated when development imposes negative externalities on the protected area, and understated when it generates positive externalities for the protected area. Ignoring the negative externality imposed on national parks from air pollution caused by power plants could result in too many power plants. Conversely, ignoring the positive externality to gateway communities from the aesthetic and scenic amenities of national parks could result in too little federal support for national parks.

It is possible to accept weak utilitarianism and reject strong utilitarianism. In this case, one accepts that human satisfaction is the source of all value and is concerned about how different social groups are impacted by resource use. The latter attitude negates strong utilitarianism. In evaluating proposals to create a new protected area, strong utilitarianism is rejected when more weight is given to the income benefits and cost burdens of low-income families than of high-income families.

Intrinsic Value

Intrinsic value emanates from an ecocentric philosophy that gives equal rights to all living objects regardless of their human benefits. An *intrinsic rights approach* implies that all biotic elements in a protected area have equal moral status. Animal rights activists support elements of the intrinsic rights philosophy by arguing that animals and plants have a right to exist apart from their value to humans (Singer 1975). The deep ecology movement (Naess 1989; Seed 1989; Sessions 1995) embraces intrinsic rights. Deep ecology emphasizes the intrinsic value of the natural world and the notion that preserving natural ecosystem health is in the best interests of humans (Snyder 1995). Since humans are not a dominant species under the intrinsic rights philosophy, the latter implies that it is unethical to exploit living things for human gain. This position is the antithesis of the anthropocentric view that undergirds utilitarianism.

One of the ironies of the intrinsic rights approach is that nonhuman rights can be judged and articulated only from a human perspective. Nonhuman species have no way of articulating and defending their rights to life and protection. This is why organizations like Defenders of Wildlife have intervened in human-wildlife controversies, such as the introduction of wolves into Yellowstone National Park. In that case, Defenders of Wildlife agreed to compensate ranchers for livestock mortality caused by wolves. Since the rights accorded nonhumans will always be subject to human interpretation and manipulation, an ecocentric approach is subject to the same foibles as anthropocentrism.

Kantian justice is another rights-based approach that is similar to intrinsic rights. It purports that one should ignore personal gains or losses from a particular action in judging its worth (Kant 1963). Kantian justice is essentially anthropocentric because it does not confer moral status to nonhumans. Interestingly enough, values do not appear central to the intrinsic rights approach and Kantian justice (Goulder and Kennedy 1997). Whereas the intrinsic rights philosophy centers on whether nonhuman rights are being violated by a proposed course of action, Kantian justice concerns itself with making decisions in the absence of self-interest.

Spiritual and Ethical Values

Spiritual values and *ethical values* influence individual and social behavior. The relationship that American Indians have with nature and one another has a spiritual basis. They consider biotic and nonbiotic elements of nature to be their brothers and sisters. This view requires that nature be treated with respect. For this reason, Indians typically limited their harvesting of plants and animals to the amounts needed to sustain their life and culture.

As discussed in chapter 1, spiritual values were endemic in the philosophies of Emerson, Thoreau, and Muir. Emerson held to the anthropocentric-utilitarian view that nature had value because it supplied humans with goods and services. He believed that God's existence could be confirmed by nature; it did not require faith in Scripture. Thoreau made a clean break from Emerson's anthropocentric-utilitarian view by espousing the position that nature "exists in and of itself," which is ecocentric. He questioned the Emersonian teaching that nature is sufficient to prove God's existence. Muir was a biocentrist; he did not elevate humans above other creatures. His religion was the love of nature. Muir espoused the sacredness of creation and viewed God and creation as one and the same thing (pantheism).

Christian values have been charged with motivating resource exploitation and abuse of nature (White 1967). This position is rationalized based on certain verses in Genesis, the first book of the Holy Bible. In Genesis 1:28, God instructs humans to "be fruitful and increase in number; fill the earth and subdue it. Rule over the fish of the sea and the birds of the air and over every living creature that moves on the ground." Justifying human exploitation of nature based on this passage is inconsistent with other scriptures on creation, particularly Genesis 1:10, 18, 21, and 25. Every one of these verses says, "And God saw that it [creation] was good." Genesis 1:31 captures it best: "God saw all that he made, and it was very good." The NIV commentary on this verse says, "As God's representative in the creaturely realm [since man was created in God's image, Genesis 1: 26], he is steward of God's creatures. He is not to exploit, waste or despoil them, but to care for them in the service of God and man" (*The NIV Study Bible* 1995). This interpretation is corroborated by God's charge to Adam and Eve. "The Lord God took the man and put him in the Garden of Eden to work it and take care of it" (Genesis 2:15). The NIV commentary on this verse states: "Man is now charged to govern the earth responsibly under God's sovereignty" (*The NIV Study Bible* 1995).

Total Economic Value

Total economic value of a good or service is defined as the sum of the use and nonuse values enumerated above, namely:

Total economic value = Direct use value + Indirect use value + Option value
+ Quasi-option value + Existence value + Bequest value

Total economic value is the standard economic measure of the total utilitarian worth of a good or service. Measurement of total economic value is fraught with difficulty, especially for public goods, such as protected areas. Since many of the ecological goods and services provided by protected areas are unpriced (no markets exist for them), estimation of their monetary value is done using nonmarket valuation methods (see the sidebar). Direct use values for activities such as hunting, fishing, hiking, and wildlife viewing are measured using the travel cost, hedonic pricing, and/or hedonic travel cost methods. Each of these methods is described below.

Valuing Nonmarketed Ecological Goods and Services

Natural landscapes provide a wide range of ecological goods and services. Ecological goods include timber, forage (grazing), seafood, biomass fuels (coal, oil, and natural gas), natural fiber, pharmaceuticals, industrial products (waxes, rubber, and oils), and others. Ecological services include purification of air and water, generation and renewal of soil and soil fertility, conservation of biodiversity, partial stabilization of climate, and many others. The value of ecological goods can be determined using market prices. Since markets do not exist for ecological services, their values need to be determined using other methods. Several techniques are available for valuing the ecological services provided by protected areas. Contingent valuation, hedonic pricing, and travel cost are described in this chapter. Other ways to estimate the value of ecological services include change in productivity, loss (or gain) of earnings, opportunity cost, and replacement cost.

- The change in productivity method estimates the value of altering the quantity or quality of an ecological good or service by changes in the productivity of affected resources that can be valued in the marketplace. For example, suppose global climate change is expected to cause drier and warmer conditions in a mountain national park that is an important source of irrigation water for an agricultural valley. Expected declines in the future supply of irrigation water due to climate change is likely to reduce agricultural crop yields in the valley. The market value of the expected loss in crop yields is an estimate of the water supply impacts of global climate change in the park.
- The loss (or gain) of earnings method evaluates the change in human productivity resulting from degradation of a natural area. For example, suppose that current levels of snowmobile use in a national park result in levels of air pollution that are detrimental to the health of park attendants responsible for admitting snowmobile users to the park. If the adverse health effects increase absenteeism for park attendants and absences result in lost wages, then the value of lost earnings is a cost of the current snowmobile use.
- The opportunity cost method estimates the economic value of a protected area as the forgone net income (opportunity cost) from not using the area in its next best use, such as mining, oil and gas development, logging, and resort development. If the total value of all ecological goods and services provided by a park exceed the opportunity cost, which should be the case for protected areas, then protection is the best use of the area.

- The replacement cost method estimates negative impacts on a protected area in terms of how much it costs to eliminate those impacts, assuming this is possible. For example, if overuse of a particular trail and/or backcountry campsite results in ecological damages that can be alleviated by a restoration program that costs $100,000, then the replacement cost of the impacts is $100,000. If the cost of avoiding the damage is less than the replacement cost, then action should be taken to avoid the damage. Certain negative impacts on protected areas, such as the melting of glaciers due to global warming or air pollution in the park that originates in another part of the world, are very costly to avoid. In these cases, replacement cost is likely to exceed the cost of avoiding the damages, in which case mitigation of impacts is not economically efficient.

Not all nonmarket valuation methods discussed in this chapter are equally suitable for estimating impacts on protected areas. The job of the analyst is to determine which estimation method is best suited to a particular impact.

Sources: Daily (1997) and Task Force on Economic Benefits of Protected Areas of the World Commission on Protected Areas (1998).

In the *travel cost method,* surveys are used to determine the number of visits made to a recreation site by individuals residing in concentric distance zones around the site. Average number of visits per person for a zone is calculated by dividing total number of visits by the number of visitors for that zone. Similarly, average travel cost for a zone is calculated by multiplying average travel cost per mile by the average round-trip mileage between the zone and the site. Average values for number of visits, travel cost, household income, and socioeconomic characteristics of households in all zones are used to statistically estimate a travel cost demand curve for the site (Prato 1998). This method has been used to estimate demand curves for several national parks (Clawson 1959). A hypothetical travel cost demand curve for a recreational site is illustrated in figure 4.3. It shows that visitor days increase (decrease) as price per visit decreases (increases). If the site access fee is P, then visitor days are Q. Total use value is the sum of area a and area b, access fees equal area b, and consumer surplus is area a.

The *hedonic pricing method* uses the difference in the prices of marketed goods to estimate the marginal value of an attribute of that good. For example, the incremental value of being in a cabin closer to a lake versus one farther away is estimated by statistically analyzing how rental rates for cabins vary with distance from the lake. In this example, the incremental value (hedonic price) of being

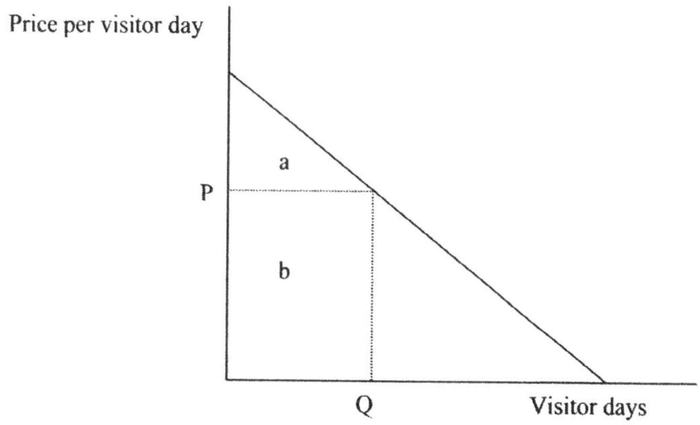

FIG. 4.3. Hypothetical travel cost demand curve for a recreational site and consumer surplus (area a) for an access fee of P.

closer to the lake is the increase in cabin rental rates as the distance to the lake decreases on a per unit basis (e.g., $5 per meter). The statistical procedures used to estimate hedonic prices account for differences in other attributes of the good. In the cabin example, other attributes of the cabin that need to be taken into account include the number of bedrooms and bathrooms, presence or absence of a fireplace, and household amenities, such as telephone, television, and washing machine.

The *hedonic travel cost method* is a variant of the hedonic pricing method. It is used to estimate marginal values of attributes of recreational sites and other nonmarketed goods (Brown and Mendelsohn 1974; Rosen 1974). Survey data are used to estimate travel cost functions for recreation sites having different levels of attributes. The marginal value of an attribute is estimated by the change in travel cost associated with a change in the level of the attribute. Both hedonic methods have their share of technical and other difficulties (Randall 1994; Sagoff 1994).

Indirect use value and nonuse values (option, quasi-option, existence, and bequest) are estimated using various methods, including contingent valuation, conjoint analysis, choice experiments, and fuzzy contingent valuation. *Contingent valuation* is the most popular method. It employs surveys or experimental settings to query individuals regarding how much they value incremental changes (increases or decreases) in the quantity and/or quality of an unpriced good or service (Cummings et al. 1986). For example, a survey of visitors to Yellowstone National Park indicated they were willing to pay an annual fee of $11 per year to help fund a program to reduce lake trout populations in Yellowstone Lake (Cherry and Shogren 2001). Introduction of lake trout, which is a nonnative species in Yellowstone Lake, has reduced populations of native Yellowstone cutthroat trout. The estimated willingness to pay by all park visitors substantially exceeded the costs of the program, which indicates the program has a positive net benefit.

Private vs. External Costs

The *private cost* of producing a good or service, such as timber or oil, is the sum of the costs of raw materials, supplies, equipment, labor, and interest on borrowed capital. The *social cost* of producing a good or service equals private cost plus external cost of production. External cost is positive when there are negative externalities and negative (an external benefit) when there are positive externalities. For example, if clear-cutting stands of timber in a forest adjacent to a national park degrades wildlife habitat for species protected by the park, then clear-cutting imposes a negative externality on the park. A negative externality is depicted in figure 4.4. The private marginal cost curve for the timber industry is the same as the supply curve for timber and is positively sloped. It indicates that the private cost of harvesting an additional unit of timber increases with the amount harvested. Social marginal cost equals private marginal cost plus *external marginal cost*. The latter equals the cost of the loss in wildlife habitat from harvesting an additional unit of timber. Figure 4.4 illustrates the case where external marginal cost increases as timber harvest increases.

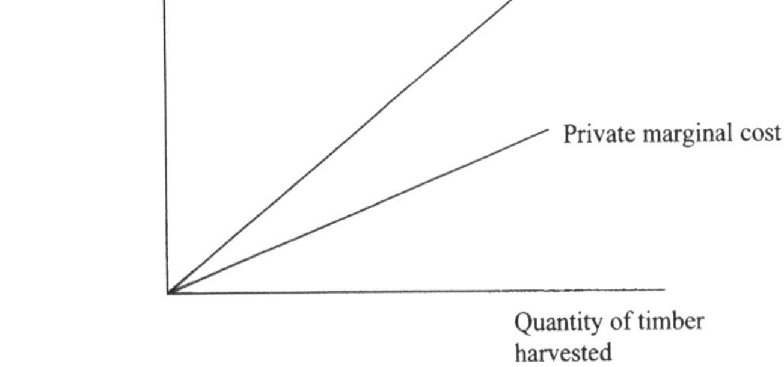

FIG. 4.4. Private and social marginal cost of timber harvesting. The difference between private marginal cost and social marginal cost is due to a negative externality.

Social Value

Social value is defined as the net worth of a good or service to society. It is the primary utilitarian metric used to determine the socially optimal level of production of a good or service and to evaluate public policies aimed at reducing external costs. Social value equals total economic value minus total social cost. Two examples of social value are presented. The first example deals with the social value of recreation in a protected area and the second example illustrates the discrepancy between the private and socially optimal levels of timber harvesting and the loss in social value caused by a negative externality.

The first example is based on the recreational demand and supply curves shown in figure 4.5. The demand curve indicates that visitor days decrease (increase) as price per visitor day increases (decreases). The supply curve indicates that the marginal cost of providing additional visitor days increases with the number of visitor days. When there is no externality, as in the example, private marginal cost (supply curve) equals social marginal cost. Q visitor days are demanded when the access fee is P. Social value equals total economic value (area a + area b + area c) minus total cost (area c), which is equivalent to the sum of consumer surplus (area a) and producer surplus (area b). *Consumer surplus* is the surplus value of the site to users above and beyond what they pay in access fees, namely total economic value (area a + area b + area c) minus total access fees (area b + area c). *Producer surplus* is the surplus value that accrues to the owners of the

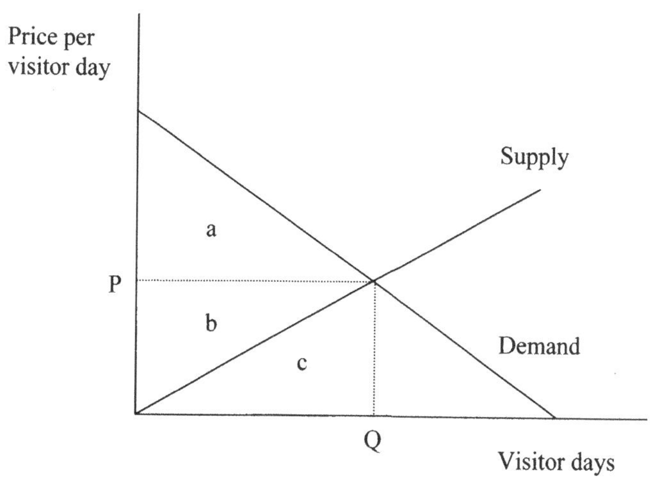

FIG. 4.5. Social value for a recreational site is the sum of consumer surplus (area a) and producer surplus (area b)

site above and beyond the cost of providing visitor services. It equals total access fee (area b + area c) minus total cost of providing visitor services (area c), namely, area b.

The second example illustrates the private and socially optimal timber harvesting on private land adjacent to a national park. The example assumes that timber harvesting imposes an external cost on the park in the form of losses in wildlife habitat. The negative externality of losses in wildlife habitat causes the social marginal cost to exceed the private marginal cost (supply curve) of harvested timber as illustrated in figure 4.4. Private and socially optimal amounts of harvested timber are determined by combining the private and social marginal cost curves (given in fig. 4.4) with the marginal revenue curve for timber harvesting shown in figure 4.6. The marginal revenue curve is derived from the demand curve for timber. It measures the increase in total timber revenue (price times quantity) from selling another unit of timber.

The privately optimal amount of harvested timber maximizes profit. This occurs at Q_r where private marginal cost equals marginal revenue. Harvest amounts higher or lower than Q_r result in less profit. The socially optimal amount of harvested timber occurs at Q_s where social marginal cost equals private marginal revenue. Social value is maximized at Q_s and is equal to area a. The privately optimal harvest amount exceeds the socially optimal harvest amount by Q_r - Q_s, which from society's point of view implies too much timber is being harvested. Compared to Q_s, Q_r imposes an external cost on society given by area b + area c.

Social value is useful in evaluating public policies for alleviating externalities. Suppose Q_s can be achieved by policies such as imposing a tax on each unit of timber harvested, limiting the total area harvested, and/or designating areas where timber harvesting is not allowed, for example, near streams and lakes. A

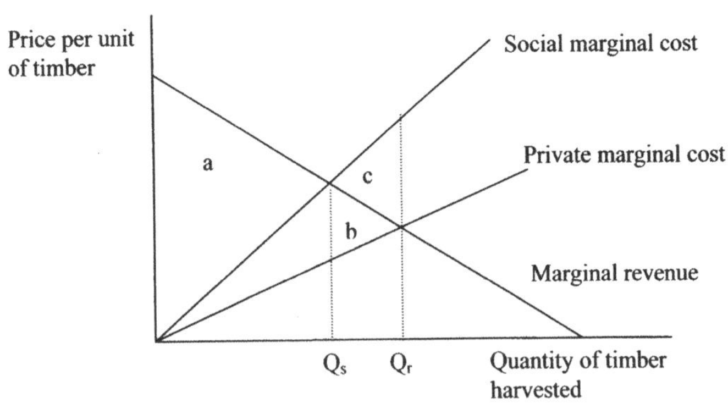

FIG. 4.6. Private and socially optimal amounts of timber harvesting with a negative externality. Q_r is the privately optimal amount of timber harvested and Q_s is the socially optimal amount of timber harvested. Social value at Q_s equals area a.

policy is cost effective when the cost of implementing it is less than the increase in social value, assuming the policy increases social value.

The preceding treatment of social value is static because the demand and supply curves are constant. Social value is dynamic because demand and supply curves change over time in response to variation in tastes and preferences, population, prices, income, and technology. Dynamics are explained in terms of whether or not to improve roads in a protected area. The proposed road improvements are expected to increase the demand for the area from Demand-B to Demand-A, and shift the supply curve from Supply-B to Supply-A as illustrated in figure 4.7. "B" indicates conditions before the improvements and "A" indicates expected conditions after the improvements. The supply curve shifts upward due to the cost of improving the roads.

The demand and supply curves in figure 4.7 show the proposed road improvements are expected to increase visitor days from Q_B to Q_A. That is to say, Q_B is the equilibrium between supply and demand (quantity demanded equals quantity supplied) before the improvements, and Q_A is the expected equilibrium after the improvements. Since social value equals area e + area f before the improvements and area d + area e after the improvements, the expected increase in social value as a result of the road improvements is area d – area f. Since area d exceeds area f, the improvements are expected to increase social value, which means they should be made. Whether social value rises or falls with road improvements depends on how the demand and supply curves shift relative to one another. Certain shifts (e.g., upward shift of supply curve exceeds downward shift

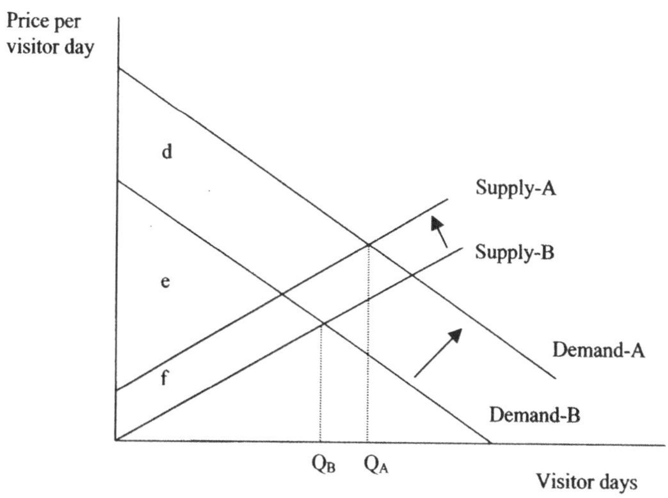

FIG. 4.7. Changes in visitor use and social value for proposed road improvements in a protected area. "A" indicates the conditions before the improvements and "B" indicates the expected conditions after the improvements.

of demand curve) result in a decrease in social value, which implies that the road improvements should not be made.

A comprehensive assessment of the proposed road improvements requires consideration of how it affects recreational demands and social values at other protected areas. For example, if the social value of other protected areas is expected to decrease as a result of the road improvements, then that decrease must be subtracted from area d – area f. If the change in social value is still positive after making the subtraction, then the improvements should be undertaken. Conversely, if the change in social value is negative after the subtraction, then the improvements should not be undertaken. When road improvements increase social value, the improvements are said to be economically efficient. Conversely, when road improvements decrease social value, the road improvements are said to be economically inefficient.

Distributional Equity and Ecological Sustainability

Social value is an efficiency-based criterion that is grounded in utilitarianism. When resources are allocated in a manner that maximizes social value, then economic efficiency is maximized. Any other allocation of resources lowers social value and is economically inefficient. While maximizing social value is desirable from an economic efficiency viewpoint, it is by no means the only goal worth pursuing for protected areas. Two other important goals are *distributional equity* and *ecological sustainability* (Daly 1992; Costanza and Folke 1997). While the goals of economic efficiency, distributional equity, and ecological sustainability are somewhat independent (Arrow and Raynaud 1986), management actions that are economically efficient (maximize social value) are likely to differ from ones that attempt to balance the three goals (Bishop 1993; Perrings 1994).

Achieving distributional equity requires consideration of how resources and property rights are distributed among different social groups in the same generation (intragenerational equity), among groups in different generations (intergenerational equity), and between humans and nonhumans (interspecies equity). The goal of distributional equity stands in stark contrast to the strong utilitarian-based goal of maximizing social value (social benefits minus social costs). Since strong utilitarianism ignores how benefits and costs are distributed among individuals and groups, it is possible for a set of management actions to be economically efficient but inequitable from an intragenerational, intergenerational, and/or interspecies viewpoint.

The goal of distributional equity can be pursued using an intrinsic rights approach, Kantian justice, or Rawlsian decision making. The intrinsic rights approach accords all living objects the same rights regardless of their benefits to humans. *Kantian justice*, as mentioned earlier, requires individuals to ignore personal gain or loss in judging alternatives. *Rawlsian decision making* invokes a "veil of ignorance" that requires individuals to evaluate alternatives as though they had no knowledge of how those alternatives might influence their personal welfare (Rawls 1971).

Moving from one management regime to another for a protected area is likely to involve trade-offs among the three goals. A trade-off between economic efficiency and distributional equity occurred when protected areas were established in the homeland of American Indians. Indians harvested natural resources in order to support their subsistence lifestyle. This still occurs today in places like Alaska and Canada's northern territories in North America and Cape York in northeastern Australia. Establishing a protected area in Indian homeland might maximize social value in terms of recreation and tourism, but it reduces the welfare of Indians when indigenous resource extraction is disallowed. This action is inequitable because the recreation/tourism industry benefits at the expense of Indian culture. Such an inequity is likely to cause Indians and others concerned about Indian welfare to oppose establishment of the protected area, unless provisions are made to accommodate subsistence living. This example illustrates a situation in which the goals of economic efficiency and distributional equity are in conflict. It may be possible to resolve the conflict by allowing the Indians to utilize the protected area for subsistence-based resource extraction as long as it does not diminish the recreational/tourist qualities of the area. This was achieved in several protected areas whose creation involved expropriation of Indian homelands. Another way to attain distributional equity is for individuals to vote their preferences for alternative management actions as a member of the community (cloaked in Rawls's veil of ignorance), or to engage in community-wide discussions aimed at reaching consensus on management actions that are equitable to various stakeholders (Costanza 2000).

The third goal (ecological sustainability) considers whether direct use of natural resources in the production of consumer goods and services, and indirect use in the production of ecological services are ecologically sustainable. Ecosystems are being used in an ecologically sustainable manner when their long-term functioning and resilience (productivity) are not impaired. For example, losses in native biodiversity reduce the functioning and resilience of protected area ecosystems, which, in the long run, decrease the supply of ecological goods and services, and human welfare (Barbier et al. 1995).

Striving to achieve ecological sustainability requires individuals to discuss and vote or reach consensus on their preferences for management actions. The decision framework for implementing ecological sustainability differs from the one used for economic efficiency, which is based on individual preferences for consumer goods and services, and the one used for distributional equity, which incorporates preferences for the distribution of resources and property rights.

Several countries have devised common property rules and other institutional arrangements for integrating economic efficiency, distributional equity, and ecological sustainability and the values underlying those goals. These include the development of Spanish *huertas* for irrigation, Swiss grazing commons (Ostrom 1990), Oceania's marine resource tenure systems (Johannes 1978), and Turkey's Mediterranean coastal fishery (Berkes and Folke 1994). Chapter 8 discusses a method for balancing the three goals.

Value Conflicts

The previous section indicates that changing management of protected areas is likely to entail trade-offs among the goals of economic efficiency, distributional equity, and ecological sustainability. When trade-offs do not exist, pursuing one goal either has no affect on attainment of the other goals, which implies independence of goals, or increases the attainment of other goals, which implies complementarity of goals. This section examines in greater depth the nature of conflicts between multiple goals or values for protected areas. Chapter 6 discusses approaches for resolving such conflicts.

Consider a national park that has three management goals: ecological integrity (EI), quality of visitor experiences (VX), and regional economic benefits (RE). Ecological integrity reflects ecological values, quality of visitor experiences embodies social values, and regional economic benefits encompass economic values. The three goals are appropriate for protected areas that are managed based on ecosystem management (Franklin 1997; Schowalter et al. 1997; Thomas 1997). Those goals are likely to be in conflict with one another. For example, high levels of visitation would increase economic benefits to gateway communities but possibly decrease ecological integrity and quality of visitor experiences.

Ecologically sustainable management of protected area ecosystems requires maintaining ecological integrity. Ecological integrity refers to the capacity of an ecosystem to provide ecological services and is measured in terms of soil, water, and air quality; vegetative cover; species richness and diversity; connectivity of landscape elements; and other indicators. Sustainability is achieved when:

1. All native plants and animals are well represented and distributed within a region
2. All ecosystems are intact
3. All evolutionary processes are functioning
4. All natural disturbances are allowed to function
5. The ecosystem as a whole is resilient to short-term and long-term environmental change (Yellowstone to Yukon Conservation Initiative, undated)

Several attributes influence the quality of visitor experiences, including aesthetically pleasing landscapes, wildlife abundance, congestion, cultural nationalism, emotional and spiritual renewal, and others (Runte 1987). Attributes of the quality of visitor experiences are experiential in nature, vary widely among visitors and types of recreational activities, and are not easy to evaluate (Box 1999).

Regional economic benefits take the form of income, employment, and tax revenues generated by recreation and tourism in protected areas (see the "Money Generation Model" sidebar). These benefits arise in connection with visitor expenditures at restaurants, service stations, motels, and other businesses; expenditures in support of visitor facilities and services, such as visitor centers, tour guides, horse, boat, and equipment rental, and educational institutes; and construction and maintenance of roads, campgrounds, trails, and administrative facilities.

Even though recreation and tourism are nonconsumptive uses of protected areas, high visitation rates can degrade natural resources and hence diminish the quality of visitor experiences. In addition, some recreational activities, notably use of personal watercraft, snowmobiles, sightseeing aircraft, and off-road vehicles, can seriously degrade the quality of natural resources and/or visitor experiences.

Money Generation Model

The original version of the Money Generation Model (MGM), developed in 1990 by Dr. Ken Hornback of the Denver Statistical Office of the National Park Service, is designed to estimate economic benefits to local economies of tourism expenditures in national parks. Benefits are estimated in terms of (1) increased purchases of goods and services, (2) increased sales and income tax revenues, and (3) new jobs. MGM provides estimates of economic benefits that are useful for planning, concessions management, budget justifications, tourism marketing, assessing the contributions of parks to gateway communities, justifying the upgrading of privately owned travel and tourism support facilities, and others. Four features of the MGM are noteworthy. First, it estimates only local economic benefits, not the benefits to the region or state in which the park is located. Second, local economic benefits pertain to visitor and park expenditures, not the benefits attributable to higher real estate values, improved recreational and cultural opportunities for local residents, enhanced community services, and other benefits generated by the park. Third, economic benefits are estimated using average values for taxable income ratios, indirect sales multipliers, and other factors. More accurate economic benefits can be obtained by using economic values that are specific to the impacted communities. Fourth, the MGM is designed to estimate benefits quickly and at very low cost.

In 2000, Daniel Stynes and Dennis Propst of Michigan State University expanded the MGM to include two additional sources of economic benefits, namely federal expenditures for National Park Service salaries, park construction projects, and other park-related activities, and other outside expenditures, such as state expenditures on park access roads and nonpark expenditures for marinas, motels, restaurants, and other park-related capital development projects. The expanded MGM, referred to as MGM2, is more useful in evaluating management, policy and marketing alternatives whether originating inside and/or outside the park. Use of MGM2 fosters partnerships within gateway communities, increases external support for park policies and interests, and clarifies the roles of national parks, local communities, and tourist-related businesses in attracting and serving visitors.

MGM2 has the following important features:

- The model is programmed as an Excel 7.0 workbook using macros and other tools to automate the calculations and present the user with a simple yet flexible interface.
- The model is available in three forms: a pencil and paper version, a "short form" Excel spreadsheet implementation of this, and a full-featured Excel workbook version. The short form represents a modest upgrade of MGM, while the full-featured MGM2 model adds extensive detail and flexibility for a modest increase in input data requirements. MGM2 produces much more complete estimates of impacts by economic sector and visitor segment. For example, MGM2 can estimate the number of jobs in the hotel sector resulting from park visitors staying overnight outside the park or the impacts that park campers have on local restaurants. The additional detail in MGM2 spending data and multipliers allows users to better customize the model to fit a particular park, local region, and particular management alternative to be evaluated.
- MGM2 encourages analysts to divide park visitors into segments with distinct spending patterns and possibly different responses to management alternatives. In particular, local visitors should be distinguished from tourists to the area, day visitors from overnight visitors, and campers and backcountry users from visitors staying in motels, cabins, and lodges. Each type of visitor has a distinct spending pattern and different impacts on the park, the local community and the region's economy. Visitors may also be segmented by activity, trip purposes, trans-

portation mode, or any other variables that help to explain spending patterns.

- MGM2 captures spending in the local area by all park visitors. Spending can be itemized in up to 12 spending categories (e.g., motel, restaurant, groceries, and souvenirs). Multipliers and economic ratios for each major tourism sector are used to estimate direct sales, income, value added, and jobs. Sector-specific multipliers capture differences in the direct and secondary economic impacts of spending in different sectors of the economy.
- For applications with limited visitor spending or local economic data, MGM2 provides sets of default or "generic" values that can be tailored to a particular application. MGM2 spending data are based on recent NPS visitor surveys so they represent park visitors, rather than general travelers. MGM2 multipliers are based on IMPLAN Pro 2.0 input-output models for local regions around park units.
- In selecting spending and multipliers for a particular area, MGM2 model users may choose from sets of defaults or generic values that represent types of parks or regions, they may adjust these to fit the application and retrieve existing data sets for other parks and regions, or they may enter their own data, when available.
- MGM2 establishes NPS spending and multiplier databases that can be directly used within the MGM2 model. These databases can be extended and updated over time. MGM2 developers have provided recommended visitor survey questions for measuring spending and utilities to extract multipliers for local areas around NPS units from the latest IMPLAN models.

MGM2 was used to estimate the following economic benefits for units of the National Park System in 2001:

- NPS visitors spent $10.6 billion in the local regions around national parks.
- Direct effects of visitor spending on gateway communities are $3.1 billion in wages and salaries (including payroll benefits) and 212,000 jobs in local tourism-related businesses.
- Direct and secondary impacts on gateway communities of spending by park visitors are $4.6 billion in value added, $4.5 billion in personal income, and 267,000 jobs.
- Park visitors who stayed overnight in commercial lodging in or near the park account for 56 percent of visitor spending. Day trips from outside

the local area account for 30 percent of spending and local residents and campers account for 7 percent each.

- Most of the direct jobs are in eating and drinking establishments (77,000), lodging establishments (59,000), retail trade (35,000), and amusements and entertainment (31,000).

These estimates do not cover all expenditures made by park visitors. They do not reflect receipts and operations of the National Park Service and visitor purchases at home or en route to national parks beyond a 20–67 km (30–100 mile) radius of the park. If visiting a park was not the primary purpose of a trip, then only the additional costs of the park visit are counted. Nevertheless, the economic benefits listed above are a good indicator of the economic losses that gateway communities would incur should park units be closed.

Source: National Park Service (2003a, 2003b).

Protected area managers need to understand the factors influencing achievement of the three goals (EI, VX, and RE) and potential trade-offs among them. Factors influencing achievement of the three goals include number of visitors, type of visitors, level and type of management, extent and type of visitor facilities, environmental quality, climate, and others. These factors affect goals in different ways. Increases in the number of visitors beyond some threshold level are likely to decrease the quality of visitor experiences and ecological integrity. Similarly, expanding and improving visitor services could increase the quality of visitor experiences (at least for some visitors) and regional economic benefits, but decrease ecological integrity. If climate change extends the summer tourist season and increases the number of visitors, then regional economic benefits would increase but ecological integrity could be adversely affected by more visitors. Climate change could adversely affect the quality of visitor experiences in protected areas located in northern latitudes by reducing the presence of certain large mammals (their ranges would move farther north), even though it might increase the number of visitors by extending the tourist season, which increases regional economic benefits. Extending the length of time that park concessionaire hotels are open makes it more attractive for banks to loan money to concessionaires for hotel renovations. While this change would increase regional economic benefits and the quality of visitor services (for those using the hotels), it could adversely affect ecological integrity by extending visitor impacts over a longer period of time.

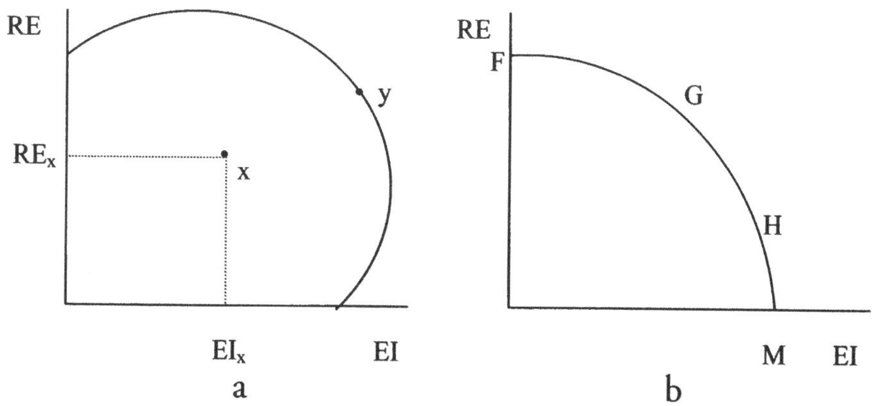

FIG. 4.8. Hypothetical trade-off curves between regional economic benefits (RE) and ecological integrity (EI).

Trade-offs among the three goals (EI, VX, and RE) can be illustrated using a three-dimensional trade-off surface (e.g., Prato 2000). Three-dimensional surfaces are difficult to understand. Therefore, this section uses a simpler two-dimensional trade-off curve shown in figure 4.8 to illustrate the trade-offs between RE and EI. A similar trade-off curve exists between RE and VX and between EI and VX. The trade-off curve between RE and EI in figure 4.8a has two positively sloped segments, one at each end of the curve. These segments indicate a complementary (win-win) relationship between RE and EI in which it is possible to obtain more of both goals. The remainder of the trade-off curve in figure 4.8a plus the entire trade-off curve in figure 4.8b is negatively sloped. This implies a competitive relationship between the two goals, that is, more RE results in less EI and vice versa.

The trade-off curve is based on two simplifying assumptions. First, it assumes that each combination of goals, such as x in panel a, results from a unique management action. This assumption is valid when there is no uncertainty regarding the outcomes (goals) achieved by management actions. Chapter 7 considers decision-making when there is uncertainty regarding the outcomes of management actions. Second, the continuous nature of the trade-off curve stems from the assumption that there are an infinite number of management actions. In practice, there are only a finite number of management actions and, hence, the trade-off curve is discontinuous or simply a series of points. Since it is easier to work with a continuous trade-off curve, the assumption is made of an infinite number of management actions.

The exact shape of the trade-off curve for a particular protected area is determined by the levels of human and financial capital, environmental conditions, and the state of science and technology. A small, resource-limited protected area

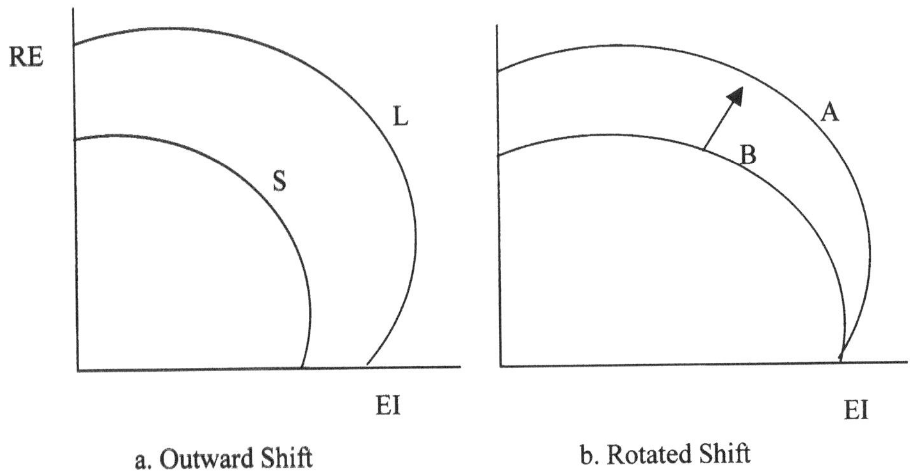

FIG. 4.9. Outward shift (panel a) and rotated shift (panel b) in trade-off curves between regional economic benefits (RE) and ecological integrity (EI). In panel a, S is a trade-off curve for a small, resource-limited protected area that employs outdated science and technology. L is the tradeoff curve for a large, well-funded protected area that utilizes up-to-date science and modern technology. In panel b, the trade-off curve rotates outward from B to A.

that employs outdated science and technology and generates lower levels of regional economic benefits and ecological integrity has a trade-off curve like S in figure 4.9a. In comparison, a large, financially well-funded protected area that employs up-to-date science and modern technology has a trade-off curve like L in figure 4.9a. What is important to note is the relative position of the trade-off curves, namely curve S is below curve L. Different protected areas have different trade-off curves, and the trade-off curve for the same protected area is likely to shift over time. Increases in the size of a protected area and the amount of human and financial capital available to it, and improvements in science and technology shift the trade-off curve for a protected area outward. On the other hand, global climate change can shift the trade-off curve toward the origin. Not all shifts in the trade-off curve are uniformly outward. For example, an improvement in technology that allows a protected area to attract more visitors without sacrificing ecological integrity (e.g., state-of-the-art mass transportation) causes the trade-off curve to rotate outward from B to A as illustrated in figure 4.9b.

A major management implication of the trade-off curve is that some combinations of goals and associated management actions are efficient and others are inefficient (Prato 1999, 2000). Consider the combination of goals given by point x in figure 4.8a. Combination x provides RE_x and EI_x, which is inefficient because it is possible to achieve more RE and/or EI by moving to a combination of goals on the trade-off curve, such as y. Therefore, any management action that leads to a combination of RE and EI below the trade-off curve is inefficient and should be rejected. By this reasoning, all management actions that achieve combinations of goals on the trade-off curve are efficient and those that do not are inefficient. Combinations of goals beyond the trade-off curve are not achievable under current conditions.

The first task of a protected area manager is to eliminate from further consideration management actions that are inefficient, namely those that yield combinations of goals below the trade-off curve. The first task requires only identification of the trade-off curve. The second task requires the manager to select an efficient combination of goals and associated management action. This task requires consideration of stakeholders' preferences for goals.

Not all combinations of goals on the trade-off curve are equally preferred by stakeholders. For example, consider the combinations along the segments of the trade-off curve labeled FG and HM in figure 4.8b. Combinations along segment FG are likely to be unacceptable to protected area managers and environmental interests because they do not provide enough EI. Conversely, combinations along the segment HM are likely to be unacceptable to economic development interests because they provide insufficient levels of RE.

To illustrate this point more clearly, suppose a business person's preferences for RE and EI are given by the indifference map shown in figure 4.10a. An indifference map consists of a collection of indifference curves that represent an individual's preferences for the goods or services shown on the axes. Combinations of RE

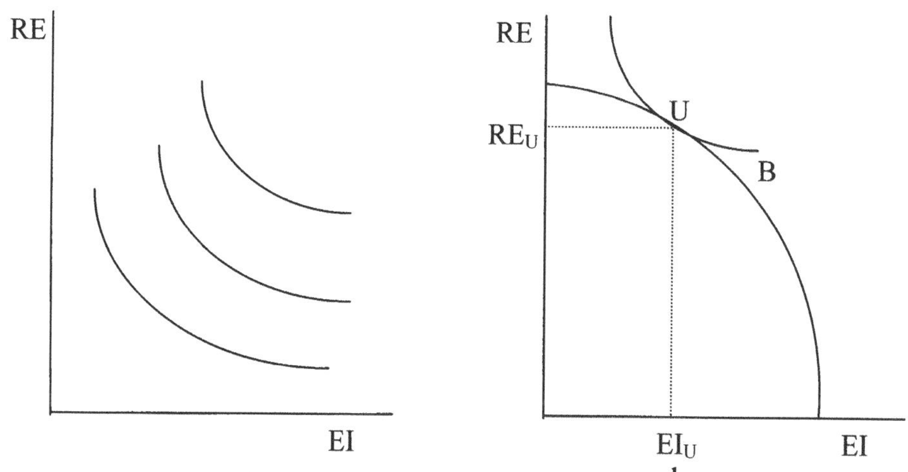

FIG. 4.10. Indifference map (panel a) and most preferred combination of regional economic benefits (RE) and ecological integrity (EI) for a business person (panel b). U designates the most preferred combination of goals for the business person.

and EI along the same indifference curve provide the same level of satisfaction or utility. Higher indifference curves contain more preferred combinations of RE and EI. Hence, moving from a lower indifference curve to a higher indifference curve increases an individual's satisfaction. The most preferred combination of goals for the businessperson is indicated by combination U in figure 4.10b, which provides RE_U and EI_U. Combination U is on the highest indifference curve (labeled B) that can be attained by the businessperson. Hence, U provides greater satisfaction to the businessperson than any other attainable combination and, therefore, the businessperson's most preferred management action is the one that provides U.

Consider a second stakeholder who has strong environmental interests. The environmental person is likely to have different preferences for RE and EI and hence a different indifference map than the businessperson. The highest indifference curve that the environmental person can achieve is labeled B in figure 4.11 (the indifference map for the environmental person is not shown). The highest indifference curve the environmental person can attain is labeled V. Hence, the most preferred combination of goals for the environmental person is given by N, which provides RE_N and EI_N. Comparing the combinations of goals provided by U and N shows that the environmental person prefers more EI relative to RE, whereas the businessperson prefers more RE relative to EI. Since points U and N provide different combinations of RE and EI, they imply different management actions (recall that a management action results in a unique combination of goals). Therefore, the businessperson prefers the management action that pro-

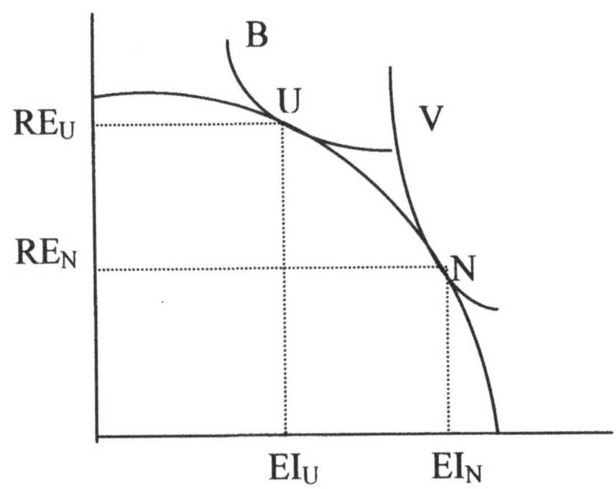

FIG. 4.11. The most preferred combination of regional economic benefits (RE) and ecological integrity (EI) for business (B) environmental (V) persons. N designates the most preferred combination of goals for the environmental person and U designates the most preferred combination of goals for the business person.

vides RE_U and EI_U and the environmental person prefers the management action that provides RE_N and EI_N.

Determining the trade-off curve is important because it identifies management actions that provide efficient combinations of RE and EI. Protected area managers should only consider management actions that provide combinations of RE and EI on the trade-off curve. Identifying stakeholder preferences for RE and EI is important because this information is needed to determine the most preferred combinations of RE and EI for different stakeholders. If a protected area manager decides to select a preferred management action based only on the trade-off curve and the indifference curves for stakeholders (the manager's preferences for goals are not considered), then a compromise management action would need to be selected that achieves a combination of goals between U and N. However, if the protected area manager decides to select a preferred management action based on both the stakeholders' preferences and the manager's preferences for goals, then determining the compromise management action requires consideration of three sets of preferences for goals. The situation is more complicated than the hypothetical example described here because, in practice, there are likely to be more than three goals and three sets of preferences to consider. Finding a compromise management action under these conditions is facilitated by using mathematical optimization methods.

Summary

Utilitarian, intrinsic, spiritual, and ethical values influence how people view national parks and protected areas. The value of ecological goods, such as timber, oil, and gas, are determined by market prices. Ecological services of protected areas, such as recreation, conservation of biodiversity, and purification of air and water, are public goods. Such goods have the properties of nonrivalry and nonexcludability and do not have market prices to guide their allocation and use.

Utilitarianism has its philosophical roots in anthropocentrism. Utilitarian theory implies that goods and services have value only if they satisfy human wants. Weak utilitarianism states that the value of a good or service to an individual depends solely on how much personal satisfaction (utility) he or she receives from it. Strong utilitarianism, which is the theoretical basis for cost-benefit analysis, implies that the value of a good or service is determined by summing the satisfaction it provides to all individuals that use it.

There are two forms of utilitarian value, use and nonuse. Use value can be either direct or indirect. Direct use can be consumptive or nonconsumptive. Nonuse values include option, quasi-option, bequest, and existence. Intrinsic value emanates from the ecocentric philosophy that all living objects have equal rights regardless of their value to humans. Intrinsic rights imply that all biotic elements of protected areas have equal moral status. Spiritual values of protected areas permeate the writings of Emerson, Thoreau, and Muir and are best exemplified by the American Indian culture.

Total economic value is the sum of direct use, indirect use, option, quasi-option, existence, and bequest values. It is the standard measure of total economic worth of a good or service. Direct use value can be measured using either market prices or, for nonmarketed goods, the travel cost, hedonic pricing, or hedonic travel cost methods. Nonuse values can be measured using contingent valuation, conjoint analysis, choice experiments, and fuzzy contingent valuation.

When production of a good or service results in a negative externality, it imposes an external cost on someone else. Social cost equals private cost plus external cost. A divergence between social cost and private cost occurs in the presence of negative externalities and causes the privately optimal level of production to exceed the socially optimal level of production of a good or service. Too much of the good/service is produced when there are negative externalities.

Maximizing social value results in economically efficient production of goods and services. Economically efficient production does not necessarily achieve the goals of distributional equity or ecological sustainability. For this reason, protected area managers should evaluate management actions in terms of their impacts on the three goals of economic efficiency, distributional equity, and ecological sustainability and identify management actions that provide efficient combinations of those goals. The satisfaction provided by various combinations of the three goals depends on stakeholders' preferences for goals. Since stakeholders typically have different preferences for the three goals, selection of a preferred management action for a protected area requires striking a compromise among stakeholders.

References

Arrow, K. J., and A. C. Fisher. 1974. Environmental preservation, uncertainty and irreversibility. *Quarterly Journal of Economics* 88: 312–319.

Arrow, K. J., and H. Raynaud. 1986. *Social choice and multicriterion decision-making.* Cambridge, MA: MIT Press.

Barbier, E. B., J. C. Burgess, and C. Folke. 1995. *Paradise lost? The ecological economics of biodiversity.* London, England: Earthscan Publications, Ltd.

Berkes, F., and C. Folke. 1994. Investing in cultural capital for a sustainable use of natural capital. In *Investing in natural capital: The ecological economics approach to sustainability,* ed. A. M. Jansson, M. Hammer, C. Folke, and R. Costanza. Washington, DC: ISEE/Island Press.

Bishop, R. C. 1993. Economic efficiency, sustainability and biodiversity. *Ambio* 22: 69–73.

Bodian, S. 1995. Simple in means, rich in ends: An interview with Arne Naess. In *Deep ecology for the 21st century,* ed. G. Sessions, 26–36. Boston, MA: Shambhala Publishers.

Box, P. 1999. *A GIS/ABM-based evolutionary algorithm for determining recreational anchorage carrying capacity using subjective information.* Book of abstracts, 1999 Congress on Recreation and Resource Capacity, Nov. 29–Dec. 2, Snowmass Village, Aspen, CO.

Brown, G., and R. Mendelsohn. 1974. The hedonic travel cost method. *Review of Economics and Statistics* 66: 427–433.
Capra, F. 1995. Deep ecology: A new paradigm. In *Deep ecology for the 21st century*, ed. G. Sessions, 19–25. Boston, MA: Shambhala Publishers.
Cherry, T. L., and J. L. Shogren. 2001. Invasive species management for the Yellowstone Lake ecosystem: What do visitors think? *Yellowstone Science* 9: 10–15.
Clawson, M. 1959. *Methods of measuring the demand for and value of outdoor recreation.* Washington, DC: Resources for the Future.
Costanza, R. 2000. Social goals and the valuation of ecosystem services. *Ecosystems* 3: 4–10.
Costanza, R., and C. Folke. 1997. Valuing ecosystem services with efficiency, fairness and sustainability as goals. In *Nature's services: Societal dependence on natural ecosystem,* ed. G. C. Daily, 49–70. Washington, DC: Island Press.
Costanza, R., et al. 1997. The value of the world's ecosystem services and natural capital. *Nature* 387: 253–260.
Cummings, R. G., D. S. Brookshire, and W. D. Schulze (Eds.). 1986. *Valuing environmental goods: An assessment of the contingent valuation method.* Totowa, NJ: Rowman and Allanheld.
Daily, G. C. 1997. Introduction: What are ecosystem services? *In Nature's services: Societal dependence on natural ecosystems,* ed. G. C. Daily, 3–6. Washington, DC: Island Press.
Daly, H. E. 1992. Allocation, distribution, and scale: Towards an economics that is efficient, just, and sustainable. *Ecological Economics* 6: 185–193.
Exxon Valdez Oil Spill Council. 2003. *Oil spill facts.* http://www.oilspill.state.ak.us/facts/index.html (accessed August 29, 2003).
Franklin, J. F. 1997. Ecosystem management: An overview. In *Ecosystem management: Applications for sustainable forest and wildlife resources,* ed. M. S. Boyce and A. Haney, 21–53. New Haven, CT: Yale University Press.
Goulder, L. H., and D. Kennedy. 1997. Valuing ecosystem services: Philosophical bases and empirical methods. In *Nature's services: Societal dependence on natural ecosystem,* ed. G. C. Daily, 23–47. Washington, DC: Island Press.
Henry, C. 1974. Option values in the economics of irreplaceable assets. *The Review of Economic Studies: Symposium on the Economics of Exhaustible Resources* 89–104.
Johannes, R. E. 1978. Traditional marine conservation methods in Oceania and their demise. *Annual Review of Ecology and Systematics* 9: 349–364.
Kant, I. 1963. Duties to animals and spirits. In *Lectures on ethics,* translated by Louis Infield. New York: Harper and Row.
Krutilla, J. V. 1967. Conservation reconsidered. *American Economic Review* 57: 777–786.
Naess, A. 1989. *Community and lifestyle: Outline of an ecosophy.* Cambridge, England: Cambridge University Press.
National Park Service. 2003a. *Money generation model.* http://planning.den.nps.gov/mgm/ (accessed April 3, 2003).
———. 2003b. *Money generation model—version 2. Systemwide impact estimates.* http://www.prr.msu.edu/mgm2/MGM2web.htm (accessed April 3, 2003).
Newmark, W. D. 1995. Extinction of mammal populations in western North American national parks. *Conservation Biology* 9: 512–526.
NIV study Bible. 1995. Grand Rapids, MI: Zondervan Publishing House.
Ostrom, E. 1990. *Governing the commons: The evolution of institutions for collective action.* Cambridge, England: Cambridge University Press.
Pearce, D. W. 1993. *Economic values and the natural world.* Cambridge, MA: The MIT Press.
Perrings, C. A. 1994. Biotic diversity, sustainable development, and natural capital. In *Investing in natural capital: The ecological economics approach to sustainability,* ed. A. M. Jansson, M. Hammer, C. Folke, and R. Costanza, 92–112. Washington, DC: Island Press.
Prato, T. 1998. *Natural resource and environmental economics.* Ames: Iowa State University Press.

———. 1999. Risk-based multiattribute decision making in property and watershed management. *Natural Resource Modeling* 12: 307–334.
———. 2000. Multiple attribute evaluation of landscape management. *Journal of Environmental Management* 60: 325–337
Randall, A. 1994. A difficulty with the travel cost method. *Land Economics* 70: 88–96.
Randall, A., and J. Stoll. 1983. Existence values in a total valuation framework. In *Managing air quality and scenic resources at national parks and wilderness areas,* ed. R. D. Row and L. G. Chestnut. Boulder, CO: Westview Press.
Rawls, J. 1971. *A theory of justice.* Oxford, England: Oxford University Press.
Rosen, S. 1974. Hedonic prices and implicit markets: Product differentiation in pure competition. *Journal of Political Economy* 82: 34–55.
Runte, A. 1987. *National parks: The American experience.* 3d ed. Lincoln: University of Nebraska Press.
Sagoff, M. 1994. Should preferences count. *Land Economics* 70: 127–144.
Schowalter, T., E. Hansen, R. Molina, and Y. Zhang. 1997. Integrating the ecological roles of phytophagous insects, plant pathogens, and mycorrihzae in managed forests. In *Creating forestry for the 21st century: The science of ecosystem management,* ed. K. A. Kohm and J. F. Franklin, 171–189. Washington, DC: Island Press.
Schultz, W., et al. 1983. The Economic benefits of preserving visibility in the national parklands. *Natural Resources Journal* 23 (January).
Seed, J. 1989. *Thinking like a mountain: Towards a council of all beings.* Gabriola Island, British Columbia, Canada: New Society Publishers.
Sessions, G. 1995. *Deep ecology for the 21st century.* Boston, MA: Shambhala Publishers.
Singer, P. 1975. *Animal liberation.* New York: Random House.
Synder, G. 1995. Cultured or crabbed. In *Deep ecology for the 21st century,* ed. G. Sessions, 47–50. Boston, MA: Shambhala Publishers.
Task Force on Economic Benefits of Protected Areas of the World Commission on Protected Areas (WCPA) of IUCN. 1998. *A new vision for protected areas, economic values of protected areas: Guidelines for protected areas management,* 11–24. Gland, Switzerland, and Cambridge, UK: IUCN.
Thomas, J. W. 1997. Foreword. In *Creating forestry for the 21st century: The science of ecosystem management,* ed. K. A. Kohm and J. F. Franklin, ix–xii. Washington, DC: Island Press.
Watson, R. A. 1983. A critique of anti-anthropocentric biocentrism. *Environmental Ethics* 5: 245–256.
Weisbrod, B. A. 1964. Collective-consumption services of individualized-consumption goods. *Quarterly Journal of Economics* 78: 471–477.
White, L. 1967. The historical roots of our ecological crisis. *Science* 155: 1203–1207.
Yellowstone to Yukon Conservation Initiative. Undated. *Preamble.* http://www.rockies.ca/y2y/default.htm.

CHAPTER

5

Institutional Environment

A sound understanding of a country's institutional environment and of the dynamics of institutional change can make...conservation...more effective.

—Stephanie Presber James

Institutions are the "set of rules actually used (the working rules or rule-in-use) by a set of individuals to organize repetitive activities that produce outcomes affecting those individuals and potentially affecting others" (Ostrom 1992). Rules are important because they influence human and organizational behavior and performance. This chapter addresses the institutional environment for national parks and protected areas. Specific topics include theory of institutions, new resource economics, institutional arrangements for managing different protected area systems, environmental statutes and rules for protected areas, the functions of courts and executive orders, and the role of interagency/interdisciplinary agreements, and environmental, scientific, educational, and professional organizations.

Theory of Institutions

The theory of institutions posits cause-and-effect relationships among institutions, human/institutional behaviors, and performance (outcomes) as shown in figure 5.1. Institutions (rules of the game) create incentives (awards) or disincentives (penalties) in the form of opportunities and constraints that influence human and organizational behaviors. These behaviors, in turn, influence the performance (outcomes) of people and organizations (North 1990; James 2001). Favorable performance (outcomes) reinforces or favors current behaviors and institutions (positive feedback) and unfavorable performance (outcomes) tends to stimulate behavioral and institutional change (negative feedback).

There are three primary elements of institutional theory: the nature of the institutional environment in which individuals and organizations operate, institutional incentives created by human and organizational behaviors, and institutional change. Each of these elements is discussed below.

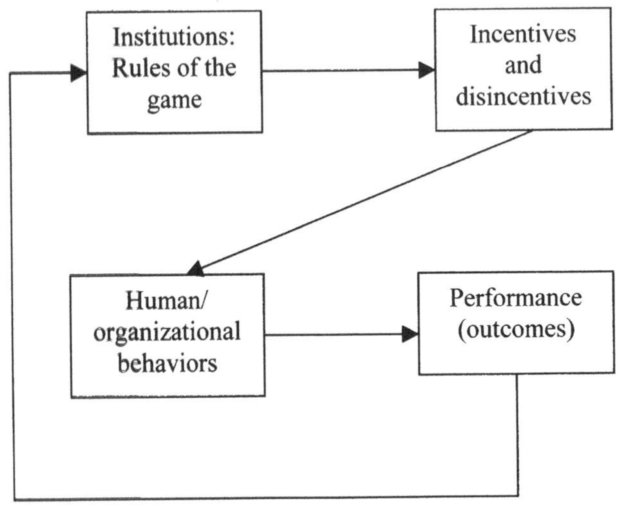

FIG. 5.1. Elements of institutional theory.

Institutional Environment

There are three interrelated determinants of the institutional environment: formal institutions, informal institutions, and enforcement. Formal institutions consist of written legally enforceable laws, regulations, policies, and property rights that control the socioeconomic and political interactions among people and organizations. For instance, the organic act for a national park is a formal institution that determines the size, location, and management objectives for a park. Organic acts vary with protected areas. For example, hunting is not allowed in most national parks. However, to accommodate traditional lifestyles, subsistence hunting, trapping, and fishing are permitted in Denali National Park and Preserve in Alaska (Denali National Park and Preserve 1999). Laws that assign property rights for wildlife to the state in which the wildlife are located are state institutions. The section on institutional arrangements discusses specific formal institutions affecting protected areas.

Informal institutions consist of unwritten rules that encompass cultural norms, social conventions, mores, beliefs, etiquette, traditions, and taboos (James 2001). By establishing expectations for individual and social behaviors, informal institutions increase the predictability of human behavior and, hence, social order. For example, backcountry etiquette sets a standard that backcountry users "leave only footprints."

Formal and informal institutions are related. Changes in formal institutions can alter expectations regarding individual and social behavior. Public policies to control invasive species on public lands often stimulate private landowners to control invasive species on their own land. Similarly, changes in informal institutions can be a precursor to changes in formal institutions. Voluntary "good citizen" efforts to keep public roadways free of litter eventually became institutionalized in litter laws that established fines for littering public roadways.

Informal institutions, such as social beliefs regarding the validity and role of formal institutions, influence acceptance of and compliance with formal institutions. For example, social attitudes toward subsistence hunting influence the extent to which poaching occurs in protected areas. If law enforcement officials are sympathetic to subsistence hunting in a protected area (informal institution), then violators are less likely to be apprehended and prosecuted even though poaching is illegal (formal institution). In this case, lax enforcement of poaching laws for subsistence hunters reduces the expected penalty for poaching, which increases the incentive for subsistence poaching.

Enforcement determines the ultimate success of formal and informal institutions. The meaning of enforcement differs between formal and informal institutions. When the behavior of individuals and organizations complies with formal institutions, there are typically no negative consequences. On the other hand, if a violation occurs and is detected, then legal action can be taken against the offending party. Even though informal institutions impose expectations regarding individual and organizational behaviors, poor performance with respect to those expectations is not illegal. Yet, violation of those expectations can have negative consequences, such as public humiliation, boycotts, mistrust, loss of social status, and others. Leaving trash in a backcountry campsite might not be illegal, but can be humiliating for parties caught doing it. Hence, there is an incentive for individuals and organizations to comply with informal institutions, namely, avoidance of the stigma and duress caused by noncompliance.

Institutional Incentives and Behavior

A fundamental premise of institutional theory is that the behavior of individuals and organizations is influenced by incentives (rewards) or disincentives (penalties) determined by the institutional environment and financial and technological constraints (North 1994). For purposes of this chapter, behavior refers to the decisions or choices that individuals or organizations make with respect to the selection, delineation, management, and use of national parks and protected areas. Rational behavior implies that individuals and organizations make decisions in a manner that maximizes personal or organizational gains subject to relevant constraints. Realized outcomes, in turn, influence individual or organizational performance. In short, incentives/disincentives influence decisions and decisions influence performance.

Behaving in what appears to be a rational manner does not always lead to desirable outcomes. The decision of the superintendent of Bandelier National Monument in New Mexico to do a prescribed burn in 2000 was consistent with National Park Service policy on wildland fire. A prescribed burn is a deliberate ignition of an area under predefined conditions for the purpose of achieving a management objective (Bunting 1996). A board of inquiry subsequently determined that questionable judgments were made in carrying out the prescribed burn policy that resulted in the Cerro Grande fire (National Park Service 2001a). Weather conditions caused the Cerro Grande fire to burn well beyond the target area causing the destruction of two hundred homes in Los Alamos, New Mexico.

The link between incentives/disincentives and behavior is illustrated in terms of compliance with a law (formal institution). Degree of compliance with a law depends on the likelihood that a violation is detected and the expected financial and other losses from being found guilty should the violator be detected and prosecuted. Consider, for example, the decision to poach in a particular game reserve by two groups of poachers, namely, commercial poachers that sell what they poach in the black market and subsistence poachers that use what they poach for survival. The disincentive to poach is clear; poaching violates the law and guilty parties are subject to fines and jail sentences. In reserves where poaching is a problem and the number of law enforcement officers is insufficient, there is a high likelihood that poachers will not be detected. On the other hand, the incentive for commercial poaching is high particularly if the reserve contains species whose parts have high market values. Hence, from a commercial poacher's perspective, the expected net gain from poaching is high and therefore poaching occurs. A different incentive exists for subsistence poachers. Even though poaching is illegal, there is a high incentive for subsistence poaching, namely, survival. Enforcement, however, is likely to be lax for subsistence poachers especially when law enforcement officers live in the same communities as the poachers. In other words, law enforcement officers have a disincentive to apprehend subsistence poachers, namely, chastisement from fellow residents. Under these conditions, subsistence poaching is more likely to be tolerated than commercial poaching. This example illustrates that two distinct incentives/disincentives can coexist for poaching on the same reserve, one for commercial poachers and another for subsistence poachers. It is not uncommon for incentives to vary across groups.

Incentives/disincentives are generated by many kinds of institutions. For example, a national park must comply with its organic act, environmental regulations for protecting air and water quality, local comprehensive planning ordinances, and views toward property rights. Interaction of diverse incentives/disincentives creates a unique incentives/disincentives structure for each national park and protected area and, hence, unique individual and organizational behaviors and performance.

Institutional Change

Feedback loops between institutions, incentives/disincentives, behavior, and performance (see fig. 5.1) cause change in one element to stimulate change in the other elements. This dynamic brings about institutional change. When performance is unacceptable, pressure mounts to change the underlying institutions (formal and informal rules of the game) in an effort to improve performance. A management challenge facing federal land management agencies is that what one group of stakeholders deems as acceptable performance another group views as unacceptable. Furthermore, institutional inertia (propensity of individuals and organizations to resist change) can delay or prevent institutional change.

Consider the following examples of institutional change. Prior to 1996, federally managed recreational lands in the United States, including national parks, were required to return all entrance fees to the headquarters of the federal agency overseeing the management of those lands. Under this institutional arrangement, there was a financial disincentive for a national park to attract and retain tourists because more visitors add to the workload but do not provide additional funds. In 1996, Congress authorized the Recreational Fee Demonstration Program, which allowed up to one hundred units managed by the National Park Service, Bureau of Land Management, Fish and Wildlife Service, and Forest Service to retain 80 percent of total entrance fees. Retained fees were earmarked for paying operating and maintenance costs for recreation areas. One hundred units in the National Park System are in the program (National Park Service 2001c). This demonstration program essentially gave participating units property rights to 80 percent of the revenues they generated and, consequently, changed both the incentives and constraints for those units. Specifically, it increased the incentive to attract and cater to visitors, and increased the budget for managing recreational lands.

Individuals and organizations external to a management agency can initiate institutional change. Yellowstone National Park's natural fire program adopted in 1972 allowed lightning-caused fires to burn as long as they did not threaten human life, property, historic and cultural sites, special natural features, and threatened and endangered species (Varley and Schullery 1991). The Yellowstone fires of 1988 burned 45 percent of the acreage in the park (400,000 ha, 988,440 acres) and an additional 170,000 ha (437,385 acres) outside the park (Bunting 1996). While most scientists agreed that the Yellowstone fires were ecologically beneficial, regional residents and the general public strongly objected to letting the fires burn for fear they would destroy the regional tourist industry. The tremendous negative reaction to the fires created a strong incentive for the National Park Service to fight the fires, with little effect in terms of controlling their extent, and change its natural fire program (Glick and Alexander 2000). Specifically, the NPS fire management policy was changed to allow fires only in a few large wilderness areas and isolated portions of national parks (Bunting 1996). A similar response occurred following a prescribed burn in Bandelier National Monument that resulted in the Cerro Grande fire. This episode prompted the National Park Service to review and revise its policy for prescribed burns.

Environmental interest groups often make negative statements and, in some cases, file lawsuits against federal land management agencies for what they consider to be poor performance or undesirable outcomes. Groups that engage in this behavior are known as institutional entrepreneurs; they find it advantageous to advance their cause by initiating institutional change (North 1994).

Another example of institutional dynamics is the evolution of the natural regulation policy in the National Park Service and Yellowstone National Park in particular. Prior to 1968, the National Park Service took a heavy-handed approach to natural resource management in the park. Species thought to be at odds with the goals of the park were extirpated. Thousands of elk, deer, and large predators were legally killed and poached; other species (lynx, wolf, and cougar) almost became extinct (Craighead 1991); and forest fires were suppressed.

The *Leopold Report on Wildlife Management in National Parks* (Leopold et al. 1963) recommended that natural processes be allowed to operate and dominate in national parks, and that parks be managed as vignettes of primitive America (Keiter 1996). This recommendation did not imply total hands-off management, but rather active management using the most advanced techniques (Craighead 1991). The National Park Service generally ignored the recommendations made in the Leopold Report for many years, but in 1968 it adopted a natural regulation policy, which was a major institutional change affecting park management. In this case, writers of the Leopold Report were the institutional entrepreneurs and the National Park Service exhibited institutional inertia.

Institutional environments that generate incentives for institutional change exhibit adaptive efficiency (North 1990). Rewarding creativity, promoting innovation, and encouraging decentralized decision making are ways to achieve adaptive efficiency. Higher adaptive efficiency means society responds to new challenges in a positive and proactive way (James 2001). When institutional inertia is high, adaptive efficiency is low, and it is more difficult for institutional entrepreneurs to implement change. Since formation of institutions involves high start-up costs and considerable investment of social capital, institutional change is threatening to individuals and organizations. Consequently, entrepreneurial activity aimed at changing institutions is often resisted. When this takes place, institutional change is slow to occur (North 1990).

New Resource Economics

A classic criticism of institutional performance for national parks is *Playing God in Yellowstone* (Chase 1987). In this book, Chase argues that the history of management of Yellowstone National Park demonstrates mistakes, inefficiencies, poor judgment, and politically motivated decisions. Examples of mismanagement cited by Chase include the natural regulation policy that allowed the 1988 Yellowstone fires to burn unabated for several weeks and overgrazing by elk and bison in the park's northern range. Chase's arguments have been criticized for being ecologically flawed (Boyce 1991).

Public opinion strongly favors having federal agencies, such as the National Park Service, manage public land. Many oppose the privatization of public land (transferring ownership and management to a private entity) for fear it would result in resource exploitation and degradation. The tension between Chase's arguments that the National Park Service mismanaged Yellowstone National Park and public advocacy of federal management of national parks and protected areas raises the following issue. New resource economics (NRE) poses the question: Can public land be managed in a way that preserves the values embodied in the public trust doctrine while capturing some of the benefits that result from allocating natural resources using market-based incentives? NRE answers this question in the affirmative by arguing that mismanagement of public land can be alleviated by establishing well-defined property rights, institutions, and private incentives. An underlying premise of NRE is that federal mismanagement of public land results because of flawed institutions and inappropriate incentives, not because federal managers are bad people. NRE maintains that individual and organizational performance in public land management would be improved by forming institutions that link the self-interest of managers with the desires of stakeholders (Copeland 1990; Stroup 1990).

NRE utilizes public choice theory to explain why individual and organizational behaviors result in undesirable performance. Public choice theory is especially applicable to public land management, regardless of whether or not the land is protected. It offers the following motivations and behaviors of federal managers, agencies, and executive officers having authority for the management of public land (adapted from Copeland 1990):

- Federal managers are motivated by self-interest and the desire for personal advancement in salary, rank, and prestige. Their primary objective is to increase the number of people they supervise and the size of their budget (Niskanen 1971). In contrast, the primary goal of private managers is to maximize profit and perhaps other objectives.
- There is no incentive for federal managers to monitor program costs, and there may be an incentive to increase costs when budget justification is based on costs. Regarding the latter, if a federal agency does not spend its entire annual budget, then there is a high likelihood that the following year's budget allocation will be reduced. In contrast, private companies have an incentive to reduce costs in order to increase profits.
- Federal managers cannot determine the value of the goods and services provided by the units they manage because access fees are administratively determined rather than being market based. Therefore, the zero or low fee charged

for access to public land is not a good indicator of value. This is not the case for marketed goods and services whose prices adequately reflect social value in the absence of negative externalities. Higher relative prices for marketed goods and services signal private companies to allocate more resources to those goods and services. Since access fees for public land are politically determined, they cannot serve the resource allocation function of market prices. For example, public grazing fees are lower than private grazing fees for livestock. This creates an incentive for livestock producers to overgraze public grazing land, other things equal. Critics claim that low fees result in ecological degradation of public grazing land.

- Federal managers face a perverse set of incentives, which causes resources to be allocated to less-productive or unproductive activities. For instance, some have argued that a major weakness of the Endangered Species Act is that it does not allocate human and financial resources to species protection until the species is endangered or threatened. A corollary of this argument is that allocating limited human and financial resources to protecting species before they become threatened or endangered would increase conservation benefits per dollar of expenditure.
- Federal managers have authority, but they are not necessarily held accountable for their decisions. In other words, there are often no adverse economic consequences of making a wrong management decision for public land. This occurs because civil service laws insulate federal managers from negative consequences of poor management decisions. For example, federal managers could not be held financially responsible for the economic losses that would occur if bison leaving Yellowstone National Park infected domestic cattle with brucellosis.
- Special interest groups have a disproportionate influence on federal managers' decisions. An individual's cost of being informed about public land management issues is large relative to the benefits. The converse is true for special interest groups, which explains their strong and effective lobbying campaigns to influence public land management decisions. Furthermore, federal managers understand that catering to special interest groups increases their chances of rewards in the long run.

  For example, in an effort to comply with the Endangered Species Act (formal institution strongly supported by conservationists) and gain political support, President Clinton implemented the Northwest Forest Plan, which terminated logging in national forests in the Pacific Northwest in an effort to protect late successional and old-growth forest related species within the range of the northern spotted owl. It was estimated that the plan would reduce timber harvesting in the region by more than 75 percent relative to 1980 levels (Johnson 1997). Environmental groups strongly supported and timber interests strongly opposed the plan. Supporters had the Endangered Species Act on their side. To appease the timber industry, the federal government established and funded a program to retrain displaced workers.

- Federal managers often make decisions that engender support from special interest groups. The argument here is that federal agencies pay particular attention to politically powerful, well-informed, and well-organized environmental interest groups, but only pay lip service to the less informed and poorly organized general public (Stroup 1990). For example, Yellowstone National Park managers are very sensitive to the opinions of the Greater Yellowstone Coalition, which is a special interest group that works to preserve the ecological integrity of the Greater Yellowstone Ecosystem.

In addition to the above motivations and behaviors, the personal values of protected area managers strongly influence the way they make certain decisions.

Implementing New Resource Economics

The theory of institutions, public choice theory, and NRE are based on the fundamental tenet that incentives drive behavior and influence the performance of people and organizations. Prescriptions offered by NRE for improving individual and organizational performance of federal land managers and agencies involve changing the incentives they face. NRE argues that the performance of federal managers is likely to be low when the full social benefits and costs of decisions are not internalized in the decision-making process. This is the same reason why negative externalities occur in production and consumption of private goods and services (see chapter 4).

Inattention to full social costs and benefits does not occur because federal managers are malevolent, but because they do not have adequate information about the social costs and benefits of alternative management actions. If better information existed concerning the social costs and benefits, then federal managers would make decisions that would be more likely to maximize net social benefit (social benefits minus social costs).

For example, the National Park Service's initial decision to let the fires burn in Yellowstone National Park in 1988 was based on the natural regulation policy in place at the time. This policy was predicated on the fact that fire is a natural disturbance and has ecological benefits. However, the policy did not consider the full social cost of the fires in terms of society's perception that fires seriously damage park resources and private lands adjacent to the park. Public outrage at watching Yellowstone National Park burning on the nightly news precipitated an eleventh hour decision to fight the fires. A complicating factor in responding to fire and other forms of natural disturbance is the inherent uncertainty surrounding the timing, extent, and consequences of disturbance events. Yellowstone Park managers did not know how much of the park would burn when the fires started, although fuel load conditions and weather were primed for a major conflagration. Similarly, the superintendent for Bandelier National Monument had no idea that the prescribed burn authorized for that area would spread to Los Alamos and destroy two hundred homes.

Unfortunately, it is difficult for federal managers to compare the social benefits and costs of alternative management actions due to incomplete markets or no markets for ecological goods and services. This situation occurs because efficient property rights have not been established for ecological goods and services provided by public land. Property rights are institutional rules that govern and facilitate the use and exchange of resources and commodities (Prato 1998).

Consider, for example, the importance of efficient property rights in facilitating the market transactions for bread. First, bread cannot be sold unless it is owned; the grocer owns the bread. This characteristic of property rights is known as ownership. Second, bread cannot be bought unless there is a way to transfer property rights; the buyer pays legal tender to the grocer and the grocer transfers ownership to the buyer. This characteristic of property rights is called transferability. Third, a mutual understanding must exist regarding the bundle of rights that apply to the item being sold, which determines what can and cannot be done with the good or service. Buyer and seller both understand that selling the bread does not give the buyer a right to litter the grocery store with bread. This act violates the grocer's property rights, which are not transferred by the sale of bread. This characteristic of property rights is known as specificity. Fourth, the transaction of exchanging legal tender for bread would be meaningless unless property rights are enforced. Taking bread out of the grocery store without paying for it constitutes theft, which is punishable by law. Lack of law enforcement creates an incentive to steal bread. This characteristic of property rights is referred to as enforceability.

Well-defined, enforced, and transferable property rights for ecological goods and services promote practical and efficient decisions, environmental protection, and sound resource stewardship (Copeland 1990). Efficient property rights for ecological goods and services would permit managers to evaluate the social benefits and social costs of alternative management actions and should result in more socially optimal decisions. Defining property rights is one way to resolve conflicts between use of snow machines and other winter activities in Yellowstone and Grand Teton National Parks (see sidebar).

As mentioned earlier, the benefits of establishing efficient property rights include promoting practical and efficient actions, environmental protection, and sound resource stewardship. Efficient property rights should only be established when the expected benefits exceed expected costs of establishing them. When benefits exceed transaction costs, then creating efficient property rights increases social welfare. Conversely, when transaction costs exceed benefits, creating efficient property rights decreases social welfare. Transaction costs are the costs of negotiating efficient property rights (Prato 1998). Such costs are generally low relative to the benefits for goods and services that are routinely purchased by large numbers of individuals and where the parties to exchange are not in adversarial positions. This is the case for marketed goods and services. However, when numerous parties are involved and there is competition among resource uses, as is the case for snow-machine use in Yellowstone National Park, transaction costs are very high. Under these conditions, establishment of property rights is not an efficient way to resolve resource use conflicts. Another different approach is needed.

Defining Property Rights for Winter Recreation in Yellowstone and Grand Teton National Parks

Well-defined property rights can be used to resolve use conflicts in national parks. Consider winter recreation in Yellowstone and Grand Teton National Parks.

In 2000, the National Park Service issued a record of decision that snow machine use in both parks would be discontinued by 2003. The National Park Service decided to phase out snow machine use because it adversely affects wildlife populations, creates significant air and noise pollution, and competes with other types of winter recreation, such as snowshoeing and cross-country skiing. Snow machine users claimed there was insufficient evidence that snow machines have negative impacts on wildlife and they had a right to recreate in the park. Many private businesses in West Yellowstone opposed the Park Service's decision on grounds that it would significantly reduce income and employment in their community.

As a result of a lawsuit brought by the snow machine industry, the National Park Service did a supplemental environmental impact statement (SEIS) to solicit further public comments on the earlier decision. The framework proposed in the SEIS would allow snow machines with strict limitations, provide for cleaner and quieter snow machines and snowcoaches using best available technology, and utilize an adaptive management approach to adjust snow machine numbers in the park.

In order to establish efficient property rights for winter recreation, wildlife, and air quality protection, and economic development, it would be necessary to first determine whether (1) wildlife has a right to be protected or individuals have a right to pursue various forms of winter recreation, (2) all park visitors have a right to clean air and peace and quiet or snow machine users have a right to pollute the air and disturb peace and quiet, and (3) businesses in West Yellowstone have a right to the income and employment benefits generated by snow machine use in the park or the park has a right to deny those benefits in order to protect natural resources and preserve other forms of winter recreation in the park. For simplicity, property rights could be used to resolve conflicts between snowmobile use and wildlife values. If wildlife were granted a well-defined property right for protection, then snow machine users would be required to provide compensation for wildlife losses associated with snow machine use. Conversely, if snow machine users were assigned property rights to winter recreation in the park, then they would need to be compensated for losses associated with restrictions imposed on snow machine use to protect wildlife.

Assigning well-defined property rights to all affected parties forces federal managers to consider those rights in making management decisions. Assignment of property rights is more difficult to accomplish when the impacts of snow machine use on air pollution, other forms of winter recreation, and West Yellowstone are taken into account.

Source: Yellowstone Center for Resources (2003).

Park Endowment Board

As long as the general public is unwilling to transfer ownership of public land to the private sector (privatization), federal agencies, such as the National Park Service, will continue to manage public land. A park endowment board (PEB) has been proposed as a way of retaining public ownership and public values of national parks and protected areas while reaping some of the benefits of improving resource allocation by using market-based incentives (Stroup 1990). A PEB would have the authority and responsibility to manage a park so as to enhance the stated values determined by Congress. For example, parks could be managed for values, such as the provision of visitor services, educational and research opportunities, and resource protection. If authorized by Congress, the PEB could buy new land and/or lease, sell, and rent land under its jurisdiction. Land sales would be subject to congressional approval. Selling or leasing of mineral, grazing, and other property rights would be allowed provided it is not expressly prohibited in the park's organic act and/or inconsistent with stated values of the park. Board members would be appointed by the president and confirmed by Congress. If a park has more than one board, then the boards would be required to collaborate in making decisions. Each board would receive an initial financial endowment and be expected to supplement that endowment by charging user fees that cover the cost of operating the park and engaging in other revenue-generating activities, such as buying and selling of land and leasing exploration and development rights for minerals.

In theory, a PEB would manage the park according to a narrowly defined mission established by Congress, and board members would be legally accountable to Congress for properly executing that mission. Proponents of NRE argue that a PEB would perform better than management by the National Park Service. They argue that the Park Service is influenced by a set of incentives and constraints (listed above) that are not consistent with achieving the public interest and is not held accountable for its actions. Purportedly, these conditions result in management actions that are inequitable, inefficient, and sometimes environmentally destructive (Stroup and Baden 1983).

Proponents of strict federal management of national parks and protected areas argue just as vociferously that even if creating a PEB is a more efficient way to manage a national park, it is likely to lead to greater resource exploitation and environmental degradation than federal management, which is unacceptable. This criticism is not entirely valid because Congress would still have the authority to define the park's mission (the stated values that guide management) and that mission could disallow resource extraction activities, such as oil and gas drilling. However, Congress can permit resource extraction. Since the mission established for each park is open to public scrutiny (it must be approved by Congress), antagonists of resource development in national parks could lobby Congress to support their positions.

One of the responsibilities of a PEB is to evaluate trade-offs between allowable and competing uses of park resources and determine the best mix of uses. For example, if the mission allows energy mineral leasing rights in the park, then the PEB must weigh the potential revenue, income, and employment benefits of leasing energy mineral rights to private companies against the potential costs in terms of reductions in visitation rates, the quality of visitor experiences, and environmental quality. Based on a comparison of benefits and costs, the PEB would decide whether or not to allow resource exploration and development in the park and, if so, under what conditions. In all cases, the PEB would be held responsible to Congress for decisions that adversely affect the park. A benefit of allowing resource extraction on parklands is that it would generate revenues that the PEB can use to protect and enhance the park's ecological integrity and biodiversity, such as purchasing additional land for the park (boundary expansion), buying development rights to timber and mineral extraction on lands adjacent to the park, buying private inholdings in the park when leases run out, and/or financing land exchanges. PEB has not been established for any national park. Yet it affords an alternative way to manage national parks and protected areas that relies, in part, on market-based incentives to manage public lands as promoted by NRE.

Institutions for Protected Areas

Institutional arrangements prescribe, shape, and limit the behavior and responsibilities of individuals and organizations that use and manage protected areas. In the United States and other countries, institutional arrangements take the form of statutes, regulations, executive orders, court decrees, agreements, and organizations. Congress enacts laws or statutes creating federal agencies and individual protected areas and authorizes and appropriates funds for those agencies. Numerous statutes determine and influence a country's protected-area system. These laws include organic acts designating the federal agencies responsible for managing protected areas; organic acts establishing individual protected areas; and environmental laws to protect air, water, declining species, and other natural resources. Executive agencies write regulations describing how particular laws are implemented. The president issues executive orders and has authority to designate national monuments. Federal courts decide whether laws have been violated and, if so, require a federal agency to take remedial action to comply with the law. The nature of court decrees varies over time in response to changes in public values, tastes and preferences, new knowledge, political opinions, and turnover in federal judges (Wagner et al. 1995). Executive agencies translate organic acts and enabling legislation for individual units into regulations that guide planning and management actions. This section describes institutional arrangements pertinent to national parks and protected areas in the United States.

The United States has several protected area systems that are governed by various statutes and regulations. Statutes and other institutions that govern national parks and other protected areas are described in the remainder of this section.

National Park System

Passage of the Yellowstone National Park Act (16 U.S.C. §§ 21–22) in 1872 created the world's first national park. Land in Yellowstone National Park was "reserved and withdrawn from settlement, occupancy, or sale under the laws of the United States, and dedicated and set apart as a public park or pleasuring-ground. The act required that "the awesome manifestations of nature" in the upper watershed for the Yellowstone River be preserved "for the benefit and enjoyment of the people." In 1916, the National Park Service Organic Act (16 U.S.C. §§ 431–33) established the National Park Service (NPS) in the U.S. Department of the Interior and charged it with managing Yellowstone National Park and most of the other national parks, monuments, and reservations that existed at the time. According to the National Park Organic Act, "the fundamental purpose of said parks is to conserve the scenery and the natural and historic objects and the wildlife therein and to provide for the enjoyment of the same in such manner and by such means as will leave them unimpaired for the enjoyment of future generations" (Rockwood 1988). Canada's National Parks Act of 1930 similarly requires that national parks be "dedicated to the people of Canada for their benefit, education and enjoyment...and shall be maintained and made use of so as to leave them unimpaired for the enjoyment of future generations" (Wilkinson 2000). In addition, starting in 1979, Canada established the maintenance of ecological integrity as a prerequisite to use and, in 1998, Parliament amended the National Parks Act to require that "[m]aintenance of ecological integrity through the protection of natural resources shall be the first priority when park zoning and visitor use are in a management plan" (Canada 1988; Dearden and Rollins 2002). Over the years, there have been dramatic increases in visitation and external threats to the ecological integrity of national park units in the form of natural resource extraction and rural residential development. These changes have brought into question the feasibility of achieving the dual mandate of conserving scenery, natural and historic objects, and wildlife, and providing for public use and enjoyment in perpetuity.

The NPS Organic Act established natural, historical, and recreational areas. Natural area designations are national park, national monument, and national preserve. Historical site designations are national historic site, national historic park, national memorial, national battlefield, national parkway, and national cemetery. Recreational designations are national recreation area, national seashore, national lakeshore, national river, and national trail. Table 5.1 describes these designations in detail.

In 1970, the National Park System General Authorities Act stipulated that all areas administered by the National Park Service be combined in one National Park System defined as "any area of land and water now or hereafter administered by the Secretary of the Interior through the National Park Service for park, monument, historic, parkway, or other purposes." The Redwood National Park Act of 1978 (16 U.S.C. §§ 1a-1, et seq.) amended the General Authorities Act of 1970 and reaffirmed that "protection, management, and administration of these areas [in the National Park System] shall be conducted in light of the high public value and integrity of the National Park System and shall not be exercised in derogation of the values and purposes for which these various areas have been established."

As of this writing, the National Park System contains 386 units covering more than 33.2 million ha (83 million acres). Units range in size from 5.3 million ha (13.2 million acres) (Wrangell-St. Elias National Park and Preserve, Alaska) to 0.003 hectares (0.02 acres) (Thaddeus Kosciuszko National Memorial, Pennsylvania) (National Park Service 2001d).

The Organic Act has undergone several amendments. Section 1a-5 was added to the Organic Act in 1976. It requires the secretary of the interior to evaluate areas that could be added to the National Park System, such as areas adjacent to the

Table 5.1. Description of National Park System Units

National Park	Generally large natural places having a wide variety of attributes, at times including significant historic assets. Hunting, mining, and consumptive activities are not authorized.
National Monument	Areas established by the president under authority of the Antiquities Act of 1906. Designates as national landmarks, structures, and other objects of historic or scientific interest situated on lands owned or controlled by the government.
National Preserve	Areas having characteristics associated with national parks, but in which Congress has permitted continued public hunting, trapping, oil/gas exploration, and extraction. Many existing national preserves, without sport hunting, would qualify for national park designation.
National Historic Site	Usually contains a single historical feature that was directly associated with its subject. Derived from the Historic Sites Act of 1935. Secretaries of the interior established a number of historic sites, but most have been authorized by acts of Congress.
National Historical Park	Generally applies to historic parks that extend beyond single properties or buildings.
National Memorial	Commemorates a historic person or episode; it need not occupy a site historically connected with its subject.
National Battlefield	Includes national battlefield, national battlefield park, national battlefield site, and national military park. In 1958, an NPS committee recommended national battlefield as the single title for all such parklands.
National Parkway	Refers to a roadway and the parkland paralleling the roadway. All were intended for scenic motoring along a protected corridor and often connect cultural sites.
National Cemetery	There are presently 14 national cemeteries in the National Park System, all of which are administered in conjunction with an associated unit and are not accounted for separately.
National Recreation Area	Twelve NRAs are centered on large reservoirs and emphasize water-based recreation; five other NRAs are urban parks located near major population centers.
National Seashore	Ten national seashores have been established on the Atlantic, Gulf, and Pacific coasts; some are developed and some are relatively primitive. Hunting is allowed at many of these sites.
National River	Includes national river and recreation area, national scenic river, wild river, etc. The first was authorized in 1964 and others were established following passage of the Wild and Scenic Rivers Act of 1968.
National Trail	Includes over 5,456 km (3,600 mi) in national scenic trails and national historic trails; authorized under the National Trails System Act of 1968.
Affiliated Areas	Comprise a variety of locations in the United States and Canada that preserve significant properties outside the National Park System.
Other Designations	Units that bear unique titles or combinations of titles, like the White House and Prince William Forest Park.

Source: Adapted from U.S. Department of the Interior (2000).

park that would enhance the park's ecological integrity, and to submit an annual report to Congress listing these areas. Expanding park boundaries requires congressional approval. In 1978, sections 1a-1 and 1a-7 of the Organic Act were amended to clarify the secretary's responsibility for complying with the Organic Act. Amended section 1a-1 mandates protection of park units from internal and external threats and section 1a-7 requires that general management plans for parks include measures to preserve park resources and indications of potential boundary modifications (Keiter 1985). Following these amendments, in 1984, the secretary promulgated regulations to control the use of boats, aircraft, and backcountry areas in park units (Keiter 1985).

Individual units of the National Park System are established by park-specific enabling legislation, except for the eight units that entered the system via the Cooperation Agreement Act of 1946 (16 U.S.C. §§ 17). Enabling legislation for individual units of the National Park System is important because they establish the boundaries of each unit and the purposes for which it is to be managed.

The importance of enabling legislation is illustrated with respect to boundary decisions (see chapter 6 for more details). Boundary decisions ultimately influence the long-term capacity to protect the ecological integrity of a national park from internal and external threats. Since external threats are typically not an issue when national parks and protected areas are established, the fact that a park does not encompass a complete ecosystem has little immediate consequence at the time of establishment. However, over time, visitation increases and rural residential and resource development grow in areas adjacent to parks. These changes magnify the adverse ecosystem impacts of the original boundaries. For example, the boundaries of Glacier National Park excluded half of the North and Middle Fork Drainages and the riverine and upland habitat areas for wildlife contained in these drainages (Keiter 1985). As a result of these omissions, the park's wildlife is more vulnerable to residential and resource development in areas adjacent to the park.

National Forest System

The Forest Reservation Creation Act of 1891 (16 U.S.C. §§ 2) authorized development of national forest reserves in the United States The act gave the president authority to designate lands withdrawn from the public domain as forest reserves. The Organic Administration Act of 1897 (30 Stat. 34) specified the purposes for establishing forest reserves, namely, timber harvesting and water supply, provided for the protection and management of those reserves and created the National Forest System of Lands. The Transfer Act of 1905 (33 Stat. 628) authorized the secretary of agriculture to designate and administer forest reserves. The Bureau of Forestry was first called the Forest Service in 1905 (U.S. Forest Service Archives Index 2001). The Transfer Act requires the Forest Service to develop integrated land management plans for forests based on multiple use management. Plans must "ensure consideration of the economic and environmental aspects of various systems of renewable resource management, including the related systems of silviculture and protection of forest resources, to provide for outdoor recreation (including wilderness), range, timber, watershed, wildlife, and fish; provide for diversity of plant and animal communities" while avoiding "substantial and permanent impairment of the productivity of the land" (Keiter 1985; Keiter et al. 1999; World Conservation Monitoring Centre 1999).

The Forest and Rangeland Renewable Resources Planning Act of 1974 as amended by the National Forest Management Act of 1976 (16 U.S.C. §§ 1601–1614) created the National Forest System. Forested lands in the system are to be "maintained in appropriate forest cover with species of trees, degree of stocking, rate of growth, and conditions of stand designed to secure the maximum benefits of multiple use sustained yield management in accordance with land management plans" (U.S. Forest Service 2001). It also requires the U.S. Forest Service to protect, manage, and develop units of the system in a manner that complies with the Multiple Use-Sustained Yield Act of 1960 (16 U.S.C. §§ 528–531) and the National Environmental Policy Act of 1969 (42 U.S.C. §§ 4321–4370) (Keiter 1985; World Conservation Monitoring Centre 1999). The Multiple Use-Sustained Yield Act (16 U.S.C. §§ 528–531) calls for national forests to be managed for multiple objectives including "outdoor recreation, range, timber, watershed and wildlife and fish and a sustained yield of products and services."

The National Forest Management Act requires the secretary of agriculture to develop, maintain, and revise Land and Resource Management Plans for units of the National Forest System. While the act emphasizes the economic benefits of managing forests, it allows for protection of soil, water, and air resources. The 77.4 million ha (191 million acres) in the National Forest System include forests, rangelands, game refuges, game preserves, wilderness areas, national recreation areas, research natural areas, national wild and scenic rivers, and national monuments (Learning Network 2001).

Wilderness Preservation System

The Wilderness Act of 1964 (16 U.S.C. §§ 1131–1136) established the National Wilderness Preservation System and standards for managing wilderness areas. A primary goal of the wilderness system is to "secure for the American people of present and future generations the benefits of an enduring resource of wilderness." The act defines wilderness as "an area where earth and its community of life are untrammeled by man, where man himself is a visitor who does not remain." Wilderness areas must retain their "primeval character and influence, without permanent improvements or human habitation," and provide "outstanding opportunities for solitude or a primitive and unconfined type of recreation." To this end, roads, permanent improvements, human habitation, motorized vehicles, and machines are not allowed in wilderness areas. Primary uses of wilderness include fishing, hiking, camping, skiing, mountaineering, and horsepacking. Hunting is allowed in wilderness areas not located in national parks.

The original National Wilderness Preservation System included 3.6 million ha (9.1 million acres). Following a ten-year study of primitive areas and roadless areas in national parks and wildlife refuges, an additional 38 million ha (96 million acres) were added to the system (Scott 2001). Currently, the system contains 42 million ha (106 million acres) (4.6 percent of the total land area of the United

States), 55 percent of which is in Alaska. Of the 42 million ha (106 million acres) in the system, 41 percent is managed by the National Park Service, 33 percent by the Forest Service, 20 percent by the Fish and Wildlife Service, and 6 percent by the Bureau of Land Management (Pacific Biodiversity Institute 2001). The Wilderness Act does not cover public lands managed by other agencies.

National parks and wilderness areas share common purposes. Management of both requires preservation of natural resources and public use and enjoyment. In particular, the Wilderness Act requires that wilderness areas be managed for "use and enjoyment of the American people" and provide opportunities for "recreational, scenic, scientific, educational, conservation and historical" uses. However, the Wilderness Act prohibits permanent roads, commercial enterprises, structures or installation of structures, use of motor vehicles, motorized equipment and motor boats, and other forms of mechanical transport, all of which are present and allowed in national parks. Under current federal law, wilderness designation affords the highest degree of protection from the impacts of human activities (Keiter 1985). Wilderness areas often buffer national parks from commercial, resource, and residential development on adjacent lands.

Bureau of Land Management Lands

The Organic Act for the Bureau of Land Management (BLM) does not provide specific statutory authority for managing protected areas. It does maintain lands designated as national monuments by executive orders issued by the president.

While the BLM manages less than 1 percent of the acreage in the National Wilderness Preservation System, it has responsibility for 52 percent of the federal nonwilderness roadless areas in the United States (Pacific Biodiversity Institute 2001). Management of these nonwilderness roadless areas is important in buffering protected areas from human disturbance.

The origin of the BLM can be traced to early laws, namely, the Land Ordinance of 1785 and the Northwest Ordinance of 1787 (*Journals of the Continental Congress*, XXVIII). These laws promoted surveying and settlement of lands ceded to the federal government by the original 13 colonies after the War of Independence. The U.S. Department of Treasury's General Land Office was established in 1812 to dispose of federal lands acquired from Spain, France, and other countries. As the public domain increased and spread west, Congress enacted several laws, including the Homestead Act of 1862 (16 U.S.C. §§ 608 to 611, 611a; 25 U.S.C. §§ 495; 43 U.S.C. §§ 663, 946 to 949), General Mining Law of 1872 (R.S U.S.C. §§ 2319 to 2328, et seq.; 30 U.S.C. §§ 22 to 24, et seq.), and Desert Lands Act of 1877 (43 U.S.C. §§ 321 to 323, et seq.), to encourage land settlement (U.S. Bureau of Land Management 2001). The period of land disposition lasted from 1862 to 1890 (Keiter 1999). The most notable anomaly during this period was the creation of Yellowstone National Park in 1872. In 1890, the western frontier was declared closed (Turner 1963).

From 1891 to 1934, the motivation for federal land management shifted from land disposition to land reservation (Keiter 1999). During this period, considerable amounts of the public domain were reserved in national parks, forest reserves, and wildlife refuges, and various laws were passed that allowed private resource development on public lands. This period was punctuated by passage of the Mineral Leasing Act of 1920 (30 U.S.C. §§ 22, et seq.), which allowed leasing of public lands to private interests for exploration and development of coal, oil, gas, and sodium. In addition, the Taylor Grazing Act of 1934 (43 U.S.C. §§ 315, et seq.) terminated homesteading and created the U.S. Grazing Service with the charge of regulating the grazing of livestock on 32 million ha (80 million acres) of public lands, mostly in the western United States. BLM was created in the U.S. Department of the Interior in 1946 through merger of the Grazing Service and General Land Office.

The trend of land reservation continued from 1934 to the present and was accompanied by intensive natural resource planning and management. In an effort to resolve conflicting laws governing public land management and provide BLM with unified authority for land management, Congress enacted the Federal Land Policy and Management Act of 1976 (43 U.S.C. §§ 1701–1784) (Bureau of Land Management 2001). The act stipulates that land be managed for multiple uses in a manner that balances scientific, historical, scenic, fish and wildlife, and water resource values and protects the land and environmental quality. The act defines multiple use as "management of the public lands and their various resource values so that they are utilized in the combination that will best meet the present and future needs of the American people." In addition, the act requires BLM to conduct comprehensive, interdisciplinary land use planning that takes into account federal, state, and local land management plans and complies with the National Environmental Protection Act and other federal land management statutes (Keiter et al. 1999). Under the act, BLM was required to review and inventory the suitability of its roadless areas of 2,025 ha (5,000 acres) or more and roadless islands in the lower 48 states for preservation as a wilderness area.

Today, BLM manages 106 million ha (264 million acres) of biological diverse grasslands, forests, high mountains, arctic tundra, and deserts; leases subsurface mineral rights on an additional 120 million ha (300 million acres); and provides wildfire management and suppression on 135 million ha (388 million acres) (Bureau of Land Management 2001). Considering that most of the land and mineral resources managed by the BLM are located in the western United States and 10 of the 12 western States with significant BLM-managed lands have the fastest rates of population growth in the United States, BLM faces monumental management challenges.

National Wildlife Refuge System

In 1869, Congress formally enacted legislation to set aside federally owned land to protect the Pribilof Islands in Alaska as a reserve for the northern fur seal. In 1903, President Roosevelt established Pelican Island National Wildlife Refuge as the nation's first wildlife refuge. Several other islands and parcels of land and water were subsequently set aside for protection of various species of colonial nesting birds, including Breton, Louisiana (1904), Passage Key, Florida (1905), Shell Keys, Louisiana (1907), and Key West, Florida (1908). During the Roosevelt administration, 51 executive orders were issued to establish wildlife reservations in 17 states and three territories. Other reserves established include the Wichita Mountains Forest and Game Preserve in 1905, the National Bison Range in 1908, and the National Elk Refuge in 1912. Legislation to establish waterfowl refuges includes the Upper Mississippi River Wildlife and Fish Refuge in 1924 and the Bear River Migratory Bird Refuge in 1928 (U.S. Fish and Wildlife Service 2001c).

The first statute establishing a refuge system was the Migratory Bird Conservation Act of 1929. The Migratory Bird Hunting and Conservation Stamp Act of 1934 (16 U.S.C. §§ 718, et seq.) authorized revenues needed to purchase migratory bird habitat. That same year, the Fish and Wildlife Coordination Act (16 U.S.C. §§ 661 to 666c) was passed. Amendments to this act over the next 31 years made it possible to purchase lands for fish and wildlife habitat to offset losses caused by federal water projects. The Fish and Wildlife Act of 1956 (15 U.S.C. §§ 713c-3 note, et seq. and 16 U.S.C. §§ 742a to 742j) created a comprehensive national fish and wildlife policy and broadened authority for acquiring and developing refuges. Lack of funds for implementation of this act led to congressional passage of the Duck Stamp Act in 1958 (16 U.S.C. §§ 718b, 718d) and the Wetlands Loan Act of 1961 (16 U.S.C. §§ 715k-3 to 715k-5); subsequent amendments authorized funds for purchasing refuge lands (U.S. Fish and Wildlife Service 2001c).

Passage of the National Wildlife Refuge System Administration Act in 1966 (16 U.S.C. §§ 668dd-668ee) and amendments in the form of the National Wildlife Refuge System Improvement Act of 1997 (16 U.S.C. §§ 668dd, 668ee) introduced guidelines and directives for administration and management of refuges. The amended act states that the mission of the refuge system is "to administer a national network of lands and waters for the conservation, management, and where appropriate, restoration of the fish, wildlife, and plant resources and their habitats within the United States for the benefit of present and future generations of Americans." According to the act, the refuge system includes "wildlife refuges, areas for the protection and conservation of fish and wildlife that are threatened with extinction, wildlife ranges, game ranges, wildlife management areas, and waterfowl production areas." Current designations in the refuge system are national wildlife refuge, waterfowl production area, coordination area, marine protected area, wetland of international importance, wilderness area, research natural area, wild and scenic river, and national natural landmark (U.S. Fish and Wildlife Service 2001c). Currently, the refuge system encompasses 547 units on 37 million ha (92 million acres) (National Geographic 2001).

Wild and Scenic Rivers System

The Wild and Scenic Rivers Act of 1968 (16 U.S.C. §§ 1271–1287) authorized the Wild and Scenic Rivers System. This act prescribes "that certain selected rivers of the Nation which, with their immediate environments, possess outstanding remarkable scenic, recreational, geologic, fish and wildlife, historic, cultural, or other similar values, be preserved in free-flowing conditions, and that they and their immediate environments shall be protected for the benefit and enjoyment of present and future generations." Threats to rivers envisioned by the act include water pollution and/or impairment of scenic values from timber harvesting, road construction, mining activities and claims, and similar activities. Since many rivers flow through several political jurisdictions, the act allows Congress and state legislatures to nominate rivers for designation as wild and scenic. The act applies to rivers that are already free flowing or have been restored to free-flowing conditions (Keiter 1985, World Conservation Monitoring Centre 1999).

The act permits three river designations. A *wild river* designation is applied to rivers or river reaches that are not impounded and located in watersheds or shorelines that are essentially primitive and have unpolluted waters. A *scenic river* label is for rivers or river reaches that are not impounded and located in watersheds or shorelines that are still largely undeveloped and have unpolluted waters. A *recreational river* covers rivers or river reaches that are readily accessible by road or railroad, and have some development along their shorelines and past impoundment or diversion (The Nature Conservancy 1975). Wild and scenic rivers must be managed to "protect and enhance the values which caused it to be included in said system."

Marine Protected Area System

The IUCN-World Conservation Union defines a marine protected area (MPA) as "any area of the intertidal or subtidal terrain, together with its overlying water and associated flora, fauna, and cultural features, which has been reserved by law or other effective means to protect part or all of the enclosed environment" (IUCN 1988; Kelleher 1999). Executive Order 13158 defines an MPA as "any area of the marine environment that has been reserved by Federal, State, territorial, tribal, or local laws or regulations to provide lasting protection for part or all of the natural and cultural resources therein" (*Federal Register* 2000). Resolution 17.38 adopted by the General Assembly of the IUCN in 1990 establishes the following goal relative to marine areas: "To provide for the protection, restoration, and wise use, understanding and enjoyment of the marine heritage of the world in perpetuity through the creation of a global, representative system of marine protected areas and through the management in accordance with the principles of the World Conservation Strategy of human activities that use or affect the marine environment" (Kelleher and Kenchigton 1992).

The Marine Protection, Research, and Sanctuaries Act of 1972, as amended (16 U.S.C. § 1431 et seq.) created the National Marine Sanctuary Program in the United States.The act authorized the secretary of commerce to establish and manage National Marine Sanctuaries in the ocean and coastal environments. Sanctuaries must have nationally significant aesthetic, ecological, historical, or recreational values. The goal of the act is to protect marine ecosystems or objects, such as coral reefs, sunken historical vessels, or unique habitats, while allowing compatible public and private uses. The National Oceanic and Atmospheric Administration (NOAA) within the U.S. Department of Commerce develops and implements approved management plans for national sanctuaries. The act indicates three reasons for designating a site as a national marine sanctuary:

1. It has special national significance.
2. Existing state and federal authorities provide inadequate protection and conservation of the area.
3. The area would benefit by coordinated and comprehensive management.

NOAA can establish two kinds of protected areas: national marine sanctuaries and national estuarine research reserves (NOAA 2001; IUCN 1991).

The Coastal Zone Management Act of 1972, as amended (16 U.S.C. § 1451 et seq.), established the National Estuarine Research Reserve System. Units of this system are managed as "natural laboratory" areas for estuarine research (IUCN 1991).

Executive Order 13158, issued by President Clinton on May 26, 2000, significantly advanced the protection of U.S. ocean and coastal resources. It directs the Departments of Commerce and the Interior, and other federal agencies, to work closely with state, local, and nongovernmental partners to create a comprehensive network of marine protected areas "representing diverse U.S. marine ecosystems, and the Nation's natural and cultural resources." The network is likely to include MPA sites established under existing management authorities, such as the Marine Protection, Research, and Sanctuaries Act. Other institutional arrangements influencing the establishment and management of MPAs includes the Presidential Proclamation 5928 of December 27, 1998, on the Territorial Sea of the United States of America; Proclamation 5030 of March 10, 1983, on the Exclusive Economic Zone of the United States of America; and Presidential Proclamation 7219 of September 2, 1999, on the Contiguous Zone of the United States (U.S. Department of Commerce and U.S. Department of the Interior 2001).

The United Nations Convention on the Law of the Sea provides a legal framework for establishing marine protected areas in international waters by allocating rights to territorial seas. Objectives of the convention are to:

1. Establish a comprehensive legal order to facilitate international communication and promote peaceful uses of the oceans and seas, equitable and efficient utilization of their resources, conservation of their living resources, and the study and protection and preservation of the marine environment.
2. Integrate and balance the right to exploit natural resources with the duty to manage and conserve such resources and to protect and preserve the marine environment.
3. Provide a comprehensive legal framework for the protection and preservation of the marine environment to be complemented and developed by further legal rules at the global or regional level and national measures.

Chapter 17 of Agenda 21, which was developed at the 1992 United Nations Conference on Environment and Development (Earth Summit), states that the Law of the Sea Convention provides the international basis upon which to pursue the protection and sustainable development of the marine and coastal environment and its resources (Fridtjof Nansen Institute 2001).

Environmental Statutes

In addition to the organic acts, which stipulate the responsibilities of federal agencies for managing protected areas, and enabling legislation, which creates individual protected areas, numerous environmental statutes have an important bearing on the management of protected areas. Environmental statutes falling in this category include the National Environmental Policy Act, Endangered Species Act, Clean Air Act, Clean Water Act, Antiquities Act, and state-specific legislation such as the Alaska National Interest Lands Conservation Act. This section discusses the roles of these statutes in protected area management.

National Environmental Policy Act

The National Environmental Policy Act (NEPA) of 1969 (42 U.S.C. §§ 4321 et seq.) established procedural requirements for federal agencies when implementing proposed actions "significantly affecting the quality of the human environment." Council of Environmental Quality regulations implementing NEPA require federal agencies to prepare an environmental impact statement (EIS) that considers the direct and indirect effects of proposed actions "on natural resources and on the components, structures, and functioning of affected ecosystems." The EIS must address environmental impacts of the proposed action, adverse environmental effects resulting from its implementation, and alternatives to the proposed action. Steps in the EIS process include preparation of a draft EIS, consultation with affected agencies regarding the proposed action, public involvement through scheduling workshops and hearings, and preparation of a final EIS. At the end of the EIS process, the agency issues a record of decision that announces the agency's final decision (chosen alternative). The EIS requirement is waived if, based on an environmental assessment, a federal agency concludes that the proposed action is not expected to have a significant impact on the environment (Keiter 1985).

NEPA influences the management of protected areas in several ways. Two examples are given. First, the National Parks and Recreation Act of 1978 (16 U.S.C. §§ 1a-5, et seq.) requires that each park unit develop a general management plan every 10 to 15 years, or sooner if conditions change more rapidly. The National Park Service requires that an EIS be prepared for all general management plans (National Park Service 2001b). Second, NEPA requires agencies proposing a federal action to (1) consult with federal and nonfederal agencies that have "jurisdiction by law or special expertise with respect to any environmental impact involved"; (2) afford an opportunity for affected agencies to comment on the proposed action; and (3) discuss possible conflicts with land use plans for federal, regional, state, and local entities (Keiter 1990, 1999). Hence, a national forest supervisor could consult with a national park superintendent regarding potentially adverse environmental effects on the park of a timber sale or oil and gas development in a national forest near a park. Even if the forest supervisor did not consult with the park superintendent, the latter can submit comments regarding environmental impacts of the proposed action on the park. Interagency disagreements regarding environmental impacts can be referred to the Council on Environmental Quality for resolution.

While implementing regulations for NEPA require federal agencies to consider the cumulative environmental impacts of proposed actions on nonfederal lands (Keiter 1999), they do not require nonfederal property owners to consider the environmental impacts of their actions on federal lands. For example, habitat fragmentation caused by economic and resource development on nonfederal lands adjacent to protected areas is not subject to NEPA. Habitat fragmentation results in the subdivision of continuous habitat into smaller pieces (Muradian 2001). When habitat fragmentation occurs on nonfederal lands adjacent to protected areas, it is very likely to have an adverse impact on species and is a major cause of species extinction and biodiversity loss (Tilman et al. 1994a, 1994b).

Endangered Species Act

The Endangered Species Act of 1973 (16 U.S.C. §§ 1531–1543) is the premier federal statute for protecting threatened, endangered, and declining species. The purposes of the act are "to provide a means whereby the ecosystems upon which endangered species and threatened species depend may be conserved, [and] to provide a program for the conservation of such endangered species and threatened species." Congress deemed species conservation to be important because

> various species of fish, wildlife, and plants in the U.S. have been rendered extinct as a consequence of economic growth and development untempered by adequate concern and conservation; other species of fish, wildlife, and plants have been so depleted in numbers that they are in danger of or threatened with extinction; and these species of fish, wildlife, and plants are of aesthetic, ecological, educational, historical, recreational, and scientific value to the Nation and its people.

Endangered species are species in danger of extinction throughout all or a significant portion of their range, and threatened species are species likely to become endangered within the foreseeable future. As of April 9, 2004, there were 1,263 listed species (U.S. Fish and Wildlife Service 2004).

Responsibility for administering the act rests with the Fish and Wildlife Service in the U.S. Department of the Interior and the National Marine Fisheries Service in the U.S. Department of Commerce. The Fish and Wildlife Service has primary responsibility for terrestrial and freshwater organisms, and the National Marine Fisheries Service has primary responsibility for marine species. Species listings must take into consideration the best scientific and commercial data available. Listings are based solely on biological considerations; economic factors cannot be considered in determining whether or not to list a species (U.S. Fish and Wildlife Service 2001b).

The ultimate goal of the act is to recover species through development and implementation of a recovery plan in consultation with public and private agencies and stakeholders. A recovery plan typically identifies critical habitat for a listed species. When a species is delisted (removed from the federal list of endangered and threatened species), management of the species is transferred from the federal government to state government(s). Federal agencies are required to consult with the Fish and Wildlife Service to ensure that the actions they authorize, fund, or carry out do not jeopardize a threatened or endangered species. To reduce adverse effects (jeopardy) of a federal action on a listed species, the Fish and Wildlife Service can write a biological opinion containing "reasonable and prudent alternatives" for modifying the proposed action of another federal agency. Other laws that protect declining species and their habitats include the Marine Mammal Protection Act (16 U.S.C. §§ 1361, et seq.), the Migratory Bird Treaty Act (16 U.S.C. §§ 703 to 708, et seq.), Anadromous Fish Conservation Act (16 U.S.C. §§ 757a to 757g), and Lacey Act (16 U.S.C. §§ 667e, 701; 18 U.S.C. §§ 42 to 44) (U.S. Fish and Wildlife Service 2001b).

The Endangered Species Act has been criticized on several grounds (Keiter 1999). First, it only focuses on a single species rather than an ecosystem and only when a species has declined to the point of being eligible for listing (Doremus 1991; Grumbine 1992). Second, the act favors mammals over plants (Coggins and Harris 1987). Third, the process used to implement the act has been criticized. Finally, the act is often viewed as infringing on the property rights of private landowners by restricting current and future land use decisions (Keystone Center 1996).

Section 10 of the act was amended to allow private landowners developing land that contains listed species to design and implement a Habitat Conservation Plan (HCP), which minimizes and mitigates harm to those species during development (Defenders of Wildlife 2001). Landowners that develop and implement an HCP for their property are allowed to have "incidental takes" of listed species without reprisals. This provision is significant because an estimated 25 percent of listed species populations in the United States occur on privately owned land. Defenders of Wildlife have criticized HCPs because (1) they often result in net loss of habitat; (2) they are not consistent with species recovery; (3) the No Surprises Rule, which assures that landowners will not have to increase their funding if the HCP is determined to have detrimental consequences, constrains the ability to improve HCPs and avoid species declines; (4) HCPs are not always based on adequate scientific assessments; and (5) the public does not have adequate opportunity to provide input for HCPs (Defenders of Wildlife 2001).

Clean Air Act

The Clean Air Act as amended in 1990 (42 U.S.C. §§ 7401–7642) has as its goal "to protect and enhance the quality of the Nation's air resources so as to promote the public health and welfare and the productive capacity of its population." The 1970 version of the act established emission areas that were permitted to establish their own emission standards and methods for achieving those standards. Areas with dirtier air had to establish more restrictive standards than areas with cleaner air. In response to the act, dirty industries, like coal-burning power plants, moved from dirtier areas to cleaner areas in terms of air quality. As a result, air pollution and haze increased and air quality decreased in areas such as the Colorado Plateau. Visibility in Grand Canyon National Park, which had been increasing in the 1990s, began to deteriorate. Lower visibility in national parks reduces visitor satisfaction (O'Brien 1999).

The 1977 amendments to the act established a national goal of "prevention of any future, and the remedying of any existing, impairment of visibility in mandatory class I Federal areas which impairment results from manmade pollution" (National Parks Conservation Association 2001b). Since all national parks and wilderness areas in excess of 2,428 ha (6,000 acres) were designated as Class I areas, the act had the potential of reducing air pollution in national parks and wilderness areas. Unfortunately, pollution control benefits were not realized because the provision was not enforced except in the Grand Canyon. Furthermore, older power plants (those built prior to 1985) were exempted from the 1977 amendments with the expectation they would be retired from service. Many older plants are still operating.

As a result of this loophole and growth in power generation, air pollution increased in the Shenandoah and Great Smoky Mountains National Parks in North Carolina and Tennessee. Air quality in Shenandoah National Park has been adversely affected by a coal-fired power plant in West Virginia and polluting sources in the Ohio Valley. Visibility in the park has decreased 60 percent on a year-round basis and 80 percent during the summer months (National Parks Conservation Association 2001b). Pollutants from 11 coal-fired power plants owned by the Tennessee Valley Authority have significantly degraded air quality in Great Smoky Mountains National Park; these plants were exempted from the pollution control requirements of the 1977 amendments because they were built before 1985 (Fordney 2001). Older power plants emit large amounts of sulfur dioxide and nitrogen oxide. Sulfur dioxide contributes to acid deposition and nitrogen oxide contributes to ground-level ozone problems.

Ground-level ozone is an air pollutant that occurs in Shenandoah and Great Smoky Mountains National Parks. This invisible gas occurs when nitrogen oxides and hydrocarbon emissions from motor vehicles react in sunlight. Ground-level ozone can irritate eyes and throats and cause respiratory problems and irregular heart beats. Acid deposition and ground-level ozone also damage plants. Unfortunately, ozone concentrations increase with distance from the source (Agee 1996). This explains why ozone originating in the San Francisco Bay area damages plants in national parks in California's Sierra Nevada, and ozone originating in the Seattle-Tacoma area damages plants in Mt. Rainier National Park in Washington.

In an effort to redress the loophole in the 1977 amendments, EPA issued final rules in 1999 that require certain grandfathered air pollution sources to adopt "best available retrofit technology" to reduce haze in 156 national parks and wilderness areas in Class I areas (Fordney 2001). In addition, the Clean Air Act of 1990 mandated a permanent 10 million ton reduction in sulfur dioxide emissions from 1980 levels and required EPA to issue regulations for phase out of nitrogen oxides emissions from boilers.

Air pollution problems occur in other national parks. Temperature inversions in Yosemite National Park in California turn auto-exhaust emissions and smoke from wood-burning stoves into air pollutants. Sequoia and Kings Canyon National Parks in California have the worst air quality of any parks in the western United States; haze in Big Bend National Park in Texas has been attributed to refineries, power plants, and factories as far away as Houston, Dallas, and Fort Worth (Fordney 2001).

Clean Water Act

The Clean Water Act of 1977, which amended the Federal Water Pollution Control Act of 1972 (33 U.S.C. §§ 1251–1387), has as its goal "to restore and maintain the chemical, physical, and biological integrity of the Nation's waters" by achieving zero discharges of pollutants into navigable waters; attaining water quality standards that support protection and propagation of fish, shellfish, and wildlife for recreation in and on water; and eliminating discharges of toxic pollutants. The act gave the U.S. Environmental Protection Agency (EPA) authority to set technology-based effluent limits for industries, continued the requirement of setting surface water quality standards for all contaminants, and required persons to obtain permits for discharging pollutants from point sources into navigable waters. Point source pollution is caused by effluent from discernable, confined, and discrete conveyances, such as a pipeline, ditch, channel, tunnel, conduit, discrete fissure, or container, as well as vessels, floating craft, and confined animal feeding operations. National Pollution Discharge Elimination System permits establish effluent limits and require the polluter to monitor and report discharges (Ditschman 2001).

In an effort to deal with water pollution problems, EPA and many other federal and state agencies have adopted an integrated watershed management (IWM) approach (U.S. Environmental Protection Agency 1996; Hooper 1997; Prato and Herath 2000). IWM entails "coordinated management of land and water resources within a region, with the objectives of controlling and/or conserving the water resource, ensuring biodiversity, minimizing land degradation, and achieving specified and agreed land and water management and social objectives" (AACM and Centre for Water Policy Research 1995). Implementation of IWM involves helping landowners understand the sources and causes of water pollution in their watershed, and using educational and financial assistance programs to motivate landowner adoption of best management practices for improving water quality. IWM is germane to protected areas because they play a significant role in watersheds in terms of being a primary source of water supplies, in the case of mountain protected areas, or feeling the brunt of water pollution, in the case of protected areas located in low-lying areas of a watershed. The following sidebar explains how IWM is used to protect water quality.

Use of Integrated Watershed Management in Water Quality Protection

Application of IWM to water quality protection has two critical elements.

First, states are required to identify watersheds that do not satisfy designated beneficial uses of water, so-called impaired watersheds. Impaired watersheds do not comply with water quality standards determined by states, territories, and tribes , and total maximum daily loads (TMDLs) programs. A TMDL is the maximum amount of a particular pollutant that a waterbody can receive and still meet water quality standards, and an allocation of that amount to pollution sources.

Second, TMDLs or watershed protection plans must be developed and implemented for impaired watersheds. Nonpoint source (NPS) pollutants enter receiving waterbodies via runoff from agricultural land, small and medium-sized animal feeding operations, construction sites, and other disturbed areas. The most common NPS pollutants are sediment, nutrients, pesticides, pathogens (bacteria and viruses), salts, grease, oil, toxic chemicals, and heavy metals. Agriculture accounts for 60 percent of the impaired river length and half of the impaired lake acreage surveyed by states, territories, and tribes. The process of establishing TMDLs involves determining the reductions in pollutant loads needed to meet water quality standards and the allocation of pollutant loads among multiple sources in a watershed. States have the responsibility for establishing TMDLs for waterbodies in their borders; EPA has authority for implementation.

In order to achieve TMDLs for a watershed, pollution sources need to adjust operations to comply with their load allocations. Several strategies can be used to achieve TMDLs, including reducing the intensity of resource development and use and/or adopting best management practices for reducing water pollution. For example, a logging operation can achieve its allocated TMDL for sediment by removing roads and/or maintaining adequate vegetative cover in riparian areas. Maintenance on septic systems for existing residential units could be increased. Septic systems for new residential developments can be required to meet more stringent water quality standards. Crop farmers could reduce fertilizer application rates, install vegetative buffer strips along waterbodies, and/or increase crop fertilizer use efficiency. Livestock producers could reduce stocking density and/or keep livestock away from watercourses. To maximize success with IWM, as many pollution sources as possible should participate in the planning and implementation of best management practices.

IWM has different implications for different protected areas. Protected areas that contribute little or nothing to water pollution in their watersheds will not have to change their operations. On the other hand, protected areas that contribute significantly to water pollution in their watersheds will most likely need to implement best management practices for pollution reduction. For example, protected areas that have old and decrepit wastewater treatment facilities and high visitation rates might need to modernize those facilities.

Source: EPA (2000a, 2001b); Ditschman (2001).

Protected areas in IUCN category I (wilderness areas) and category II (national parks) are subject to the most stringent water quality standards permitted by the Clean Water Act. Complying with stringent water quality standards in wilderness areas and national parks is relatively easy in upland areas, provided levels of human disturbance are relatively low. Stringent water quality standards are more difficult to achieve in waterbodies located in lowland regions of protected areas. For example, the North Fork of the Flathead River, which flows along the western border of Glacier National Park in Montana, is subject to pollution from residential development on adjacent private land west of the river. Similarly, logging and energy development and associated road building in the portions of the North and Middle Fork drainage outside the park are potential sources of water pollution (Keiter 1985).

The Clean Water Act has increased awareness and action in dealing with water pollution threats to national parks and protected areas. Several cases are noteworthy. The proposed Cabin Creek coal mine in British Columbia, Canada, would have discharged wastewater laden with phosphorus, dissolved nitrates, and sediment into the North Fork of the Flathead River (Keiter 1985). This river originates in British Columbia and forms the western border of Glacier National Park. In this case, the source of water pollution is in another country. Hence, the Clean Water Act is not applicable. Fortunately for Glacier National Park and downriver areas, the Cabin Creek coal mine was not built.

Ecological integrity of Everglades National Park in Florida has been severely impaired by rapid conversion of natural areas to agricultural and residential uses and major decreases in the volume and quality of water entering the park from upland areas (Walker and Solecki 2001; Walker 2001). Water problems in the Everglades are attributed to construction of an extensive water distribution network by the U.S. Army Corps of Engineers that benefits sugarcane producers, and nitrogen and phosphorus in wastewater pumped into the system by the South Florida Water Management District (Whittlesey and Finley 2001). A 20-year, multi-billion dollar plan has been developed to reestablish south Florida's historic rainwater sheetflow, remove 115 km (76 mi) of canals and 198 km (131 mi) of levees, improve southward flow of water, restrict urban development, purchase drained land, and preserve buffer areas on the eastern boundary of the park (National Parks Conservation Association 1999).

Noranda Minerals' proposed New World Mine threatened Yellowstone National Park. Gold and silver mining at the mine site, located just 3.8 km (2.5 mi) from the northeast border of the park, would have generated enormous volumes of sulfide rock (four tons per ounce of gold). If not properly contained, leaching of mine waste would have caused acidic heavy metal pollution of nearby streams, one of which flows into the park. The UN World Heritage Commission declared that the proposed mine would endanger the park, which is a World Heritage Site. In 1996, President Clinton announced a buyout settlement with Crown Butte that compensated shareholders for losses incurred by ceasing all exploration and mining activities in the area (O'Brien 1999).

Protected areas located near marine environments and oil pipelines, such as Kenai Fjords National Park in Alaska, are particularly vulnerable to oil spills. River water entering Kruger National Park in South Africa is characteristically low in volume and polluted by residential, silvicultural, agricultural, and energy developments in upland areas of the watershed.

Alaska National Interest Lands Conservation Act

The Alaska National Interest Lands Conservation Act (ANILCA) of 1980 (16 U.S.C. §§ 410hh 1 to 410hh 5, et seq.) made Alaska the top-ranked state in the United States for protected areas. ANILCA expanded the U.S. system of protected areas by more than 40 million ha (100 million acres). It doubled the size of the U.S. National Park and Refuge Systems by adding over 16 million ha (37.5 million acres) in 10 new national parks and preserves, expanding by 2.2 million ha (5.6 million acres) the area of 3 existing park units, and tripled the amount of U.S. land designated as wilderness causing Alaska to contain over half of the land in the National Wilderness Preservation System (National Parks Conservation Association 2001a). The Wilderness Act of 1964 did not designate wilderness areas in Alaska. ANILCA was significant for other reasons; it established new protected areas that would be unencumbered by the problems faced by protected areas in other parts of the world, filled a void caused by underrepresentation of certain kinds of natural areas in the units of the National Park System in Alaska, and contained model provisions for preserving subsistence resource uses in national parks and preserves (Wright 1996). Many conservationists view ANILCA as the premier land conservation statute in the United States and hold it up as a model for how to create protected areas that balance development and conservation interests.

ANILCA was a legal consequence of the Alaska Native Claims Settlement Act (ANCSA) of 1971 (43 U.S.C. §§ 1601 to 1624). ANCSA created 12 Native-owned regional corporations, granted $962 million in seed money, and authorized Native corporations to select 18 million ha (44 million acres) of federal lands in Alaska. Section 17 (d)(2) of the ANCSA directed the secretary of the interior to withdraw from development 32 million ha (80 million acres) of unreserved federal lands (so-called d-2 lands) and study their suitability for designation as national parks, refuges, wild and scenic rivers, or national forests. Congress had until 1978 to make such designations. Failure to do so would prohibit future withdrawals, thus allowing the lands to be developed. Congress did not act in time, prompting the Carter administration to withdraw over 40 million ha (100 million acres) of federal lands from development. Sixteen million ha (40 million acres) were placed under the authority of the secretary of the interior, and 22 million ha (56 million acres) were designated as national monuments using the presidential authority under the Antiquities Act. In late 1980, Congress passed ANILCA, which institutionalized how the 40 million-plus ha (100 million-plus acres) were to be used.

Antiquities Act

The Antiquities Act of 1906 (16 U.S.C. §§ 431–433) gave the president of the United States discretionary authority to "declare by public proclamation historic landmarks, historic and prehistoric structures, other objects of historic or scientific interest that are situated upon the lands owned or controlled by the Government of the United States to be national monuments." In bills preceding the act, the size of national monuments was limited to 128 ha (320 acres) or 256 ha (640 acres). The bill approved by Congress indicated the size of national monuments should "be confined to the smallest area compatible with the proper care and management of the objects to be protected." This stipulation gave the president flexibility to select much larger areas (Wilderness Society 2001b). Fort Matanzas in Florida is the smallest national monument at 0.4 ha (one acre) and Grand Staircase Escalante National Monument in Utah is the largest national monument at 800,000 ha (1.9 million acres).

Theodore Roosevelt was the first president to establish a national monument under the act. On September 24, 1906, he created Devils Tower National Monument, an impressive geologic formation in Crook County, Wyoming. Roosevelt designated a total of 18 national monuments in nine states. The most significant and substantial designations under the act occurred in 1978 when President Jimmy Carter set aside 22 million ha (56 million acres) of Alaska as national monuments (Lee 2001).

While the act gave the president substantial powers to set aside lands for conservation purposes, Congress retained authority to designate monuments, change monument boundaries, authorize and appropriate funds for the management of monuments, convert national monuments to national parks, and abolish monuments. For example, from 1929 to 1969, special acts of Congress authorized the creation of 28 national monuments (Lee 2001). Several national monuments were subsequently designated as national parks by Congress, including Bryce Canyon, Death Valley, Denali, Glacier Bay, Grand Canyon, Joshua Tree, and Olympic. The act has been used by 14 of the 17 presidents that have served in office since 1906 (Wilderness Society 2001b). There were 96 national monuments covering 32.6 million ha (81.6 million acres) as of February 2002 (Wilderness Society 2001a). National monuments created in 2000 and 2001 are listed in table 5.2.

The act is not without its detractors. Opponents claim it vests too much power in the president to designate conservation lands, allows public land to be locked up, infringes on private property rights, and does not require sufficient input from state and local governments. Attempts to limit presidential power under the act have been unsuccessful due to considerable public support (Wilderness Society 2001b).

States and provinces have authority to enforce federal environmental laws that influence the management and operation of national parks and protected areas

Table 5.2. U.S. National Monuments Created in 2000 and 2001

Agua Fria, AZ (January 11, 2000)
Anderson Cottage, Washington, DC (July 7, 2000)
California Coastal (January 11, 2000)
Canyons of the Ancients, adjacent to Mesa Verde National Park, CO (June 9, 2000)
Cascade-Siskiyou, south-central Oregon (June 9, 2000)
Craters of the Moon, ID (expansion; November 9, 2000)
Giant Sequoia National Monument, CA (April 15, 2000)
Grand Canyon-Parashant National Monument, AZ (January 11, 2000)
Hanford Reach, southeast Washington (June 9, 2000)
Ironwood Forest, near Tucson, AZ (June 9, 2000)
Pinnacles National Monument, CA (January 11, 2000—expansion)
Vermilion Cliffs, AZ (November 9, 2000)
Buck Island National Monument (U.S. Virgin Islands/expansion—January 17, 2001)
Carrizo Plain National Monument, CA (January 17, 2001)
Minidoka Internment National Monument, ID (January 17, 2001)
Pompeys Pillar National Monument, MT (January 17, 2001)
Kasha-Katuwe Tent Rocks National Monument, NM (January 17, 2001)
Santa Rosa and San Jacinto Mountains National Monument , CA (October 24, 2000)
Sonoran Desert National Monument, AZ (January 17, 2001)
Upper Missouri Breaks National Monument, MT (January 17, 2001)
U.S. Virgin Island Coral Reef National Monument (January 17, 2001)

Source: Wilderness Society (2001a). Used by permission of the Wilderness Society.

within their political jurisdictions. In addition, most states and provinces have enacted laws to establish state parks and protected areas. For example, Missouri's National Areas Systems has 168 units covering all major terrestrial and aquatic communities in the state (Missouri Departments of Conservation and Natural Resources 2001) and Missouri's State Park System consists of 80 units covering nearly 55,200 ha (138,000 acres) (Missouri Department of Conservation 2001).

Rule Making

Congress passes laws governing the establishment and management of national parks and protected areas, and protection of land, water, and air resources. The executive branch of government is responsible for implementing those laws, which it does by writing rules that have the "full force and effect of law" (Keiter 1999). The process of rule making must comply with the Administrative Procedures Act (5 U.S.C. § 553) and involves three steps. First, the federal agency responsible for enforcing the law is required to analyze the best way to implement the law. Based on this analysis, a proposed rule is developed and published in a notice that appears in the *Federal Register*. The notice informs the public about the purpose and content of the proposed rule and solicits public comments. Second, based on comments received from the public and other agencies, the proposed rule is revised and published in the *Federal Register* as a final rule. Third, final rules issued by executive departments and agencies of the federal government are incorporated in the Code of Federal Regulations (CFR). The CFR has 50 titles covering a broad range of subjects. Most environmental regulations are contained in title 40 (U.S. Environmental Protection Agency 2001a)

Development and enforcement of rules is not a simple matter and can be contentious. For example, in 1985 and 1992, EPA issued regulations implementing section 303(d) of the 1972 Clean Water Act. This section requires states, territories, authorized tribes, and EPA to establish and implement Total Maximum Daily Loads (TMDLs) for waterbodies impaired by pollution. Except on a very limited basis, impaired waterbodies were not being listed and TMDLs were not being developed and implemented as required by law. Citizen organizations in 38 states successfully challenged EPA for its lack of performance causing court orders and consent decrees to be filed against EPA. These actions were intended to accelerate the listing of impaired waterbodies and TMDL development. Implementation of TMDL regulations is highly contentious because they essentially bring all sources of water pollution under regulation (prior to the rule, nonpoint sources were not regulated), and TMDLs are very costly to develop, implement and enforce (U.S. Environmental Protection Agency 2001b).

Federal agencies can legitimize a new policy by issuing new regulations as long as they are compatible with existing statutes. Ecosystem management is a case in point. New rules for implementing ecosystem management can have several benefits, including:

1. Establishing procedures for meaningful interagency coordination
2. Clarifying the way federal land management agencies address and resolve negative impacts of development on adjoining lands
3. Illuminating how and when to address biodiversity issues in land management planning
4. Determining how to incorporate scientific information in ecosystem management decisions
5. Providing guidance in conducting risk assessments of scientific uncertainty in decision making for ecosystem management (Keiter 1999)

Court Decrees

Federal policy for the management of national parks and protected areas is changed by amending existing laws, which is the purview of Congress, altering implementing regulations, which is the purview of executive departments, or interpreting existing laws, which is the purview of the federal and other courts. Article III of the U.S. Constitution established the judicial branch and supporting federal court system. Federal courts protect the rights and liberties guaranteed by the Constitution by making fair and impartial judgments in connection with disputes regarding the interpretation and application of law. The 94 federal judicial districts in the United States have authority to hear most civil and criminal federal cases within the limits prescribed by Congress and the Constitution. To ensure that federal courts function in a manner that ensures fairness and justice for all U.S. citizens, the Constitution developed a rigid procedure for appointing and compensating federal judges (U.S. District Courts 2001).

Court cases relevant to national parks and protected areas are too numerous to mention here. Three examples are given to illustrate their influence on the management of national parks and protected areas. In *Sierra Club v. Department of the Interior*, the district court held that the secretary of the interior was statutorily responsible and legally bound to uphold the public trust doctrine as it applies to protecting national parks. The public trust doctrine for national parks was subsequently rejected. In *Sierra Club v. Andrus*, the courts rejected the public trust doctrine argument in an action challenging the secretary of the interior's refusal to develop laws to protect water quality in the Grand Canyon National Park (Keiter 1985). In *McDowell v. Collingsworth*, the Alaska Supreme Court ruled that the state violated the equal protection and common-use clause of its constitution by giving priority in subsistence resources uses to rural Alaskan residents, which violated the Alaska National Interests Lands Conservation Act. One consequence of this case is that the authority for managing subsistence resource uses for all nationally owned public land in Alaska was transferred from state to federal agencies (Caufield 1991).

Executive Orders

Executive orders are consecutively numbered official documents issued by the president of the United States that govern particular activities or operations of the federal government. The text of executive orders appears in the daily Federal Register as each executive order is signed by the president and received by the Office of the Federal Register. The first executive order (number 7316) was issued March 13, 1936. Several executive orders pertain to national parks and protected areas. During his second term, President Clinton issued four executive orders related to protected areas. Executive Order 13093 American Heritage Rivers was signed July 27, 1998; order 13089 Coral Reef Protection was signed June 11, 1998; order 13158 Marine Protected Areas was signed May 26, 2000; and order 13178 Northwestern Hawaiian Islands Coral Reef Ecosystem Reserve was signed December 4, 2000 (National Archives and Records Administration 2001).

Agreements, Organizations, and Programs

Interagency-international agreements and environmental, scientific, educational, and professional organizations comprise the institutional environment affecting the designation and performance of national parks and protected areas. Several international organizations, agreements, and programs have established goals, standards, procedures, and guidelines that affect the selection, delineation, and management of protected areas. They include the IUCN, UNESCO's Biosphere

Reserve Program, World Heritage Convention's World Heritage Sites, Ramsar Convention Wetlands, UN Law of the Sea, and Convention on Biological Diversity.

IUCN's seven categories of protected areas are used worldwide (see chapter 2, table 2.1). IUCN's World Commission on Protected Areas (WCPA) is a global organization whose mission is "to promote the establishment and effective management of a world-wide representative network of terrestrial and marine protected areas, as an integral contribution to the IUCN" (World Commission on Protected Areas 2002). Specific activities include (1) strengthening the capacity and effectiveness of protected area managers by providing management guidelines, tools, and information; (2) providing strategic advice to policy makers on integrating protected areas with sustainable development and biodiversity conservation; (3) persuading public and corporate entities to increase their investment in protected areas; and (4) collaborating with IUCN and partners to strengthen the capacity of the WCPA to implement its programs (Worboys et al. 2001). The WCPA includes some one thousand members from over 140 countries. There are five categories of membership in the WCPA: (1) managers of protected areas and field practitioners; (2) experts in protected areas; (3) academic specialists in protected areas, resource economics, biogeography wildlife management, marine conservation, and other related fields; (4) officials from relevant nongovernment organizations that deal with protected areas; and (5) ex-officio members from key partner organizations (World Commission on Protected Areas 2002). In addition, every 10 years IUCN hosts a World Parks Congress that develops strategies and philosophies for protected area management (Whitehouse 1992).

UNESCO's Man and the Biosphere Program established the Biosphere Reserve Program in 1974; the first Biosphere Reserve was established in 1976. As of July 2003, there were 440 biosphere reserves in 97 countries covering coastal, forest, desert, island, mountain, plains, polar, and tropical areas (Green and Paine 1997; UNESCO 2004a). Linked to UNESCO is the International Council on Monuments and Sites. This international organization is concerned with the conservation and evaluation of areas of cultural significance.

The 1972 World Heritage Convention established World Heritage Sites having outstanding cultural and historical values. As of July 2003, there were 754 World Heritage Sites in 129 States Parties or countries (UNESCO 2004b). North America contained 43 World Heritage Sites as of July 2003 (UNESCO 2004b).

A 1971 intergovernmental treaty called the Ramsar Convention on Wetlands established Ramsar Wetlands. Signatories to the treaty encourage designation of wetland-protected areas that benefit waterfowl and promote wise use of wetland ecosystems. Under the convention, 1,069 wetland sites are protected (Ramsar Convention on Wetlands 2001). Other international conventions that influence protected area designation/management and declining species include the UN Law of the Sea, which is discussed in the section above on marine protected areas, the International Convention for the Regulation of Whaling, the Convention on International Trade in Endangered Species of Wild Flora and Fauna, and the Convention on the Conservation of Migratory Species of Wild Animals.

Article 8 of the Convention on Biological Diversity encourages each party to purse in situ conservation by establishing and managing a system of protected areas that support natural and cultural resource conservation, sustainable use, and equitable sharing of benefits (Glowka et al. 1994).

The IUCN, World Wildlife Fund for Nature, and United Nations Environment Program jointly established the World Conservation Monitoring Centre (WCMC). The WCMC provides information services on the conservation and sustainable use of the world's living resources, helps others to develop their own information systems, and periodically publishes international inventories of national parks and protected areas (World Conservation Monitoring Centre 2002). Other global-scale nongovernmental organizations (NGOs) that work toward biodiversity conservation and ecosystem protection include Conservation International, Nature Conservancy, and United Nations Foundation.

North American environmental organizations that actively lobby for national parks and protected areas include Defenders of Wildlife, National Parks Conservation Association, National Wildlife Federation, Sierra Club, Wilderness Society, and many others. In addition, there are several NGOs and individuals that purchase private lands for the purpose of protecting species and their habitats. Examples include The Nature Conservancy, Rocky Mountain Elk Foundation, Ducks Unlimited, Pheasants Forever, and Turner Foundation, Inc.. Numerous environmental organizations in the United States inform and educate the public, monitor management activities, and raise funds for protected areas. For example, the Greater Yellowstone Coalition consists of individuals, environmental organizations, and businesses that promote ecosystem management in the Greater Yellowstone Ecosystem (Greater Yellowstone Coalition 2001). The latter encompasses Yellowstone and Grand Teton National Parks, portions of six national forests, and three national wildlife refuges (Keiter 1991). The Greater Yellowstone Coordinating Committee (GYCC) was established in 1964 by federal agencies responsible for managing land in the Greater Yellowstone Ecosystem. The U.S. National Park Service, U.S. Fish and Wildlife Service, and U.S. Forest Service are the members of the GYCC. The overall goal of the GYCC is to improve interagency cooperation and coordination by (1) providing leadership in making resource management decisions; (2) setting management priorities and assigning resources to specific objectives; (3) ensuring coordination of strategies and practices across National Park, National Forests, and National Wildlife Refuge units; (4) minimizing duplication of effort; and (5) ensuring, to the extent possible, that rules and regulations are consistent across agencies (Yellowstone Ecosystem Studies 2001).

Scientific and professional organizations publish journals and reports and conduct educational programs for national parks and protected areas that significantly enhance the capacity of protected area managers to make wise and informed decisions. Scholarly journals that publish scientific and management studies dealing with protected areas include *George Wright Forum, International Journal of Wilderness Management, Journal of Forestry, Journal of Leisure Research, Leisure Sciences, Natural Areas Journal, Parks and Recreation, Park Science, Society and Natural Resources,* and *Yellowstone Science.* Many large national parks have organizations that provide field-based educational programs for their park. For example, the Glacier Institute provides outdoor education field seminars dealing with natural and cultural values of the Crown of the Continent Ecosystem, which includes Glacier National Park (Glacier Institute 2001). The Glacier Natural Historical Association (Glacier Natural History Association 2001) supports educational, interpretive, cultural, and scientific programs for Glacier National Park through the sale of books. The Glacier Park Foundation promotes appreciation of wilderness preservation, history, and traditions in Glacier National Park and Montana (Glacier Park Foundation 2001).

Summary

Institutions consist of formal and informal "rules of the game" that influence how people and organizations behave, interact, and perform. Formal institutions encompass the laws establishing federal agencies that manage protected areas, enabling legislation for individual protected areas, environmental laws, regulations implementing laws, executive orders, and court decrees that influence the interactions among people and organizations. The National Park Service Organic Act and implementing regulations are formal institutions. Informal institutions consist of the cultural norms, social conventions, mores, beliefs, etiquette, traditions, and taboos that govern expectations regarding individual and social behaviors. "Leaving only footprints" is a slogan or cultural norm among backcountry users. The ultimate success of formal institutions depends on enforcement. While informal institutions are not enforced in a legal sense, noncompliance is deterred by humiliation, boycotts, mistrust, and loss of social status. Getting caught in the act of leaving trash in a backcountry campsite can be humiliating to the offender. Institutions are important because they create incentives or disincentives in the form of opportunities and constraints that, along with other factors, influence how people and organizations behave and ultimately perform.

Institutional theory maintains that individuals and organizations face incentives and disincentives that influence their decisions and performance. Unacceptable performance stimulates institutional change. The high incidence of poaching in a wildlife reserve (poor performance) can be reduced by passing a law (creating a formal institution) that makes it illegal to export wildlife and their parts. Enforcement of antipoaching laws is essential if poaching is to be reduced. The propensity of individuals and organizations to resist change (institutional rigidity) works against institutional change and better performance.

New resource economics explains why bureaucratic organizations generate incentives and disincentives that cause individuals in those organizations to make decisions that result in poor performance or undesirable outcomes. According to new resource economics, the performance of federal protected area managers is deficient because they do not take account of the full social benefits and costs of their decisions. Presumably, if federal managers considered full social costs and benefits, then better decisions (in terms of increasing net social benefits) would be made. Full social costs and benefits have a greater likelihood of being considered when efficient property rights have been assigned to the ecological goods and services provided by protected areas. Such rights should be established only if benefits exceed costs. This condition is unlikely to be achieved for protected areas because of the high transaction costs of establishing efficient property rights and the general unwillingness of the public to privatize the management of public land. Proponents of new resource economics argue for establishment of park endowment boards that would retain public ownership of national parks and use market-based incentives to allocate a park's natural resources among competing uses.

Institutional arrangements shape the use and management of national parks and protected areas in the United States. They include statutes that vest federal agencies with authority to manage protected areas and establish individual protected areas, regulations that implement statutes, and environmental statutes. Executive orders, court decrees, agreements, and organizations are other institutions that influence protected areas. Major protected area systems in the United States include the National Park System, National Forest System, certain lands managed by the Bureau of Land Management, National Wildlife Refuge System, Wild and Scenic Rivers System, and Marine Protected Area System. Major environmental statutes influencing the management of protected areas are the National Environmental Policy Act, Endangered Species Act, Clean Air Act, Clean Water Act, Antiquities Act, and state-specific legislation such as the Alaska National Interest Lands Conservation Act. Organizations, agreements, and programs that establish goals, standards, procedures, and guidelines for selecting, delineating, and managing protected areas include the IUCN-World Conservation Union, UNESCO's Biosphere Reserve Program, the World Heritage Convention's World Heritage Sites, Ramsar Convention wetlands, UN Law of the Sea, Convention on Biological Diversity, and others. Many national and regional scientific, educational, professional, and environmental organizations play a major role in the establishment, stewardship, and management of national parks and protected areas.

References

AACM and Centre for Water Policy Research. 1995. Enhancing the effectiveness of catchment management planning. Final report to the Department of Primary Industries and Energy. AACM International, Adelaide, Victoria, Australia.

Agee, J. K. 1996. Ecosystem management: An appropriate concept for parks? In *National parks and protected areas: Their role in environmental protection,* ed. R. G. Wright, 31–44. Cambridge, MA: Blackwell Science.

Boyce, M. S. 1991. Natural regulation or the control of nature? In *The Greater Yellowstone Ecosystem: Redefining America's wilderness heritage,* ed. R. B. Keiter and M. S. Boyce, 183–208. New Haven, CT: Yale University Press.

Bunting, S. C. 1996. The use and role of fires in natural areas. In *National parks and protected areas: Their role in environmental protection,* ed. R. G. Wright, 277–302. Cambridge, MA: Blackwell Science.

Canada. 1988. *An act to amend the national parks bill C-30.* Minister of Supply and Services, Ottawa, Canada.

Caufield, R. A. 1991. Alaska's subsistence management regimes. *Polar Record* 28: 23–32.

Chase, A. 1987. *Playing God in Yellowstone: The destruction of America's first national park.* Fort Washington, PA: Harvest Book Company.

Coggins, G., and F. A. Harris. 1987. The greening of American law? The recent evolution of federal law for preserving floral diversity. *Natural Resources Journal* 27:247–307.

Copeland, M. D. 1990. The new resource economics. *In The Yellowstone primer: Land and resource management in the greater Yellowstone ecosystem,* ed. J. A. Baden and D. Leal, 13–23. San Francisco, CA: Pacific Research Institute for Public Policy.

Craighead, J. J. 1991. Yellowstone in transition. *In The Greater Yellowstone Ecosystem: Redefining America's wilderness heritage,* ed. R. B. Keiter and M. S. Boyce, 27–39. New Haven, CT: Yale University Press.

Dearden, P., and R. Rollins. 2002. *Parks and protected areas in Canada: Planning and management.* 2d ed. Ontario, Canada: Oxford University Press,.

Defenders of Wildlife. 2004. Habitat conservation plans. http://www.defenders.org/hcp/ (accessed April 9, 2004).

Denali National Park and Preserve. 1999. Subsistence users guide: Denali National Park and Preserve, Subsistence Resource Commission, Denali Park, AK.

Ditschman, E. P. 2001. Introduction: Stormwater regulation and nonpoint source policy—complimentary or contradictory. *Water Resources IMPACT* 3: 3–4.

Doremus, H. 1991. Patching the ark: Improving legal protection of biological diversity. *Ecology Law Quarterly* 18: 265–333.

Federal Register. 2000. Presidential documents. Executive Order 13158 of May 26, 2000, May 31. Volume 65, No. 105. Washington, DC: U.S. Government Printing Office.

Fordney, C. 2001. Bad air days. *National Parks* 75: 32–35.

Fridtjof Nansen Institute. 2001. Yearbook of international co-operation on environment and development 2001/2002, 9th volume. London, England: Earthscan Publications Limited.

Glacier Institute. 2001. http://www.glacierinstitute.org/ (accessed Dec. 2001).

Glacier Natural History Association. 2001. http://www.nps.gov/glac/gnha.htm (accessed Dec. 2001).

Glacier Park Foundation. 2001. http://www.glacierparkfoundation.org/ (accessed Dec. 2001).

Glick, D., and B. Alexander. 2000. Development by default, not design: Yellowstone National Park and the Greater Yellowstone Ecosystem. In *National parks and rural development: Practice and policy in the United States,* ed. G. E. Machlis and D. R. Field, 182–205. Washington, DC: Island Press.

Glowka, L., F. Burhenne-Guilmin, H. Synge, J. A. McNeely, and L. Gundling. 1994. A guide to the convention on biological diversity. Environmental Policy and Law Paper No. 30, IUCN Environmental Law Centre, Gland, Switzerland.

Greater Yellowstone Coalition. 2001. http://www.greateryellowstone.org/ (accessed Nov. 2001).

Green, M. J. B., and J. R. Paine. 1997. State of the world's protected areas at the end of the twentieth century. Paper presented at Protected Areas in the 21st Century: From Islands to Networks. Albany, Australia, November 24–27.

Grumbine, R. E. 1992. *Ghost bears: Exploring the biodiversity crisis.* Washington, DC: Island Press.

Hooper, B. An integrated catchment management protocol for water quality management in Australia. *Water Quality International* January/February: 48–50.

International Union for the Conservation of Nature (IUCN). 1988. Resolution 17.38 of the 17th General Assembly of the IUCN. Gland, Switzerland, and Cambridge, UK.

———. 1991. *IUCN directory of protected areas in Oceania: United States of America.* http://www.wcmc.org.uk/cgi-bin/pa_paisquery.p (accessed Nov. 2001).

James, S. J. 2001. An institutional approach to protected area management performance. *In The politics and economics of park management,* ed. T. L. Anderson and A. James. Lanham, MD: Rowman & Littlefield Publishers, Inc.

Johnson, K. N. 1997. Science-based assessments of the forests of the Pacific Northwest. In *Creating forestry for the 21st century: The science of ecosystem management,* ed. K. A. Kohm and J. F. Franklin, 397–409. Washington, DC: Island Press.

Keiter, R. B. 1985. On protecting the national parks from the external threats dilemma. *Land and Water Review* 20: 355–420.

———. 1990. NEPA and the emerging concept of ecosystem management on the public lands. *Land and Water Law Review* 25: 43–60.

———. 1991. An introduction to the ecosystem management debate. In *The Greater Yellowstone ecosystem: Redefining America's wilderness heritage,* ed. R. B. Keiter and M. S. Boyce, 3–18. New Haven, CT: Yale University Press.

———. 1996. Ecosystem management: Exploring the legal-political framework. In *National parks and protected areas: Their role in environmental protection,* ed. R. G. Wright, 63–88. Cambridge, MA: Blackwell Science.

Keiter, R. B., T. Boling, and L. Milkman. 1999. Legal perspectives on ecosystem management: Legitimizing a new federal land management policy. In *Ecological stewardship: A common reference for ecosystem management,* III, ed. W. T. Sexton, A. J. Malk, R. C. Szaro, and N. C. Johnson, 10–41. Oxford, UK: Elsevier Science.

Kelleher, G. 1999. Guidelines for marine protected areas. Gland, Switzerland, and Cambridge, UK, IUCN—The World Conservation Union.

Kelleher, G., and Kenchigton, R. 1992. Guidelines for establishing marine protected areas. Gland, Switzerland: IUCN.

Keystone Center. 1996. Keystone national policy dialogue on ecosystem management: Implementing community-based approaches. Keystone, CO.

Learning Network. 2001. *National Forest System.* http://www.infoplease.com/ce6/us/A0834943.html (accessed Dec. 2001).

Lee, R. E. 2001. *The Antiquities Act of 1906.* Electronic Edition. http://www.cr.nps.gov/aad/pubs/ (accessed Nov. 2001).

Leopold, A. S., et al. (Advisory Board on Wildlife Management). 1963. Wildlife management in the national parks. *American Forests* 69: 32–35, 61–63.

Missouri Departments of Conservation and Natural Resources. 2001. *Missouri Natural Areas System.* http://www.conservation.state.mo.us/nathis/naturalareas/natdir/ (accessed Dec. 2001).

Muradian, R. 2001. Ecological thresholds: A survey. *Ecological Economics* 38: 7–24.

National Archives and Records Administration. 2001. *Executive orders disposition table.* http://www.nara.gov/fedreg/eo.html#orders (accessed Nov. 2001).

National Geographic. 2001. *U.S. National Wildlife Refuges.* http://www.nationalgeographic.com/refuges/ (accessed November 2001).

National Oceanic and Atmospheric Administration (NOAA). 2001. *The National Marine Sanctuaries Act.* http://www.sanctuaries.nos.noaa.gov/natprogram/nplegislation/nplegislation.html (accessed Nov. 2001)

National Parks Conservation Association. 1999. River of grass ready to heal. *National Parks* 7: 19.

———. 2001a. *Fact sheet on ANILCA.* http://208.226.12.12/media_center/factsheets/anilca.asp (accessed Nov. 2001).

———. 2001b. *Regional haze in national parks.* http://208.226.12.12/media_center/factsheets/haze.asp. (accessed Nov. 2001).

National Park Service. 2001a. *Cerro Grande prescribed fire: Board of inquiry final report.* http://www.nps.gov/band/Board_report-feb26final.pdf (accessed Dec. 2001).

———. 2001b. *General management planning.* http://www.nps.gov/carl/gmp_desc.htm (accessed Nov. 2001).

———. 2001c. *Recreational fee demonstration program fact sheet.* http://www.nps.gov/feedemo/facts.htm (accessed December 2001).

———. 2001d. *When did the NPS begin?* http://www.nps.gov/legacy/ (accessed Nov. 2001).

Nature Conservancy. 1975. Preserving our national heritage. Vol. 1. Federal activities. Washington, DC: The Nature Conservancy and U.S. Man and the Biosphere Program.

Niskanen, W. 1971. *Bureaucracy and representative government.* New York: Atlantic Monthly Press.

North, D. 1990. *Institutions, institutional change and economic performance.* Cambridge, England: Cambridge University Press,.

———. 1994. Constraints on institutional innovation: transaction costs, incentive compatibility, and historical considerations. In *Agriculture, environment and health: Sustainable development in the 21st century,* ed. V. W. Ruttan. Minneapolis: University of Minnesota Press.

O'Brien, B. R. 1999. *Our national parks and the search for sustainability.* Austin: University of Texas Press.

Ostrom, E. 1992. *Crafting institutions for self-governing irrigation systems.* San Francisco, CA: Institute for Contemporary Studies Press.

Pacific Biodiversity Institute. 2001. Reported in *Return of the Wild: The Future of Our Natural Lands.* Washington, DC: Island Press.

Prato, T. 1998. *Natural resource and environmental economics.* Ames: Iowa State University Press.

Prato, T., and G. Herath. 2000. Multiple-criteria decision analysis for integrated catchment management. CD Proceedings for 10th World Water Congress, March 12–17, Melbourne, Australia.

Ramsar Convention on Wetlands. 2001. http://www.ramsar.org/ (accessed June 11, 2001).

Rockwood, T. H. 1988. Organizational responses to a conflicting mandate: Legitimacy and structural adaptations in the National Park Service. MA Thesis, Washington State University, Pullman.

Scott, D. W. 2001. A brief history of wilderness time. In *Return of the wild: The future of our natural lands,* ed. T. Kerasote, 11–24. Washington, DC: Island Press.

Stroup, R. L. 1990. Rescuing Yellowstone from politics: Expanding parks while reducing conflict. In *The Yellowstone primer: Land and resource management in the greater Yellowstone ecosystem,* ed. J. A. Baden and D. Leal, 169–184. San Francisco, CA: Pacific Research Institute for Public Policy.

Stroup, R. L., and J. A. Baden. 1983. *Natural resources: Bureaucratic myths and environmental management.* Cambridge, MA: Pacific Research Institute for Public Policy and Ballinger Publishing Co.

Tilman, D., J. Downing, and D. Wedin. 1994a. Does diversity beget stability? *Nature* 371: 113–114.

Tilman, D., R. May, C. Lehman, and M. Nowak. 1994b. Habitat destruction and the extinction debt. *Nature.* 371: 65066.

Turner, F. J. 1963. In *The significance of the frontier in American history,* ed. H. P. Simonson. New York: Fredrick Unger.

U.S. Bureau of Land Management (BLM). 2001. *BLM facts.* http://www.blm.gov/nhp/facts/index.htm (accessed Nov. 2001).

U.S. Department of Commerce and U.S. Department of the Interior. 2001. *Marine Protected Areas of the United States: The national MPA initiative.* http://mpa.gov/mpadescriptive/natinitiative.html (accessed Nov. 2001).

U.S. Department of the Interior. 2000. *Designation of National Park System units.* http://www.nps.gov/legacy/nomenclature.html (accessed April 13, 2004).

U.S. District Courts. 2001. http://www.uscourts.gov/districtcourts.html (accessed Nov. 2001).

U.S. Environmental Protection Agency (EPA). 1996. *Watershed approach framework.* Washington, DC.

———. 2000a. *National water quality inventory: 1998 report to Congress.* EPA 841-R-00-001. Washington, DC: Office of Water.

———. 2000b. *Water quality conditions in the United States: A profile from the 1998 National Water Quality Inventory Report to Congress.* EPA-841-F-00-006. Washington, DC: Office of Water.

———. 2001a. *Laws and regulations.* http://www.epa.gov/epahome/rules.html#codified (accessed Nov. 2001).

———. 2001b. *Total maximum daily load (TMDL) program.* http://www.epa.gov/owow/tmdl/ (accessed Nov. 2001).

U.S. Fish and Wildlife Service. 2001a. *The Endangered Species Act of 1973.* http://endangered.fws.gov/esa.html (accessed Nov. 2001).

———. 2001b. *ESA basics.* http://endangered.fws.gov (accessed Nov. 2001).

———. 2001c. *Query of the National Wildlife Refuge System lands database.* http://realty.fws.gov/nwrs.htm (accessed Nov. 2001).

———. 2004. *Species information threatened and endangered animals and plants.* http://search.netscape.com/ns/search?query=Endangered+Species+List&fromPage=NSBoom (accessed April 9, 2004).

U.S. Forest Service. 2001a. *Archives index.* http://fs.jorge.com/archives/default.html (accessed Nov. 2001).

———. 2001b. *NEPA, NFMA, and appeals.* http://www.fs.fed.us/forum/nepa/.

UNESCO. 2000. *World Heritage list.* http://www.unesco.org/whc/nwhc/pages/sites/main.htm (accessed June 10, 2001).

———. 2004a. *The MAB Programme, world network of Biosphere Reserves.* http://www.unesco.org/mab/wnbr.htm (accessed April 9, 2004).

———. 2004b. *World Heritage List.* http://www.unesco.org/whc/heritage.htm (accessed April 9, 2004).

Varley, J. D., and P. Schullery. 1991. Reality and opportunity in the Yellowstone fires of 1988. In *The Greater Yellowstone ecosystem: Redefining America's wilderness heritage,* ed. R. B. Keiter and M. S. Boyce, 105–121. New Haven, CT: Yale University Press.

Wagner, F. H., R. Foresta, R. B. Gill, D. R. McCullough, M. R. Pelton, W. F. Porter, and H. Salwasser. 1995. Natural-resources values, goals, and policies of the system. *Wildlife Policies in the U.S. National Parks.* Washington, DC: Island Press.

Walker, R. 2001. Urban sprawl and natural areas encroachment: Linking land cover change and economic development in the Florida Everglades. *Ecological Economics* 37: 357–369.

Walker, R., and W. Solecki. 2001. South Florida: The reality of change and the prospects for sustainability. *Ecological Economics* 37: 333–337.

Whitehouse, J. F. 1992. Fourth world congress on national parks and protected areas, Caracas, Venezuela. *Australian Zoologist* 28: 39–46.

Whittlesey, L., and M. Finley. 2001. Never running from a fight: Mike Finley looks back on his Park Service career. *Yellowstone Science* 9: 9–19.

Wilderness Society. 2001a. *America's national monuments. List of presidentially designated monuments.* http://www.wilderness.org/Library/Documents/loader.cfm?url=/commonspot/security/getfile.cfm&PageID=2894 (accessed April 13, 2004).

———. 2001b. *The Antiquities Act.* http://www.wilderness.org/newsroom/pdf/antiquities.pdf (accessed Nov. 2001).

Wilkinson, P. F. 2000. Protecting for ecological integrity in Canada's national parks: Allowable and appropriate visitor use. Fourth Science and Management of Protected Areas Association Conference (SAMPAA), University of Waterloo, Waterloo, Ontario, May 14–19, 2000.

Worboys, G., M. Lockwood, and T. DeLacy. 2001. *Protected area management: Principles and practices.* South Melbourne, Australia: Oxford University Press.

World Commission on Protected Areas. 2002. World Commission on Protected Areas and the IUCN Program on Protected Areas. http://wcpa.iucn.org/ (accessed March 5, 2002).

World Conservation Monitoring Centre. 1999. *1991 IUCN directory of protected areas in Oceania.* http://www.wcmc.org.uk (accessed November 1999).

———. 2002. *About UNEP-WCMC.* http://www.unep-wcmc.org/ (accessed March 5, 2002).

Wright, R. G. 1996. Expansion of the U.S. National Park System in Alaska. In *National parks and protected areas: Their role in environmental protection,* ed. R. G. Wright, 165–171. Cambridge, MA: Blackwell Science.

Yellowstone Center for Resources. 2002. New framework for winter use of parks announced. *Yellowstone Science* 10: 23.

Yellowstone Ecosystem Studies. 2001. *Think like an ecosystem: GYCC.* http://fp2.in-tch.com/www.yellowstone.org/about/news/gycc1.html (accessed Nov. 2001).

CHAPTER

6

Selection and Delineation

Some of the more significant problems in national park management occur in relation to park boundaries, and...many of these problems can be avoided if more care was exercised in setting those boundaries.

—John J. Pigram and Paul Friesma

Wars and battles are fought throughout the world over boundary disputes and control of land. One of the most contentious boundary disputes in modern times is the Israeli-Palestinian conflict over control of the West Bank and Gaza Strip. While selection and delineation of national parks and protected areas are not typically the cause of military battles, they nonetheless cause conflicts among competing stakeholders, which sometimes fester for decades. Disagreement regarding site selection, delineation, and management of protected areas occurs due to economic development on adjacent lands; changes in technology that cause access and harm to environmentally sensitive areas; better scientific knowledge about resource conditions needed to sustain natural landscapes; and changes in human values, emotions, and perceptions (Dennerlein 1997).

A well-designed program for nature conservation involves three kinds of decisions: creation of core reserves, expanding the boundaries of existing reserves into adjacent areas, and using large natural corridors to connect existing reserves (Wilson 2002). Establishing and expanding core reserve areas involve making decisions related to four factors: how many and what types of protected areas; locations of protected areas; size, shape, and boundary of a protected area; and expanding the boundary of a protected area. These highly interrelated normative decisions are collectively referred to as selection and delineation decisions.

Selection and delineation decisions are made within a social, cultural, economic, environmental, and political setting that changes in accordance with established priorities and management objectives for protected areas. Making selection and delineation decisions is especially difficult when stakeholders have conflicting preferences for the ecological goods and services provided by protected areas, and disagree about their socioeconomic and environmental impacts.

Selection and delineation decisions occur at several jurisdictional levels. In the United States, for instance, protected areas are established at three levels: local (community), state, and national. In addition, selection and delineation decisions are influenced by whether or not countries seek or already have internationally designated protected areas, such as a Biosphere Reserve, World Heritage Site, or Ramsar Wetland.

This chapter discusses seven aspects of selection and delineation:

1. The significance of social-cultural-economic-ecological conditions
2. The importance of protected area categories
3. Decision making related to establishment of a protected area,
4. Decision making related to expanding the boundary of an existing protected area
5. National system planning
6. Successful development and implementation of national plans
7. Global system planning

Significance of Social-Cultural-Economic-Ecological Conditions

Selection and delineation of national parks and protected areas are done in a social-cultural-economic-ecological setting that shapes decision outcomes for individual protected areas and protected area systems. Consider, for example, the setting within which the world's first national park was birthed. In the early 1870s, the region to be known as Yellowstone National Park was wild and unsettled by white men and women and communication was primitive. These conditions had two important consequences.

First, during the so-called era of disposition (1862–1891), vast tracts of undeveloped public land in the western United States were made available to private interests in order to stimulate economic development (Keiter et al. 1999). Two federal incentives were proffered for this purpose. Land grants were given to railroads to entice westward expansion of rail lines (Gates 1979). Railroad land grants were instrumental in allowing the Northern Pacific Railroad to gain a monopolistic position in transporting early visitors to Yellowstone National Park. Congress enacted several laws designed to transfer public lands to private ownership. These laws included the Homestead Act of 1862, the General Mining Law of 1872, the Desert Lands Act of 1877, and the prior appropriation doctrine. At the same time that public lands were being privatized, strong interest developed in preserving unique public lands from human settlement and resource development. Designation of Yellowstone National Park and Yosemite National Park marked the beginning of the public land preservation movement. Creation of these parks was not opposed by economic and political interests because the lands they encompassed were not considered valuable for mining, ranching, farming, and homesteading (Runte 1987; Dennerlein 1997).

Second, the primitive form of communication that existed at the time made it difficult for the general public to comprehend the unique natural resources of the Yellowstone region. Sketches made by Thomas Moran and photographs taken by W. D. Jackson during the Hayden Survey convinced Congress and the general public that the wonders of Yellowstone were real. What swayed congressional and pubic opinion in favor of creating a national park in the Yellowstone region was its spectacular geothermal and other natural features (hot springs, geysers, mud pots, waterfalls, deep canyons), not its later-to-be discovered importance as a wildlife reserve. Yellowstone's spectacular geothermal features documented in the 1871 Hayden Survey were instrumental in delimiting the 0.9 million ha (2.2 million acres) included in the park. This is not surprising considering their eye appeal and Hayden's background in geology.

In summary, creation of Yellowstone National Park was supported because the area proposed for the park encompassed geologic features that were aesthetically and scenically appealing and did not contain valuable extractive resources. In addition, establishment of the park was expected to contribute to westward expansion of and higher revenues for the Northern Pacific Railway. Establishing nature-based national parks for cultural or ecological reasons did not become important until the early 1960s. During this period, Paul Ehrlich, Donnella Meadows, and Rachael Carson drew national attention to the adverse human and ecological consequences of growth in population, industrial and pesticide pollution, and consumerism. Mass media, comprehensive and widespread science education programs, and growth in environmental organizations solidified public support for the scientific, educational, and biological values of national parks and protected areas.

Importance of Protected Area Categories

Size and delineation of protected areas are influenced by the IUCN protected area categories (see table 2.1). The seven categories define the activities and functions of protected areas that fall in a particular category and have implications for the size of protected areas. For example, if a country wants to establish a protected area to commemorate a significant military battle or historic event, then category III is appropriate. This category includes areas "containing one, or more, specific natural or natural/cultural feature, which is of outstanding or unique value because of its inherent rarity, representative or aesthetic qualities or cultural significance" (IUCN 1984). Size and delineation of a protected area in category III need only encompass the natural or cultural feature being protected. In contrast, a category II national park is, among other things, suppose to "protect the ecological integrity of one or more ecosystems for present and future generations" (IUCN 1994). The amount of land needed for this purpose is ordinarily much greater than the amount of land required to protect a natural or cultural feature.

Not surprisingly, the mean size of protected areas in 1994 was 1,183 km^2 (457 mi^2) in category II and 91 km^2 (35 mi^2) in category III (IUCN 1994), a 13-fold difference. On a percentage basis, protected areas in category II made up 40 percent of the protected area acreage in North America and 30 percent of the protected acreage in the world, whereas protected areas in category III made up only 1.43 percent of the protected area acreage in North America and 1.46 percent of the protected area acreage in the world in 1997 (Green and Paine 1997).

Not all the acreage in a protected area is managed according to the primary management objectives for the category in which it falls. For example, many national parks in North America (category II) contain large backcountry areas that are de facto wilderness (category Ib). In these cases, wilderness areas are nested within national parks. A rule of thumb is that at least three-quarters and preferably more of a protected area should be managed for primary objectives with the remaining area managed so as not to conflict with the primary objectives. Based on this rule, at least three-quarters of the acreage in category II national parks should be managed for the primary objectives of species/genetic diversity, environmental services, and tourism and recreation; the remainder of the acreage should be managed for the secondary objectives of scientific research, wilderness protection, natural/cultural features, and education.

Strict application of the rule of thumb is unlikely because the same acreage of protected areas is often managed according to multiple objectives. For example, the frontcountry area of national parks in North America contain campgrounds, visitor centers, and roads that benefit tourism/recreation (a primary objective of national parks), as well as certain forms of scientific research and education (secondary objectives of national parks), such as stream monitoring and field studies for the general public. A common practice is to partition the land area in national parks into management zones, each with its own set of compatible management objectives.

Decision Making for Establishing Protected Areas

There are two general kinds of selection and delineation decisions: those dealing with establishment of a protected area and those associated with expansion of an existing protected area. This section discusses decision making regarding establishment of protected areas; the next section considers decision making related to boundary expansion.

Establishment of a protected area involves four major issues. First, how selection and delineation influence the initial and future ecological and socioeconomic values of a protected area. Second, how the establishment of a protected area affects social, economic, and cultural values in gateway communities. Third, how selection and delineation can be done so as to minimize future ecological degradation caused by economic development in gateway communities. Fourth, how international agreements influence selection and delineation. These four issues are discussed in the context of establishing an individual protected area. A gateway community is a town or city that borders public land (Howe et al. 1997). Gateway communities provide a whole host of services to visitors to national parks and protected areas.

Ecological and Socioeconomic Impacts

The type, location, size, shape, and boundary of a protected area influence the mix of ecological goods and services it provides, and hence its value to visitors, research scientists, educators, conservationists, and others. This section discusses decisions regarding the specific location of a protected area within a biome (distinct natural ecosystem) as well as its size, shape, and boundary. It is assumed that the types (with respect to IUCN categories) and general locations or spatial distribution of protected areas among biomes have been already been determined based on national system planning, which is discussed in the next section. The first part of this section focuses on how location, size, shape, and boundary decisions affect ecological values, and the second part addresses how to take account of stakeholders' tastes and preferences in making selection and delineation decisions.

Decisions regarding the location, size, shape, and boundary of a protected area have a direct bearing on its ecological value. Basing evaluations of ecological impacts on the best information/data and sound ecological principles enhances ecological values. Conversely, evaluations based on inadequate and/or faulty information/data and inadequate consideration of ecological principles are likely to result in a degradation of ecological values. Important considerations in designing individual protected areas include (adapted from Davey 1998):

- Habitat/area requirements for rare or other species and their minimum viable population sizes (especially important for large carnivores)
- Potential for increasing wildlife migration via corridors between units included in the protected area as well as proximity to other protected areas and multiple-use areas; conversely, achieving isolation to minimize transfer of disease, losses due to disease and predation, and other factors
- Perimeter/area relationships
- Linkages to natural processes, including erosion, flooding, volcanism, ocean currents, aeolian or other active geomorphic systems, and location relative to hydrologic (watershed) and political boundaries
- Ability to undertake management operations or inability to deter potentially negative impacts from activities such as poaching
- Extent of natural resource degradation and vulnerability to external threats
- Traditional uses (e.g., subsistence living) and current occupancy of area

- Cost of achieving protected area status, including the cost of acquiring land, compensation to current landowners, and cost of implementing collaborative management strategies.

Comprehensive evaluation of these factors leads to a good protected area design. Simple although somewhat extreme illustrations of poor and good locations of protected area are shown in figure 6.1.

In the human dimension, changes over time in stakeholders' tastes and preferences for ecological goods and services influence use and nonuse values of protected areas. Selection and delineation decisions should take account of both use and nonuse values at the time a protected area is established and how these values are likely to change in the future. A primary objective of many protected areas is to provide enjoyment for visitors. Visitation is important because, in the long run, a protected area will not be viable unless it provides enjoyment to visitors and meets the needs of stakeholders. The latter group includes individuals who want a protected area preserved even though they may not visit it (they have option value for the area). Therefore, it is important to consider tastes and preferences of both actual and potential visitors in making selection and delineation decisions.

Consider how visitation rates are likely to differ for protected areas having three different location and site characteristics. The first protected area is located near a heavily populated area and has average flora/fauna and scenic qualities. This protected area is likely to have high visitation rates because of its high accessibility. Shenandoah National Park in the United States is an example of this type of protected area.

The second protected area is remotely located with respect to major population centers and has outstanding flora/fauna and scenic qualities. This protected area is likely to have high visitation rates despite the relatively high travel cost because of the outstanding flora/fauna and scenic qualities. Yellowstone National Park in the United States is an example of this type of protected area.

The third protected area is located close to a major population center and has outstanding flora/fauna and scenic qualities. This protected area is likely to have high visitation rates. Yosemite National Park is an example of this type of protected area. The above comparison suggests that visitation is likely to be low for remote protected areas that have low flora/fauna and scenic qualities, and high for protected areas that are close to heavily populated areas and have outstanding flora/fauna and scenic qualities.

Selection and delineation decisions are likely to have different impacts on user groups. Consider the differences in use values between tourists and scientists. Whereas the value to tourists depends largely on aesthetic characteristics of a protected area, such as abundance of megafauna and spectacular scenery, the value to scientists is likely to depend on features that may be of little direct interest to

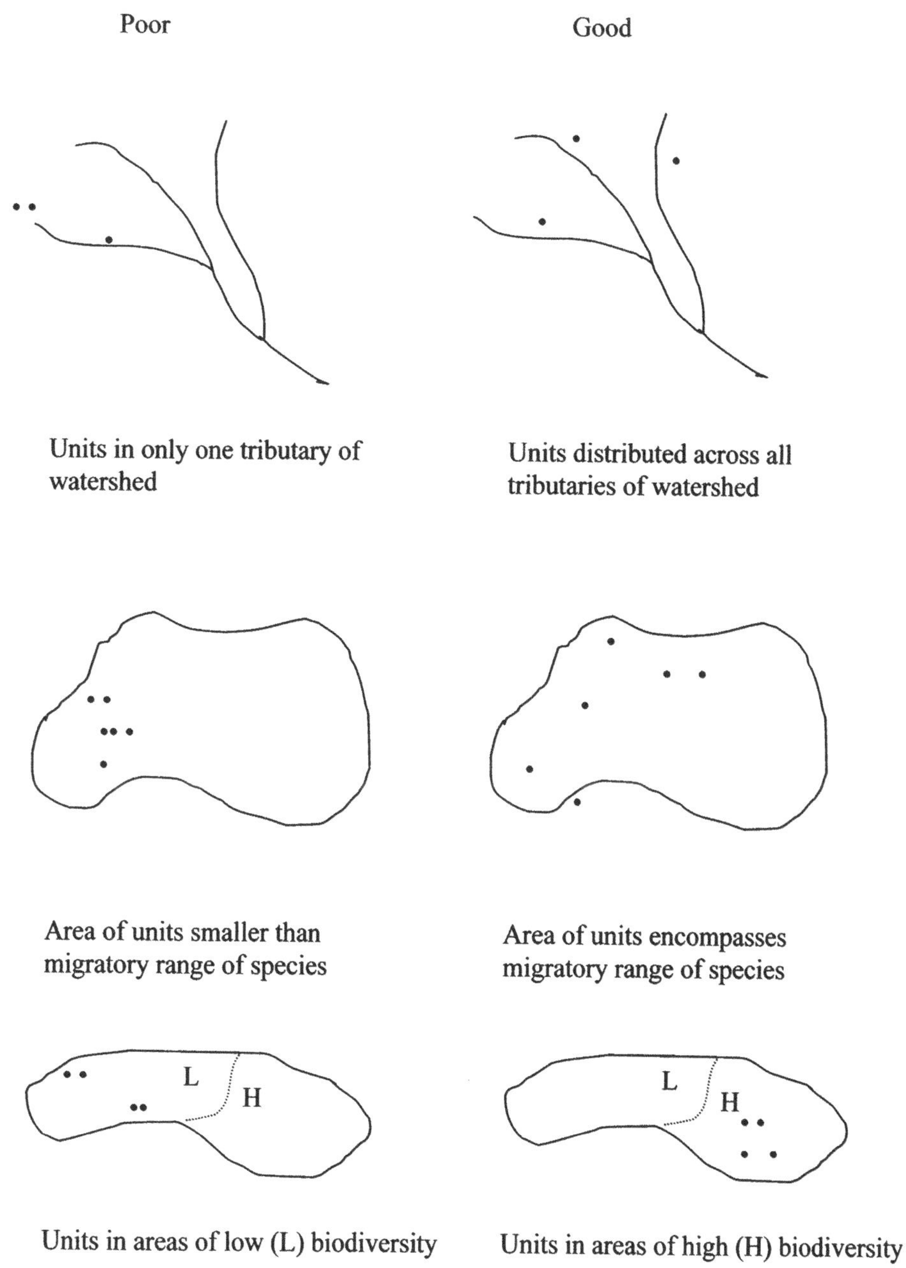

FIG. 6.1. Examples of poor and good locations for protected areas.

tourists, such as the presence of a rare species or unique geologic formation, or occurrence of a major natural disturbance, such as fire or disease. For example, an ecologist studying the impacts of glaciation on landforms or the impacts of climate change on the retrenchment of glaciers is likely to be more interested in protected areas at extreme latitudes than those closer to the equator. It is possible for tourists and scientists to have diametrically opposed preferences for the same condition. Remoteness of a protected area may be appealing to a scientist because it generally means less human impact on natural landscapes, whereas it tends to be a drawback for tourists.

Expected future changes in visitation should be assessed and taken into consideration when selecting the location, size, and boundary of a protected area. If the number of visitors is expected to increase significantly over time, then the size and boundary of the protected area should allow for growth in visitation without jeopardizing the quality of visitor experiences.

Protected areas have experienced dramatic increases in visitation. During the 1980–2000 period, for example, visitation for national parks in the United States increased twice as fast as resident population (43 percent vs. 21 percent; see fig. 2.1) and per capita visits to national parks increased 18 percent. Growth in visitation occurs because of increases in per capita income (U. S. income per capita, unadjusted for inflation, almost tripled during this period), more leisure time, improved health care (which allows for greater mobility of the elderly and increases longevity), greater use of jet travel, advertising and promotion of tourism, improvements in visitor services and transportation within and adjacent to protected areas, and development of recreational equipment and vehicles that render outdoor recreation more pleasurable and accessible. For example, ecotourism has grown dramatically due to increased visitor demand for protected areas with high biodiversity.

Decisions regarding the type, location, size, and boundary of protected areas are interrelated. Consider the creation of a new protected area whose primary purpose is to conserve biodiversity. In this case, it is important that the size and delimitation be sufficient to incorporate critical areas for protecting biodiversity. The scientific principle underlying this design strategy is island biogeography. This foundational concept in conservation biology states that biodiversity decreases and extinction rates increase as the size of an area decreases. Extinction rates appear to be higher in smaller areas because they are more vulnerable to edge effects, and invasions of exotic plants, animals, and pathogens, which accelerate species extirpation and ecosystem degradation (Soule and Noss 1998). Research shows that smaller protected areas have higher local extinction rates (Newmark 1995). Island biogeography also provides the scientific basis for regional and continental scale conservation, which has the goal of establishing corridors between relatively isolated national parks and wilderness areas in order to facilitate migration of large carnivores. The Yellowstone-to-Yukon Conservation Initiative is a prime example of regional scale conservation in North America (Chadwick 2000). Chapter 7 discusses continental scale conservation in depth.

The freedom of planners to select the type, location, size, and boundary of a protected area varies spatially even within the same country. For example, there was greater opportunity to establish relatively large size wilderness areas in Alaska under the Alaska National Interest Lands Conservation Act of 1980 than to create wilderness areas in the eastern United States because Alaskan landscapes are wilder and less populated than landscapes in the eastern United States.

The U.S. National Park Service (1978, 1988) identified several criteria for site selection (also see the sidebar):

1. National significance of a unit based on the uniqueness and quality of its scenic areas, biological communities and geologic phenomena
2. Suitability and feasibility of an area as measured by its integrity relative to physical and biological resources
3. Capacity of the area to be preserved, managed, developed, and used
4. Comprehensiveness of the proposed unit (contiguity) for effective management and accessibility
5. Whether an area is or will receive adequate protection outside the park system and is available for public appreciation and use

Criteria for U.S. National Parklands

The public, state, and local officials, Indian tribes, members of Congress, or the National Park Service (NPS) can propose adding new areas to the U.S. National Park System. Proposed areas are screened and evaluated by NPS to ensure that "only the most outstanding resources are added to the National Park System." NPS has established criteria for evaluating proposed additions to the system. Proposed areas must:

1. Possess nationally significant natural, cultural, or recreational resources
2. Have characteristics that make the area suitable and feasible for addition to the system
3. Be such that the benefits of NPS protection must exceed the benefits of protection by some other governmental agency or by the private sector

National significance of a proposed addition is evaluated in terms of the following four standards:

- It is an outstanding example of a particular type of resource.
- It possesses exceptional value of quality illustrating or interpreting the natural or cultural themes of our Nation's heritage.
- It offers superlative opportunities for recreation for public use and enjoyment, or for scientific study.

- It retains a high degree of integrity as a true, accurate, and relatively unspoiled example of the resource.

Somewhat different criteria are used to evaluate proposed additions for natural areas than for cultural areas. While considerable weight is given to an area's potential to support public use and enjoyment, its natural and/or cultural resources are also important.

If a proposed area is deemed to have national significance, then it is evaluated for suitability and feasibility. For an area to be judged suitable, it "must represent a natural or cultural theme or type of recreational resource that is not already adequately represented in the National Park System or is not comparably represented and protected for public enjoyment by another land-managing entity." Adequate representation is determined "by comparing the proposed area to other units in the National Park System for differences or similarities in the character, quality, quantity, or combination of resources, and opportunities for public enjoyment."

A proposed addition to the National Park System is deemed feasible when "an area's natural systems and/or historic settings [are]...of sufficient size and appropriate configuration to ensure long-term protection of the resources and to accommodate public use [and]...have potential for efficient administration at a reasonable cost." Feasibility is judged in terms of landownership, acquisition costs, access, threats to the resource, and staff or development requirements.

Even if a proposed addition to the National Park System is nationally significant, suitable, and feasible, it may not be appropriate for NPS to manage it. For this reason, NPS evaluates management alternatives for proposed additions including whether they should be managed "by state or local governments, Indian tribes, the private sector, or other federal agencies; technical or financial assistance from established programs or special projects; management by others as a designated national natural landmark, a national historic landmark, a national wild and scenic river, a national trail, a biosphere reserve, a state or local park, or some other specially designated and protected area." Other possible federal designations include areas of critical environmental concern, marine or estuarine sanctuaries, national conservation areas, national recreation areas, national wildlife refuges, and wilderness.

According to the National Parks Omnibus Management Act of 1998, NPS can send a list of candidate study areas to Congress. Congress can also propose study areas and is the only body that can authorize complete studies of proposed areas. Studies are done in consultation with interested federal, state, and local agencies, Indian tribes, and the public.

Final authority for deciding whether or not to add a proposed area to the National Park System rests with Congress. Typically, congressional committees hold hearings to discuss adding units to the National Park System. During the hearings, the secretary of the interior provides recommendations on the proposed additions, which are based in part on study reports prepared by NPS. Congressional approval of an addition to the National Park System comes in the form of an organic act (legislation) that authorizes a new area, explains its purpose, and outlines "any specific directions for additional planning, land acquisition, management, and operations." Congress can confer more than 20 different designations to new units of the National Park System, such as national park, seashore, lakeshore, historic site, battlefield, and recreation area. The national park designation is usually "reserved for the most spectacular natural areas with a wide variety of features." Unlike many other units of the National Park System, hunting, mining, and grazing are usually prohibited in national parks.

Boundaries are an important consideration in studying proposed additions to the National Park System. Decisions about boundaries involve "compromise between the ideal and what is practical considering costs and other factors." As part of its planning procedures for existing units of the National Park System, NPS evaluates proposals to adjust the boundaries of existing park units. Their evaluation considers the feasibility of administering proposed units taking into account size, configuration, ownership, costs, and other factors. Proposals must demonstrate that a boundary adjustment would "include significant features or opportunities for public enjoyment related to the purposes of the park...address operational and management issues such as access and boundary identification along topographic or natural features or roads, or...protect park resources critical to the park's purposes." Studies of boundary adjustments for existing units of the National Park System do not require congressional approval.

Source: National Park Service (2003).

An important concept in designing individual protected areas is bioregionalism. The aim of bioregionalism is to achieve greater harmony between human, cultural, political, and social values and natural systems (Plant 1990). Bioregionalism attempts to integrate elements of a landscape that have become separated (Aberley 1993).

While a bioregion can be defined in terms of climatology, physiography, flora and fauna, natural history, and other biophysical characteristics, it is best described and implemented in terms of the meanings that residents associate with living in a particular place or region, which are called place-based values (Hansson and Wackernagel 1999; Berg and Dasmann 1990). For example, residents of the U.S. Pacific Northwest strongly support restoration of salmon populations in the Columbia River System because salmon are symbolic of the region.

For a bioregionalist, living and planning should be done at the community level. Community encompasses the entire region potentially affected by a decision; it is not restricted to a narrow political jurisdiction, such as an incorporated city. Support for community-based decision making derives from the premise that community residents know what is best for their community, have the best understanding of community needs and aspirations, and have the most to gain or lose from land-use decisions affecting their community. Successful application of bioregionalism requires devolution of political and economic power to the community level (Hansson and Wackernagel 1999).

Bioregional planning is relevant to establishment of protected areas because they are socially, economically, culturally, and ecologically linked to adjacent communities. In terms of selecting and delineating protected areas, bioregionalism supports creating effective buffer zones for protected areas, minimizing adverse natural resource impacts of development in gateway communities, designing incentives for ecologically sustainable land use/management and ecological restoration of degraded ecosystems (Davey 1998). A major challenge in applying bioregionalism to selection and delineation of protected areas is the need to achieve a balance between the public goods nature of national parks and protected areas, which requires deference to national goals and management objectives for public lands, and the goals, needs and desires of the local communities.

Impacts on Gateway Communities

Two socioeconomic linkages exist between protected areas and gateway communities that need to be considered in selection and delineation decisions. First, establishment of protected areas has socioeconomic impacts on gateway communities. Second, future economic development in gateway communities has the potential to adversely impact ecological integrity of protected areas. The two linkages are interrelated. Increased visitation stimulates economic development in gateway communities, which in turn results in conversion and fragmentation of natural habitats and extermination of populations of vulnerable species. The first linkage is discussed in this section and the second linkage in the next section

Socioeconomic and community impacts of establishing protected areas are likely to differ between developing and developed countries. A major concern in many developing countries is "how many protected areas can we afford" (Dixon and Sherman 1997). In contrast, the issue facing developed countries is "how many protected areas do we need." This difference in perspective occurs for two reasons.

First, developing countries generally place a lower priority on establishing protected areas than on reducing environmental degradation, such as soil, air, and water pollution. Cleaner soil, air, and water decrease human illness and disease and increase labor productivity. These benefits are disproportionately borne by lower-income groups. While air and water quality benefits can be claimed for protected areas, they are, for the most part, borne by residents of the region in which the protected area is located. Consequently, programs to improve environmental quality are apt to receive greater political support and public acceptance in developing countries than creation of a protected area. In addition, environmental protection efforts are more apt to attract international development funds than additions to the protected area estate. In comparison, most developed countries have the resources to improve environmental quality and create protected areas.

Second, establishment of protected areas entails a high opportunity cost in developing countries when it reduces natural resource extraction activities, such as mining, livestock grazing, gathering of firewood and medicinal plants, water collection, fishing, hunting, and timber harvesting by indigenous cultures and eliminates or restricts subsistence uses of natural resources. Socioeconomic and cultural impacts of protected area designation are more acute when the size of the protected area is large relative to the area needed to support subsistence uses of natural resources and/or the location of the protected area infringes on areas critical for subsistence living. Under these conditions, establishment of a protected area propagates social inequity and rural poverty and is likely to be opposed by local communities in developing countries. For example, Jean Chretien took charge of Canada's national parks in 1968 with an ambitious plan to add 40 to 60 new parks to the system by 1985. Local communities and Aboriginal people strongly objected to the proposed expansion. In 1979 and again in 2000, Canada amended its national parks policy by abandoning the practice of expropriating land from indigenous cultures and private landowners. Canada's National Parks Act was amended to protect the rights of Aboriginal people to hunt, trap, and fish in certain national parks (McNamee 2002a).

Negative impacts of protected area designation on indigenous cultures and subsistence living can be alleviated two ways. First, location, size, and boundary should be selected so as to minimize conflicts with cultural values and subsistence living. Second, if subsistence resource use is significant, then the protected area should be established in a category that allows such uses, such as IUCN category VI. Unfortunately, local, indigenous groups often lack the political and economic clout to ensure fair and equitable treatment of their concerns regarding the potentially negative social, cultural, and economic impacts of designating protected areas.

The high opportunity cost of establishing protected areas in developing countries might explain why developing countries have a greater percentage of total protected area in category VI than developed countries. In 1997, the proportion of total protected acreage in category VI was 45.4 percent in the WCPA regions of North Africa/Middle East, Southeast Asia, South America, Central America, and Eastern/Southern Africa regions vs. 23.4 percent in the WCPA regions of North America, Australia/New Zealand, and Europe. The high opportunity cost of creating category II national parks in developing countries could explain why national parks in many developing countries exist only on paper (so-called paper parks).

Extractive resource industries and companies that manufacture off-road motorized recreational vehicles often oppose the establishment and expansion of protected areas when these actions eliminate or restrict motorized access to those areas. Creation of Yellowstone National Park in northwestern Wyoming (United States) was unopposed by economic interests because the area proposed for the park was considered to have little economic value for farming, ranching, and mineral extraction. In comparison, off-road vehicle users opposed expansion of Canyonlands National Park in Utah in the western United States on the grounds that it would reduce the area available for off-road recreation. Oil and gas interests, ranchers, and others have opposed the creation of protected areas by the president of the United States under the authority of the Antiquities Act of 1906. To the extent possible, the location and/or boundary for a protected area should be chosen so as to minimize conflicts between resource preservation and extraction. This strategy is difficult to implement when protected area status is being proposed in order to preserve the remnants of a biome that is being degraded rapidly by resource extraction and economic development.

Third, even when a protected area generates additional income and employment for a gateway community in a developing country, social inequities result when higher-skilled jobs, such as ranger, manager, planner, biologist, and engineer, are filled with individuals from outside the region. Filling lower-skilled jobs generated by a protected area with local residents does not reduce such inequities. Overall economic (income, employment, and tax) benefits of a protected area need to be weighted against contraction in economic activity caused by reductions in resource extraction (second objection above).

Due to differences in purpose, size, and location, the socioeconomic and community impacts of protected areas differ within the same country and even within the same IUCN category. For example, economic, environmental, and demographic differences between eastern and western regions of the United States result in substantial differences in the socioeconomic and community impacts of national parks and protected areas. National parks in the eastern United States are relatively close to heavily populated urban areas and large commercial/industrial centers and are surrounded by predominately private land. Many national parks in the western United States are located in more remote, sparsely populated rural areas surrounded by public land. Consequently, the regional economies in which western national parks are embedded are generally more dependent on income and employment generated by the parks than the regional economies for eastern national parks.

Establishment of protected areas in remote, natural-resource-based rural areas can have four negative socioeconomic effects. First, since tourism is highly labor intensive (Bull 1991) and wages for labor-intensive tourism-related jobs (service, retail, construction, real estate, transportation, and public utilities) tend to be lower than for jobs in resource extraction and manufacturing (mining, timber harvesting, wood processing, aluminum manufacturing), increasing a community's dependence on tourism at the expense of resource extraction/manufacturing reduces average wages.

Second, compared to economic reliance on the agricultural and wood products industries, greater economic reliance on tourism is likely to increase income inequality (Marcouiller and Green 2000). Specifically, income inequality rises because tourism-related jobs make rural residents subservient to wealthy tourists (Ashworth 1992), and subjects low-income families, often headed by females, to persistent poverty (Smith 1989).

Third, the remote, rural locations of many protected areas mean the income and employment they generate are highly seasonal (part-time) reaching a peak in the summer and winter seasons and a trough in the off-peak periods. Job and income seasonality are forms of income inequality.

Fourth, the highly attractive environmental and/or cultural amenities afforded by gateway communities have spurred in-migration of recreational homeowners, especially in developed countries (Moss 1994). If current demographic trends continue, the next 20 years will bring rapid growth in in-migration to gateway communities (Howe et al. 1997). While in-migration stimulates economic development and enhances property tax receipts in gateway communities, the higher income status of many recreational homeowners induces income inequality (Fritz 1982; Deller et al. 1997).

In summary, relative to developing countries, creation of protected areas in developed countries is likely to generate net benefits in terms of higher income and employment provided it does not displace significant commercial or subsistence use of natural resources or adversely affect indigenous cultures. Gains can be significant in gateway communities that have experienced or are likely to experience significant declines in traditional economic sectors, such as mining, farming, ranching, and timber harvesting. However, compared to traditional resource extraction activities, income and employment benefits generated by protected areas are highly seasonal, especially in more remote rural areas, wages are generally lower, and income inequities are greater.

Minimizing Ecological Degradation

There is mounting evidence that economic development in gateway communities has adverse ecological impacts on protected areas. In particular, construction of roads, residential housing units, and commercial/industrial centers in gateway communities result in conversion and fragmentation of natural habitats and impairment or extermination of vulnerable species. For example, economic development on lands adjacent to Florida's National Key Deer Refuge has had devastating impacts on the habitat and hence viability of the endangered dwarf key deer for which the refuge was named. Double-digit growth in residential subdivisions adjacent to the National Elk Refuge in Jackson, Wyoming, has diminished winter range for the 10 thousand elk that use the refuge and displaced corridors that the elk use to reach summer range in Yellowstone and Grand Teton National Parks (Howe et al. 1997).

While resource managers tend to be most concerned about negative ecological impacts of gateway communities after creation of a protected area, they, nonetheless, should anticipate and assess such impacts at the time of selection and delineation. Specifically, the risk of ecological degradation from economic development in gateway community should be evaluated for alternative protected area designs (sizes, locations, and boundaries). Other things equal, a design should be selected that minimizes the likelihood of negative ecological impacts. When such impacts are ignored in designing protected areas, there is greater risk of subsequent losses in ecological integrity, and protected area managers will find it more difficult to sustain ecological integrity (Connally 1982).

Ways to minimize adverse ecological impacts of future economic development include:

1. Selecting as large a size of protected area as technically, financially, and politically feasible
2. Nesting protected areas in categories I, II, and III within protected areas in categories IV, V, and VI
3. Nesting protected areas within other multiple-use public lands that limit economic development, such as a national forest
4. Establishing land-zoning regulations that increase the minimum lot size in subdivisions and/or reduce the proximity of subdivisions to protected areas through the creation of green spaces or buffer zones

While these options are likely to reduce the negative ecological impacts of future economic development on protected areas, their effectiveness is very site specific. For instance, it is much easier to implement option 1 in remote areas, option 2 in countries that are establishing multiple protected areas at the same time, option 3 in areas having a large proportion of land in public ownership, and option 4 in areas that are more receptive to land use controls.

A promising approach for preserving the ecological integrity of protected areas is ecosystem management, which attempts to balance and sustain the social, economic, and ecological values provided by large ecosystems over long time periods

(Diaz and Bell 1997; Schowalter et al. 1997; Thomas 1997). Ecosystem management is a relatively new approach to natural resource management that is replacing the more traditional resource-by-resource approach to management. Chapter 7 discusses ecosystem management in more detail.

Role of International Agreements

An important determinant of the types of protected areas in a country is whether it has signed and implemented Article 8 of the Convention on Biological Diversity (CBD). Article 8 was adopted by 181 parties, including the United States and Canada, at the 1992 Earth Summit Conference in Rio de Janeiro, Brazil. It establishes global priorities and policies for in situ conservation of biodiversity and calls for establishment and maintenance of a global system of protected areas. The CBD encourages countries to establish a system of protected areas that are managed in a manner that supports natural and cultural resource conservation, sustainable use, and equitable sharing of the benefits (Task Force 1998). Development of a national system plan is needed because several ecosystem types, notably lowlands, wetlands, streams, humid forests, and coastal areas, are underrepresented, particularly in national parks (Soule and Noss 1998). One way to alleviate this deficiency is to increase protected area designation for underrepresented ecosystems. National system planning for protected areas is discussed in greater depth below.

Decision Making for Boundary Expansion

Expanding the boundary of a protected area is typically not considered until several years after a protected area is established. Boundary expansion decisions require evaluation of the pros and cons of enlarging the boundary and size of a protected area. Enlargement becomes an issue for several reasons: (1) the original boundary constrains the protected area's capacity to accommodate growth in visitation; (2) new scientific knowledge indicates the original boundary does not adequately protect ecological processes and functions; (3) opportunities arise for transnational boundary expansion, purchasing private land (inholdings) within a protected area, or changing protected area status (designating a national monument as a national park); and (4) prospects brighten for making mutually advantageous private-federal land exchanges.

In 1998, the National Parks and Conservation Association completed a study that indicated the need for adjustments on 69 percent of the boundaries of primarily natural areas and 41 percent of the boundaries of primarily cultural areas of the National Park System in the United States (National Parks and Conservation Association 1988). Reasons given for this recommendation are that (1) resources pertinent to the theme of the park needed to be incorporated into the park, (2) resources within the boundaries and related resources outside the boundaries required buffering from adjacent activities, (3) the view from within the park needed to be preserved, and (4) park management would be improved.

Boundary adjustments are common. Through 1997 the United States made more than five hundred boundary adjustments in two hundred national park units. In addition, national legislation has been introduced and/or passed that would add or delete land to existing areas, decommission existing units, change the organic act for a unit, and transfer ownership from state to federal government or vice versa (Dennerlein 1997).

The potential benefits of expanding the boundary of a protected area are significant when human population and resource development are increasing on lands adjacent to the protected area. Under these conditions, boundary expansion reduces a protected area's vulnerability to conversion or fragmentation of natural habitats and adverse impacts on populations of vulnerable species. Unfortunately, boundary expansion is more difficult when human population and resource development in adjacent areas are increasing. Attempts to restrict land uses and management practices on private property adjacent to protected areas through zoning and other means (e.g., Endangered Species Act) have been resisted on the grounds that this constitutes a taking of private property rights. Legislators sympathetic to this argument have introduced legislation that requires compensation for takings of property rights.

Several criteria are important in making boundary decisions for parks (Wright 1985):

- The boundary of the park should encompass resources that are nationally significant and should provide a setting for those resources.
- Size and configuration of the park should contain all the habitat attributes required by faunal communities.
- Socioeconomic impacts must be taken into account to minimize existing or potential resource conflicts.
- Park areas should be sufficiently large to accommodate future expansion in visitor facilities and growth in recreational activities.

Resource managers need to be mindful of several features of protected area delineation and management, namely: (1) it is by nature a series of experiments that inform the next experiment (essentially adaptive management); (2) it is especially important when biodiversity is threatened; (3) it is urgent, but requires patience (it is a one-hundred-year not a one-year mission); (4) success requires continued development and application of advanced research and mapping techniques; (5) it demands continued diligence for improving understanding and personal communication with diverse stakeholders; (6) it is time-consuming and labor intensive; and (7) it is challenging when budgets of resource management agencies are constant or shrinking (Dennerlein 1997).

Several analytical methods are suited to evaluating boundary decisions for protected areas. They include the ABC Resource Survey Approach (Grigoriew et al. 1985), the Natural/Generated Edge Approach (Schonewald-Cox and Bayless 1986), and the General Spatial Approach (Sundell 1997). The latter involves five steps: (1) establishing study parameters, (2) defining the study area, (3) applying decision analysis methods, (4) collecting and developing data, and (5) analyzing the data to produce a series of thematic maps that identify resource areas of importance to resource managers (Sundell 1997).

National System Planning

Selection and delineation and site management plans for a protected area influence its success. However, success in achieving a broad range of protected area goals is seldom advanced by the selection and delineation decisions for a single protected area. This occurs because success in selecting, delineating, and managing a single protected area is insufficient to achieve a broad range of protected area goals. In contrast, a wide range of protected area goals can be attained by a well-designed and executed national system plan for protected areas. A plan stipulates coordinated management of protected areas included in the national system as well as coordination with other land-use planning in the country. As a way of protecting biodiversity and achieving sustainable development, the UN World Commission on Environment and Development (1987) endorsed and promoted a network of protected areas for all nations that represented each of the Earth's major ecosystems. The commission also recommended that each nation set aside 12 percent of its land in protected areas. The Canada Wilderness Charter called for "a national effort to establish at least one representative protected area in each of the national regions of Canada by the year 2000," which establishes protected areas equivalent to "at least 12 percent of the lands and waters of Canada" (McNamee 2002b).

Reasons for undertaking national system planning for protected areas are listed in the following sidebar. National system planning by itself does not guarantee a successful protected area system. The plan may be poorly conceived and/or not implemented. National system planning requires units to be managed in a manner that accounts for the complex and dynamic socioeconomic and biophysical connections among units. It takes several years to implement a long-range system plan. Hence, progress in achieving the goals of the plan can be slow.

Reasons for National System Planning for Protected Areas

National system planning for protected areas is done for many reasons, including the following:

- To relate protected areas to national priorities, and to prioritize different aspects of protected area development
- To facilitate access to international funding by defining priorities for investment in protected areas and increasing the level of confidence in the efficient use of funds and resources
- To get away from a case-by-case ad hoc approach to resource management decision making
- To target proposed additions to the protected area estate in a more rational and persuasive manner than ad hoc planning
- To facilitate integration with other relevant planning strategies, such as those for national tourism, national biodiversity conservation, or sustainable development
- To help resolve conflicts, assist in making decisions relating to trade-offs, clarify roles and responsibilities of different stakeholders, and facilitate diverse stakeholder involvement
- To provide a broader perspective for addressing site-specific issues, such as tourism management
- To enhance the effectiveness and efficiency of the way in which budgets are developed and spent
- To assist in meeting obligations under international treaties
- To assist countries to be more proactive in conservation management and in developing effective protective area systems
- To encourage consideration of a "system" that incorporates formal protected areas and areas outside of protected areas
- To provide a structured framework for a system of protected areas, ranging from areas managed for strict conservation to areas managed for a range of conservation and appropriate ecologically sound activities
- To assist protected area agencies to build political support for protected areas as a worthwhile concern
- To define a better process of decentralization and regionalization of protected area activities, resources, and responsibilities, including the involvement of NGOs and the private sector, and
- To foster transboundary collaboration

Source: Davey (1998).

National system planning and system plans for protected areas are unique to each country because they reflect social, economic, cultural, historical, and environmental conditions. Because these conditions change over time (for example, a developing country becomes a developed country), national system plans require periodic revision.

An effective national system plan should (adapted from Davey 1998):

- Articulate the national significance and priority given to protected areas
- State the goals of planning and present a strategy for achieving those goals
- Establish priorities for creating new and managing existing protected areas
- Determine the relationships between different units and categories of protected areas and between protected areas and other land uses
- Identify current and potential impacts of land use/management in gateway communities on candidate protected areas and the socioeconomic impacts of establishing new protected areas on gateway communities
- Establish criteria for screening and selecting protected areas
- Identify and discuss national policies relevant to protected areas
- Describe the roles of local, regional, and national agencies and stakeholder groups in promoting, establishing, and/or managing protected areas, as well as the relationships between these players
- Contain maps that depict the locations and enumerate the attributes and current conditions in existing protected areas
- Identify deficiencies in the current system of protected areas, including opportunities and requirements for connecting protected areas
- Present other information needed for effective planning

Design Criteria

A national system of protected areas should be designed based on a set of criteria. The World Commission on Protected Areas (WCPA) proposed five design criteria: (1) representativeness, comprehensiveness, and balance; (2) adequacy; (3) coherence and complementarity; (4) consistency; and (5) cost effectiveness, efficiency, and equity (Davey 1998). These criteria can be used to judge the effectiveness of an existing system of protected areas and the pros and cons of expanding the system. The WCPA has not prioritized these criteria for two reasons. First, the five criteria are not independent of one another. Second, relative importance of the criteria is likely to change over time. For example, if a national system is not representative, but is ranked high on the remaining four criteria, then high priority should be placed on adding units to the system that enhance representativeness.

How are the design criteria defined? Representativeness, comprehensiveness, and balance are achieved when units in the system encompass the full range of environmental types contained in a country and there is balanced sampling of environmental types. Environmental types refer to biomes, levels of biological organization (genetic, population/species, community/ecosystem, and landscape or region), landforms, and cultural landscapes. A national system of protected areas is not representative, comprehensive, and balanced, for example, when there is overrepresentation of charismatic megafauna and spectacular scenery, and underrepresentation of the full suite of plant and animal species and biomes contained in a country. In this case, the system of protected areas does not exhibit balanced sampling of the country's biodiversity (see the "*Worthless Lands Hypothesis and Protected Area System in Tasmania, Australia*" sidebar for discussion of the consequences of departing from this criterion).

On a worldwide basis, there is considerable variability in the percentage of biomes represented in protected areas. In 1997, temperate broad leaf forests/woodlands, evergreen sclerophyllous forests, warm deserts/semideserts, temperate grasslands, and lake systems had less than 5 percent of their area protected. Temperate needle-leaf forests, tropical dry forests/woodlands, cold-winter deserts, and tropical grasslands/savannas had 5–7 percent of their area protected. Tropical humid forests, tundra communities, and mixed mountain systems had 8–10 percent of their area protected. Subtropical/temperate rain forests/woodlands and mixed island systems had more than 10 percent of their area protected (Green and Paine 1997). Biomes included in a representative national system of protected areas generally differ from country to country due to spatial variability in biophysical conditions.

Gauging the representativeness of a national system of protected areas requires a classification scheme or typology for environmental types. The scheme should be scientifically defensible and appropriate to the scale of planning. Since several classification schemes and numbers of classes per scheme are possible, it is best to evaluate how sensitive a system's representativeness is to different schemes and classes per scheme. Implementation of a classification scheme requires data from resource inventories of protected areas.

There are several ways to design a classification scheme. One way to gauge representativeness is by a simple presence-or-absence rule that determines whether or not a candidate protected area contains any amount of the environmental type. Another way to gauge representativeness that complements the criterion of adequacy of protected areas (see discussion below) is to compare the actual percentage of a candidate protected area that contains an environmental type to a selected minimum threshold. If a candidate area exceeds the threshold, then it contributes to representativeness, whereas if a candidate area falls below the threshold, then it does not contribute to representativeness. Selection of the threshold is somewhat arbitrary. Ideally, evaluation of a candidate protected area should be based on the multiple design criteria listed above, not just its contribution to representativeness.

Worthless Lands Hypothesis and Protected Area System in Tasmania, Australia

One of the strategies for developing a protected area system for a country is to select areas that conserve biodiversity across ecosystem types. This strategy ensures that the system represents a wide range of flora and fauna. It has been argued rather convincingly that the first national parks were selected on the basis that they had outstanding scenic and aesthetic qualities but lacked economically valuable natural resources. In other words, the early national parks did not represent a wide range of ecosystem types. Evidence from many GAP-analysis studies indicates "striking gaps in the conservation of biodiversity and biases in reserve systems toward those areas historically perceived to have no significant economic value."

This situation has led to the so-called worthless lands hypothesis. Although it is a poor choice of terms, this hypothesis maintains that many protected area systems are relatively invaluable in terms of conserving biodiversity because many ecosystem types are underrepresented in these systems resulting in gaps in the representation of biodiversity. As pointed out in the narrative, lowlands, wetlands, streams, humid forests, and coastal areas are underrepresented in national park systems, particularly in national parks.

Consider the worthless lands hypothesis within the context of Tasmania's protected area system. Research shows that prior to the 1970s, plant communities occurring in scenic alpine and subalpine areas, which are not economically valuable, were overrepresented in the Tasmanian National Park System (TNPS). Conversely, economically valuable communities were underrepresented in TNPS. Specifically, TNPS protected less than 3 percent of the economically valuable plant communities that had more than 15 percent of their pre-European area represented in the reserve system, compared to 17 percent for plant communities considered not to be economically viable. Another way of stating this is that economically valuable communities were grossly underrepresented (<3 percent vs. 17 percent). Representativeness of noneconomically and economically valuable communities increased substantially after 1970 due to a change in the motivations for establishing reserves and the availability of analyses that quantified the extent to which plant communities were unreserved (gap analysis). In the period 1970–1992, more than 20 percent of the economically valuable plant communities and 58 percent of the economically invaluable plant communities had more than 15 percent of their pre-European area reserved.

Sources: Runte (1976, 1987); Hall (1988); Hall and Shultis (1991); Pressey (1994); Scott et al. (2001); Mendel and Kirkpatrick (2002); USGS (2004).

Adequacy of a national system of protected areas refers to the extent to which the specific locations, sizes, shapes, and boundaries of contributing units support systemwide goals. In the case of biodiversity, for example, contributing units should buttress ecological functions/processes, habitats, populations, and communities. For the most part, adequacy concerns have already been addressed in discussing the design of individual protected areas. Furthermore, representativeness considers the importance of having a network of protected areas that represent the diversity of biomes that occur in a country.

Coherence and complementarity refer to the extent to which the addition of a protected area makes a positive contribution to the system. Consistency refers to the extent to which (1) standardized procedures are used to establish protected areas, (2) procedures are compatible with the protected area policies (laws and regulations) and classification schemes established by the country, and (3) a unit is managed so as to support the management objectives it is designed to meet. Cost effectiveness and equity are useful in evaluating the efficiency and fairness of adding a particular unit to a protected area system. A protected area system is cost effective when the objectives of the system are achieved while minimizing the spatial extent of protected areas. A system is considered equitable when individuals and communities equitably share positive and negative impacts. There are likely to be trade-offs between efficiency and equity; making a system more equitable reduces efficiency.

Successful Development and Implementation of National Plans

Development of a national plan for protected areas is complex and time-consuming. Basing plan development on a systematic step-by-step procedure increases the odds of success. The six-step planning procedure outlined below provides a general framework for system planning. It is an adaptation of the planning procedure developed by Dixon and Sherman (1997).

First step: the purpose for developing a national protected area system must be clearly articulated and supported by stakeholders. Involving stakeholders in developing the plan increases public support and stakeholder acceptance of the resulting plan and the likelihood of success. Each unit of a protected area system has a specific purpose in terms of achieving particular management objectives, such as protection of threatened and endangered species and their habitats, preservation of aesthetically, culturally, and historically significant landscapes and ecosystems, provision of unique recreational opportunities, fostering opportunities for scientific research and education, and others. A national system plan does not describe how individual units in the system are managed (that is the purpose of the site management plan), but it should provide broad management guidance for crafting site-specific plans. Most countries have legislation and regulations that explicate the purpose of different categories of protected areas and guide the management of specific protected areas.

Second step: a complete inventory and assessment is made of candidate areas. The inventory and assessment describes the physical characteristics (location, size, and shape) of each candidate area, as well as current resource conditions, potential social, economic, and ecological functions and values, risk of degradation from economic development, and opportunity cost of designating a protected area in a specific location (forgone revenues from losses in resource extraction, displacement of people and communities, and negative impacts on subsistence lifestyles). Functions and values can be measured in nonmonetary or monetary terms. For example, habitat suitability for threatened and endangered species can be assessed based on total suitable habitat area for a species and degree of land fragmentation and habitat loss in the area. Opportunity cost equals any decreases in income/employment and cultural integrity, if any, associated with reductions in commercial and subsistence extraction of natural resources.

The assessment requires identifying the management objective(s) achieved by each candidate area, constraints on protected area designation, and the cost of alleviating those constraints. Constraints include lack of adequate transportation infrastructure, which is a significant issue in protected areas for which tourism is the primary objective; widespread use of the area for commercial and subsistence agriculture and forestry; and others.

Third step: a preliminary allocation of candidate areas to protected area categories is made based on the inventory/assessment and constraint analysis performed in step 2 and the protected area categories established by the country. This step matches candidate protected areas and management objectives, and allocates candidate areas to protected area categories based on those matches. For instance, if a candidate area is best suited for scientific research, then it is allocated to IUCN category Ia (Strict Nature Reserve).

Fourth step: the relative importance of management objectives for the protected area system is determined. Management objectives are defined in terms of enhancing particular values of protected areas, such as recreation, scientific research, education, conserving biodiversity, and others. Management agencies should determine the relative importance of management objectives in consultation with stakeholders. If it is determined that establishing protected areas in support of scientific research is a high priority, then adding units in IUCN category Ia (Strict Nature Reserve) contributes more to the value of a protected area system than adding units in other categories. The marginal benefit of adding units to a given category is likely to decline as the number of units in that category increases.

Fifth step: the candidate areas assigned to each category in step 3 are ranked relative to one another based on their capacity to achieve the management objectives for that category. For example, ranking protected areas in IUCN category Ia

(Strict Nature Reserve) based on efficiency would result in high rankings for candidate areas that afford outstanding opportunities for scientific research (high benefit) at a relatively low establishment and maintenance cost (low cost).

Sixth step: a final selection of protected areas is made subject to limitations on human and financial resources and other criteria, such as regional equity (achieving a balanced geographic apportionment of protected areas) and political acceptability. Since financial resources for acquiring protected areas are limited, trade-offs are inevitable. For instance, if a protected area in IUCN category Ia has a high net benefit but is very costly, then that area should be added to the system only if the net benefit exceeds the opportunity cost in terms of the loss in socioeconomic, cultural, and ecological values from having fewer protected areas in other categories. Final selection of protected areas is facilitated by using an optimization model that maximizes the resource manager's utility function subject to human, financial, and other constraints. In the simple but restrictive case of additive utility and risk neutrality, the utility function is a weighted average of the objectives established for the protected area system. Weights are based on the relative importance of protected area categories determined in step 4.

Global System Planning

Whereas a national system plan includes protected areas that balance management objectives for a nation, a global system plan includes protected areas that balance management objectives on a global scale. Even though global system planning has been endorsed by the IUCN, it has been resisted for a variety of reasons. For example, designating protected areas as Biosphere Reserves has been opposed on grounds that it infringes on private property rights and national sovereignty. Opposition can also arise when achieving global objectives for protected areas generates international inequities. For example, Thailand has lost many of its coastal mangrove forests to shrimp culture (farming). One way to offset this loss on a global scale would be to protect mangrove forests in other countries. Other countries are likely to resist such a strategy if it reduces national income and employment.

Global strategies for protecting mangrove forests and other biomes can be made more palatable internationally by offering financial and other incentives. In debt-for-nature swaps, a portion of the debt owed developed countries by developing countries is reduced provided the developing country agrees to protect an area that is rich in biodiversity. Financial incentives to protect biodiversity are provided by the Global Environment Facility. Global system planning for protected areas is fostered through international cooperation, such as the Convention on Biological Diversity and the establishment of international peace parks. The latter case requires a country's national system plan to incorporate priorities not only for establishing protected areas within its own national boundary, but also for cooperating with other nations in establishing transboundary protected areas. Besides expanding the opportunities for conserving nature, international peace parks improve international relations and facilitate the sharing of information, experience, and training capacity (Davey 1998).

Summary

Selecting and delineating national parks and protected areas involves consideration of the following:

1. Social-cultural-economic-ecological conditions in the country,
2. The importance of protected area categories
3. Decision making related to establishment of a protected area
4. Decision making related to expanding the boundary of an existing protected area
5. National system planning
6. Successful development and implementation of national plans
7. Global system planning

Establishment of a protected area (third consideration) requires making decisions about how many and what types of protected areas to establish; where to locate them; and the size, shape, and boundary of a protected area.

Social-cultural-economic-ecological conditions within a country shape the outcomes of selection and delineation decisions for individual protected areas as well as the goals attained by the resulting protected area system. The location, size, and boundary of Yellowstone National Park received support because it encompassed geologic features that were aesthetically and scenically appealing, did not lock up valuable extractive resources, and contributed to westward expansion of the Northern Pacific Railway. The IUCN protected area categories influence selection and delineation because they define the activities and functions permitted in each category. They also have implications for the size of protected areas.

Selection and delineation of protected areas should consider how overall socioeconomic, cultural, and ecological values held by individuals and gateway communities are affected, ways of reducing ecological degradation to protected areas from future economic development in gateway communities, and the implications of international designation of protected areas.

Expanding the boundary of a protected area is generally not considered until several years following establishment. Yet, considering boundary expansion at the time of site selection can reduce future adverse impacts of economic development in gateway communities.

Since selection and delineation of a single protected area is unlikely to achieve a broad range of protected area goals, many countries design and implement a national system of protected areas. An effective national system plan requires coordinated management with respect to protected areas in the system and other land-use planning efforts in the country. A national system plan is unique to a country because it is based on social, economic, cultural, historical, and environmental conditions. Plans need to be revised as conditions change.

Developing a national plan for protected areas is complex and time-consuming. A six-step procedure is presented for system planning: (1) the purpose or reason for developing a national protected area system is clearly articulated to stakeholders; (2) a complete inventory and assessment is made of candidate areas; (3) a preliminary allocation of candidate areas to protected area categories is made based on the inventory/assessment, constraint analysis, and the protected area categories established by the country; (4) the relative importance of management objectives for the protected area system is determined in consultation with stakeholders; (5) candidate areas assigned to each category are ranked relative to one another based on their capacity to achieve the management objectives for that category; and (6) a final selection of protected areas is made subject to limitations on human and financial resources and other criteria, such as regional equity and political acceptability.

Global system planning attempts to establish and maintain a global system of protected areas. Such planning has been opposed on grounds that it infringes on private property rights and national sovereignty, and can distort the international distribution of income and employment.

References

Aberley, D. 1993. *Boundaries of home: Mapping for local empowerment.* Philadelphia, PA: New Society Publishers.

Ashworth, G. I. 1992. Planning for sustainable tourism. *Town Planning Review* 63: 325–329.

Berg, P., and R. F. Dasmann. 1990. Reinhabiting California. In *Home! A bioregional reader,* ed. V. Andruss, J. Plant, and E. Wright. Philadelphia, PA: New Society Publishers.

Bull, A. 1991. *The economics of travel and tourism.* Melbourne, Australia: Pitman.

Chadwick, D. H. 2000. Yellowstone to Yukon. Washington, DC: National Geographic Society.

Connally, E. H. (Ed.). 1982. *National parks in crisis.* Washington, DC: National Parks and Conservation Association.

Davey, A. G. 1998. Protected areas categories and management objectives (appendix 2). *National System Planning for Protected Areas,* 49–55. Gland, Switzerland, and Cambridge, UK: IUCN.

Deller, S. C., D. W. Marcouiller, and G. P. Green. 1997. The influence of recreational housing development on local government finances. *Annals of Tourism Research* 24: 687–705.

Dennerlein, C. 1997. Lines to hold, lines to ignore: The paradox of boundaries in contemporary management of parks and protected areas. In *National parks and protected areas: Selection, delineation, and management,* ed. J. J. Pigram and R. C. Sundell, 45–59. Armidale, Australia: Centre for Water Policy Research, University of New England.

Diaz, N. M., and S. Bell. 1997. *Landscape analysis and design. In* Creating forestry for the 21st century: The science of ecosystem management, ed. K. A. Kohm and J. F. Franklin, 255–270. Washington, DC: Island Press.

Dixon, J. A., and P. B. Sherman. 1997. Selection of protected areas (chap. 3). *Economics of protected areas: A new look at benefits and costs.* Washington, DC: Island Press.

Fritz, R. G. 1982. Tourism, vacation home development, and residential tax burden. *American Journal of Economics and Sociology* 41: 375–385.

Gates, P. 1979. *History of public land law development.* New York: Arno Press.

Green, M. J. B., and J. R. Paine. 1997. *State of the world's protected areas at the end of the twentieth century.* Paper presented at Protected Areas in the 21st Century: From Islands to Networks. Albany, Australia, November 24–27.

Grigoriew, P., J. Theberge, and J. Nelson. 1985. *Park boundary delineation manual, the abc resource survey approach.* Occasional Paper #4. University of Waterloo/Parks Canada, Waterloo, Ontario.

Hall, C. M. 1988. The "worthless land hypothesis" and Australia's national parks and reserves. In *Australia's ever changing forests,* ed. K. J. Frawley and N. Semple, 441–456. Canberra: Australian Defense Force Academy.

Hall, C. M., and J. Shultis. 1991. Railways, tourism and worthless lands: The establishment of national parks in Australia., Canada, New Zealand, and the United States. *Australian Canadian Studies* 8: 57–74.

Hansson, C. B., and M. Wackernagel. 1999. Rediscovering place and accounting space: How to re-embed the human economy. *Ecological Economics* 29: 203–213.

Howe, J., E. McMahon, and L. Propst. 1997. *Balancing nature and commerce in gateway communities.* Washington, DC: Island Press.

IUCN. 1984. Categories, objectives, and criteria for protected areas. In *National parks, conservation, and development: The role of protected areas in sustaining society,* ed. J. A. McNeely and K. R. Miller. Washington, DC: Smithsonian Institution Press.

Keiter, R. B., T. Boling, and L. Milkman. 1999. Legal perspectives on ecosystem management: Legitimizing a new federal land management policy. In *Ecological stewardship: A common reference for ecosystem management,* vol. III, ed. W. T. Sexton, A. J. Malk, R. C. Szaro, and N. C. Johnson, 9–41. Oxford, England: Elsevier Science Ltd.

Marcouiller, D. W., and G. P. Green. 2000. Outdoor recreation and rural development. In *National parks and rural development: Practice and policy in the United States,* ed. G. W. Machlis and D. R. Field, 34–49. Washington, DC: Island Press.

McNamee, K. 2002a. *From wild places to endangered spaces: A history of Canada's national parks. In* Parks and protected areas in Canada: Planning and management, ed. D. Dearden and R. Rollins, 21–50. 2nd ed. Ontario, Canada: Oxford University Press.

———. 2002b. Protected areas in Canada: The endangered species campaign. In *Parks and protected areas in Canada: Planning and management,* ed. D. Dearden and R. Rollins, 51–68. 2nd ed. Ontario, Canada: Oxford University Press.

Mendel, L. C., and J. B. Kirkpatrick. 2002. Historical progress of biodiversity conservation in the protected-area system of Tasmania, Australia. *Conservation Biology* 16: 1520–1529.

Moss, L. A. G. 1994. Beyond tourism: The amenity migrants. In *Coherence and chaos in our uncommon futures: Visions, means and actions,* ed. M. Mannermaa, S. Inayatullah, and R. Slaughter, 121–128. Turku, Finland: Turku School of Economics and Business.

National Parks and Conservation Association. 1988. *Park boundaries: Where do we draw the line.* Investing in park futures: A blueprint for tomorrow, vol. 5. NCPA Publications.

National Park Service. 1978. *Management policies manual.* Washington, DC: U.S. Department of the Interior.

———. 1988. *Management policies.* Washington, DC: U.S. Department of the Interior.

———. 2003. National Park Service, U.S. Department of the Interior, Criteria for Parklands. www.nps.gov /legacy/criteria.html (accessed September 9, 2003).

Newmark, W. D. 1995. Extinction of mammalian populations in western North American national parks. *Conservation Biology* 9: 512–526.

Plant, C. J. 1990. *Turtle talk: Voices for a sustainable future.* Philadelphia, PA: New Society Publishers.

Pressey, R. L. 1994. Ad hoc reservations: Forward or backward steps in developing representative reserve systems? *Conservation Biology* 8: 662–668.

Runte, A. 1976. The West: Wealth, wonderland, and wilderness. In *The American environment: Perceptions and policies,* ed. J. Wreford Watson and T. O'Riordan, 47–62. London, England: John Wiley & Sons, Inc.

———. 1987. *National parks: The American experience.* Lincoln: University of Nebraska Press.

Schonewald-Cox, C. M., and J. W. Bayless. 1986. *The boundary model: A geographical analysis of design and conservation of nature reserves.* Biological Conservation *38: 305–322.*

Schowalter, T., E. Hansen, R. Molina, and Y. Zhang. 1997. *Integrating the ecological roles of phytophagous insects, plant pathogens, and mycorrihzae in managed forests. In* Creating forestry for the 21st century: The science of ecosystem management, ed. K. A. Kohm and J. F. Franklin, 171–189. Washington, DC: Island Press.

Scott, J. M., F. W. Davis, R. G. McGhie, R. G. Wright, C. Groves, and J. Estes. 2001. Nature reserves: Do they capture the full range of America's biological diversity? *Ecological Applications* 11: 999–1007.

Smith, M. 1989. *Behind the glitter: The impacts of tourism on rural women in the southeast.* Lexington, KY: Southeast Women's Employment Coalition.

Soule, M., and R. Noss. 1998. *Rewilding and biodiversity: Complementary goals for continental conservation.* Wild Earth (Fall): 19–28.

Sundell, R. C. Delineating park and nature reserve boundaries: A preliminary approach to integrating spatial analysis techniques. In *National parks and protected areas: Selection, delineation, and management,* ed. J. J. Pigram and R. C. Sundell, 135–165. Armidale, Australia: Centre for Water Policy Research, University of New England.

Task Force on Economic Benefits of Protected Areas of the World Commission on Protected Areas (WCPA) of IUCN, in collaboration with the Economics Service Unit of IUCN. 1998. *A new vision for protected areas, economic values of protected areas: Guidelines for protected areas management.* Gland, Switzerland, and Cambridge, UK: IUCN.

Thomas, J. W. 1997. Foreword. In *Creating forestry for the 21st century: The science of ecosystem management,* ed. K. A. Kohm and J. F. Franklin, ix–xii. Washington, DC: Island Press.

U.S. Geological Survey. 2004. *National GAP analysis program.* http://www.gap.uidaho.edu/ (accessed April 10, 2004).

United Nations World Commission on Environment and Development. 1987. *Our common future.* Oxford, England: Oxford University Press.

Wilson, E. O. 2002. *The future of life.* New York: Alfred A. Knopf.

Wright, R. G. 1985. Principles of new park-area planning as applied to the Wrangell-St. Elias region of Alaska. *Environmental Conservation* 12: 59–66.

CHAPTER

7

Concepts for Ecosystem Management

Even though the concept of ecosystem management has been proclaimed during the last decade as an important method to integrate parks and protected areas into their surrounding landscapes, putting the concept into practice has not been easy.

—R. Gerald Wright

Protected areas used to be considered independently of the larger landscapes in which they exist. Boundaries seemed distinct and were often viewed as encompassing an "ecosystem." Now we know differently and have begun to understand how dependent protected areas are on the processes and resources beyond their boundaries. Within this context, park and protected area managers make planning and management decisions that require balancing competing social, economic, and ecological values of protected areas. The need to balance competing values is perhaps best illustrated by the requirement to prepare an environmental impact statement for management actions that are expected to have major impacts on people and/or natural resources. For example, the development of General Management Plans for national parks in the United States requires preparation of an environmental impact statement that identifies and compares the social, economic, and environmental impacts of proposed alternatives. Even management actions that do not require preparation of an environmental impact statement, such as the decision of where to locate a new backcountry campsite or whether to close a degraded trail or campsite, often involves trade-offs among competing social, economic, and ecological values.

Chapters 7 through 10 examine ecosystem management and its application to protected areas. Concepts for ecosystem management are addressed in this chapter. Topics covered include the meaning of ecosystem management; biodiversity conservation, ecological functioning, and ecosystem resilience; ecosystem dynamics and complexity; natural regulation; carrying capacity; regional scale conservation; continental scale conservation; and ecological economics. Chapter 8 describes various approaches to ecosystem management. Case studies in protected area management that employ several of the concepts and approaches examined in chapters 7 and 8 are presented in chapters 9 and 10.

Ecosystem Management

Traditional natural resource management for public lands is, for the most part, based on production of goods and services and/or protection of species. For example, the Multiple Use Sustained Yield Act of 1960 (16 U.S.C. §§ 528–531) was the dominant management philosophy of the U.S. Forest Service in the 1950s and 1960s. It required national forests to be managed for multiple objectives including "outdoor recreation, range, timber, watershed and wildlife and fish and a sustained yield of products and services." Due to a lack of ecological knowledge for practicing integrated resource management, and the emphasis placed on timber production, livestock grazing, and game fish and wildlife, implementation of the act did not focus on long-term stewardship of natural resources (Franklin 1997). The Endangered Species Act of 1973 (16 U.S.C. §§ 1531–1543) reinforces the species orientation of public land managers by requiring developing a "means whereby the ecosystems upon which endangered species and threatened species depend may be conserved, [and] to provide a program for the conservation of such endangered species and threatened species." The ultimate goal of the act is to recover species through development and implementation of recovery plans that identify critical habitat for listed species. A prime example of the act's impact on natural resource management is the Northwest Forest Plan. The plan curtailed logging in national forests in the Pacific Northwest in an effort to protect late successional and old growth forest within the range of the northern spotted owl, which was important habitat for a suite of other species. As a result of the plan, timber harvesting in the region is expected to decline by more than 75 percent relative to 1980 levels (Johnson 1997).

Ecosystem management represents a fundamental shift in the philosophy for managing people and natural resources to incorporate larger spatial scales, longer time periods, and more variables than commodity-based resource management (Thomas 1997) and strives for sustainable productivity of the entire ecosystem (Franklin 1997; Schowalter et al. 1997; see also sidebar). The goal of ecosystem management is to manage patterns and processes over large spatial and temporal scales (Diaz and Bell 1997). Ecosystem management focuses on achieving and sustaining a balance between the production of consumer goods and services and ecological goods and services (MacKenzie 1996). Consumer goods and services include food, clothing, durable goods, such as houses and automobiles, a wide range of consumer services, such as automobile and home repair, and others. Consumer goods and services are important because their total value equals gross national product. Ecological goods include timber, biomass fuels (coal, oil, and natural gas), and natural fiber, which are used to produce intermediate goods and consumer products. Timber is used to produce wood products that are used in housing and furniture construction. Biomass fuels are refined into gasoline, heating fuel, and other petroleum-based products. Natural fiber is used in clothing.

The economic values of ecological goods are derived, in part, from the market value of products they are used to produce. Commercial timber values are derived from the value of wood products.

Ecosystem Management in the Pacific Northwest

The World Resources Institute (WRI) defines ecosystem management as an approach that broadly evaluates how human use affects the functioning and productivity of an ecosystem. According to WRI, an ecosystem approach does the following:

1. Attempts to optimize the mix of goods and services provided by one or more ecosystems
2. Looks beyond traditional jurisdictional boundaries by systematically analyzing whole ecosystems
3. Respects ecosystem processes at the micro level, but evaluates them at a variety of larger spatial and temporal scales
4. Explicitly links human needs for ecosystems to their biological capacity and attempts to fulfill those needs by integrating social and economic information with environmental information about the ecosystem
5. Views management as being successful if it preserves or increases the productivity of ecosystems
6. Considers production of ecosystem goods and services as the natural product of a healthy ecosystem, not as an end in itself

Salmon losses in the Columbia River Basin have resulted from the destruction and alteration of habitat caused by construction of 19 major dams and 60 smaller hydropower projects during the period 1930–70. The Columbia River System is the world's largest hydroelectric power system (~12,000 MWe). Hydropower development combined with losses in upland habitat due to grazing, timber harvesting, and agricultural production reduced fish runs in the basin from their historic high of 10–16 million prior to European settlement to 2.5 million in the late 1970s. In addition, 106 major Northwest salmon runs have became extinct and 214 runs in northern California and the Pacific Northwest are at risk of extinction. Reduced salmon runs caused economic losses of $500 million per year (present value losses are $13 billion) and the loss of 25,000 jobs.

The Northwest Power and Conservation Council was established by the Pacific Northwest Electric Power Planning and Conservation Act of 1980. The act requires the council, consisting of representatives from the states of

Idaho, Montana, Oregon, and Washington, to develop a program to "protect, mitigate and enhance fish and wildlife, including related spawning grounds and habitat, on the Columbia River and its tributaries...affected by the development, operation and management of [hydroelectric projects] while assuring the Pacific Northwest an adequate, efficient, economical and reliable power supply." Pursuant to the act, the council established the Columbia River Basin Fish and Wildlife Program. Development and monitoring of the program is the responsibility of the council and implementation is the responsibility of the Bonneville Power Administration, the U.S. Army Corps of Engineers, the Bureau of Reclamation, and the Federal Energy Regulatory Commission and its licensees. In addition, the council makes annual recommendations to the Bonneville Power Administration regarding funding of projects that contribute to the program.

The Columbia River Basin Fish and Wildlife Program is ecosystem management with the goal of partially restoring salmonid fish populations in the Columbia River System by artificial propagation in fish hatcheries, enhancing fish habitat, and improving fish passage at dams and fish hatcheries. In an effort to target the region's substantial fish and wildlife investment to areas and species where it yields the greatest benefits, the council initiated a fifth program revision in 2000. Unlike past programs, the latest program establishes comprehensive basinwide goals and objectives based on scientific and policy principles that apply to the entire Columbia River Basin. Future efforts call for adding to the program locally developed plans for the more than 50 tributary subbasins of the Columbia River and a plan for the main stem. Plans will be consistent with goals and objectives for the basin and 11 ecological provinces of the basin. The latter are groups of adjacent subbasins that have similar ecological features.

The three fundamental elements of the program are the "vision, which describes what the program is trying to accomplish with regard to fish and wildlife and other desired benefits from the river; biological objectives, which describe the ecological conditions needed to achieve the vision; and implementation strategies, procedures and guidelines, which guide or describe the actions leading to the desired ecological conditions." The Columbia River Basin Fish and Wildlife Program has many of the elements of an ecosystem approach as defined by WRI.

Sources: Northwest Power Planning Council (1991, 2003); Digital Studios (1997); World Resources Institute (2000).

Ecological services include air and water purification, water supplies, mitigation of floods and drought, detoxification and decomposition of wastes, generation and renewal of soil, maintenance of biodiversity and partial stabilization of climate (Daily 1997). The economic value of ecological services is more difficult to measure because they are not traded in markets. Nonmarket valuation techniques have been used to estimate the value of ecosystem services. For example, Costanza et al. (1997) estimated that the current economic value of 17 ecosystem services provided by 16 biomes is between US$16 and 54 trillion per year, with an average of US$33 trillion per year.

Traditional approaches to resource management have given way to ecosystem management for several reasons. Specifically, traditional management approaches:

1. Ignore important connectivity among geographic areas
2. Focus on short-term goals, such as commodity production at the expense of long-term ecological health, resulting in landscapes that lack economic and ecological productivity
3. Are inefficient because agencies and other groups are not aware of each other's efforts
4. Ignore the diversity of human interests in resource management
5. Result in conflict-laden impasses such as the spotted owl controversy in the Pacific Northwest

On the other hand, implementing ecosystem management is challenging because:

1. It encounters all the problems commonly associated with human relations and group decision making, the latter being particularly difficult when stakeholders disagree about management objectives and actions
2. The collaborative decision-making approach on which it is based requires agency personnel to let go of a top-down technocratic approach
3. It requires considerable time for stakeholders to understand one another and have meaningful interaction
4. It requires considerable time, money, and effort to accumulate, organize, and interpret the requisite data and information
5. Informational needs require scientists to share ideas with one another and scientists and managers to collaborate, which has been difficult to achieve in the past
6. Current policies governing budgeting and administrative procedures are not particularly supportive of ecosystem management
7. Fear that ecosystem management will infringe on private property rights

Despite these challenges, progress is being made in implementing ecosystem management (Yaffee et al. 1996).

Early definitions of ecosystems made reference to the study of organisms and their interrelations, but excluded the abiotic environment; even humans were ignored. Contemporary views of ecosystems give equal attention to biotic and abiotic elements and explicitly recognize the value of studying ecological processes at multiple scales (Franklin 1997). For example, Likens (1992) defines an ecosystem as "a spatially explicit area of the Earth that includes all organisms, along with all components of the abiotic environment within its boundaries." This definition is consistent with ecosystem management.

Ecosystem management has become popular because of the connectedness among various parts of an ecosystem and the interdependence between ecological integrity and human welfare. Many resource management agencies have changed their management philosophy from resource extraction and exploitation to ecologically sustainable management of natural resource systems, which is at the core of ecosystem management (National Research Council 1992; Williams et al. 1997). For example, management of federal lands in the United States has shifted from a focus on individual resources like timber, minerals, forage for livestock, and scenery to holistic management of multiple resources. Eighteen federal agencies have adopted or are considering adoption of ecosystem management (Haeuber and Franklin 1996).

A primary goal of ecosystem management is to manage ecosystems in a sustainable manner. Ecological sustainability attempts to maintain in perpetuity the productive capacity of terrestrial and aquatic ecosystems as measured by the quantity and quality of ecological goods and services they can produce. Ecosystems are being managed sustainably when there is no net loss in productive capacity including genetic diversity (Franklin 1993, 1997).

Management at the ecosystem scale is challenging for several reasons. First, data are imprecise and sparse. It is difficult to observe interactions between human activity and ecosystem states, and measurement of ecosystem attributes are inexact. Second, ecological theories do not allow inference much beyond observation, especially for large and unprecedented human interventions. Third, due to the limitations of ecological theory, predictions are often wrong (Lee 1993).

Ecological sustainability is consistent with sustainable development. The Brundtland Commission defines sustainable development as "development that meets the needs of the present without compromising the ability of future generations to meet their own needs" (World Commission on Environment and Development 1987). While this definition has broad appeal, it is difficult to implement because of its subjective elements. In particular how do we know when the needs of the present generation are being met and whether the ability of future generations to meet their needs is being compromised by current levels of resource extraction and use? The Brundtland Commission Report also states that, "If needs are to be met on a sustainable basis, the Earth's natural resource base must be conserved and enhanced" (World Commission on Environment and Development 1987). This interpretation is compatible with the World Conservation

Strategy's goal of maintaining "essential ecological processes and life support systems" (IUCN 1980).

The Brundtland Commission established six determinants of sustainable development, namely, (1) keeping consumption levels within ecological bounds, (2) ensuring increased economic production potential and equity, (3) protecting life support systems, (4) using renewable resources at rates below the limits of regeneration and regrowth, (5) depleting exhaustible resources at rates that minimize foreclosure of future options, and (6) minimizing adverse impacts on environmental quality. Several of these determinants are applicable to protected areas. For example, keeping consumption levels within ecological bounds requires that visitation not be allowed to increase to levels that threaten the ecological integrity of a protected area. This criterion is consistent with establishing ecological carrying capacities for protected areas, which is discussed later in this chapter. Protecting life support systems is important for protected areas because they provide ecological services that support life, such as the provision of clean air and water, and conservation of biodiversity. Minimizing adverse impacts on environmental quality requires keeping resource extraction and development in and around protected areas from degrading the environment.

Based on the foregoing discussion, maintaining the productive capacity of ecosystems is an appropriate goal for ecosystem management. Productive capacity is sustained when the total amount of *reproducible capital* and *natural capital* do not decrease over time. Reproducible capital includes human knowledge and human-made capital, such as automobiles, buildings, and equipment. It is fairly easy to value reproducible capital because there are market prices for products produced using reproducible capital. For instance, the value of the current stock of automobiles can be valued using market prices for automobiles. Natural capital includes the stock or total amounts of natural resources such as soil, forests, water, and wildlife. It is more difficult to determine the values of different forms of natural capital because there are no or limited markets for certain forms of natural capital. For example, ecological goods, such as wood products, forage, and energy minerals, have markets and prices, which can be used to value the natural capital that is used to produce these goods, namely forests, grazing land, and energy reserves, respectively. Ecological services, such as water purification, mitigation of floods and drought, maintenance of biodiversity, provision of recreational services, and others do not have markets and prices. The value of ecological services is determined using nonmarket valuation methods, such as willingness to pay (Prato 1998a).

Measuring productive capacity of an ecosystem in terms of the amount of total capital is consistent with the concept of weak sustainability (Pearce et al. 1990; Prato 1999). Under weak sustainability, a decrease in the amount of one form of capital is acceptable as long as there are compensating increases in the amounts of other forms of capital.

Strong sustainability imposes a more stringent requirement for maintaining an ecosystem's productive capacity. It requires each and every form of reproducible and natural capital to be nondecreasing over time. Strong sustainability has several advantages relative to weak sustainability. First, it does not require measuring all forms of capital in monetary terms, that is, it can be measured in physical and/or monetary units. Second, it avoids having to deal with the situation where productive capacity (as measured by the monetary values of ecological economic goods and services that can be produced over the life of that capital) is increasing while the physical amounts of capital are decreasing over time. Third, it allows a protected area manager to vary the level at which natural capital is maintained over different types of protected areas. For example, it is reasonable to establish stringent standards for conserving natural capital in national parks (category I) and relaxed standards for managed resource protected areas (category VI). Proponents of ecological economics generally favor the strong sustainability condition.

Biodiversity Conservation, Ecological Functioning, and Ecosystem Resilience

Successful ecosystem management involves making decisions that maintain or restore the integrity of ecosystems. In the case of protected areas, this requires understanding biodiversity conservation and how it influences the functioning and resilience of protected area ecosystems. Biodiversity conservation means conserving "all forms of life, including species and genetic variants within species and all ecosystems that contain and sustain those diverse forms of life" (CAST 1999). The Second World Conservation Strategy stipulates, "Biological diversity should be conserved as a matter of principle because all species deserve respect regardless of their use to humanity...and because they are all components of our life support system" (IUCN/UNEP/WWF 1991). Respecting the rights of all species regardless of their human value is based on the intrinsic value of biodiversity. It is a moral argument. The economic argument for biodiversity conservation is that species and ecosystems benefit humans by providing ecological goods and services and opportunities for scientific investigation. Whereas the traditional approach to biodiversity conservation focuses on the potential adverse impacts of species loss and extinction on human production and consumption, the new approach emphasizes the importance of biodiversity in supporting the functioning and resilience of ecosystems. In particular, biodiversity, as measured by species richness (number of species per unit area), buffers ecosystems against large disturbances, which stabilizes ecosystem processes, and enhances ecosystem productivity by processing nutrients and waste relatively faster and more efficiently (Smith 1996). The goal of the new approach is to avoid losses in biodiversity that reduce economic activity and human welfare. Under the new approach, maintaining biodiversity is important because of the role of living organisms in maintaining the functioning and resilience of ecosystems (Barbier et al. 1995).

Ecological functioning refers to "those basic processes of ecosystems, such as nutrient cycling, biological productivity, hydrology and sedimentation, as well as the ability of ecosystems to support life" (Barbier et al. 1995). In more technical terms, ecological functioning refers to the "aggregated ecosystem processes involved in the flux of biomass or energy" (Muradian 2001). Examples of ecosystem processes include community respiration, energy recycling, decomposition, nutrient retention, plant productivity, and water retention.

There are several ways to define ecosystem resilience. One definition is "the capacity of an ecosystem to recover from and thus absorb external shocks and stresses, whether they be natural (such as drought, fire or earthquakes) or human-induced (such as pollution or biomass removal)" (Barbier et al. 1995). This definition requires an ecosystem to have at least one equilibrium or stable state (Muradian 2001). A second definition of ecosystem resilience is "the ability of the system to maintain its structure and patterns of behavior in the face of disturbance, that is, the capacity to absorb perturbations and still persist" (Holling 1973, 1986; Muradian 2001). A measure of ecosystem resilience based on the second definition is the amount of disturbance an ecosystem can sustain before changing to another state. Disturbances can be natural or human. Natural disturbance includes fire, windstorms, avalanches, landslides, treefall, floods, insect epidemics, and climate variability. Human disturbance includes harvesting of biomass, pollution, habitat fragmentation, resource management, and biological invasions (Muradian 2001).

Changes in ecosystem states are likely to occur when one or more ecological thresholds are exceeded. Ecological thresholds can occur with or without discontinuities. Figure 7.1 depicts a continuous, nonlinear, inverse relationship between biodiversity and disturbance. Note that biodiversity falls precipitously below B_1^* when the level of disturbance exceeds D_1^*. Figure 7.2 illustrates a discontinuous, nonlinear relationship between biodiversity and disturbance; the discontinuity occurs at D_2^*. When the disturbance level reaches D_2^*, biodiversity ratchets downward from B_2^* to B_2^{**}. Both cases assume the threshold is known. D_1^* and D_2^* can be viewed as threshold disturbance levels for conserving biodiversity. Current ecological knowledge is insufficient to pinpoint ecological thresholds. As Muradian (2001) points out, "at the current state of knowledge, the ecological science is more able to predict the magnitude of change (the possible alternative state) than the threshold values." In other words, there is considerable uncertainty regarding ecological thresholds and, hence, the amount of biodiversity worth preserving.

When an ecosystem exceeds a threshold and moves to a less desirable state there is no guarantee the system will revert to its original state when the distur-

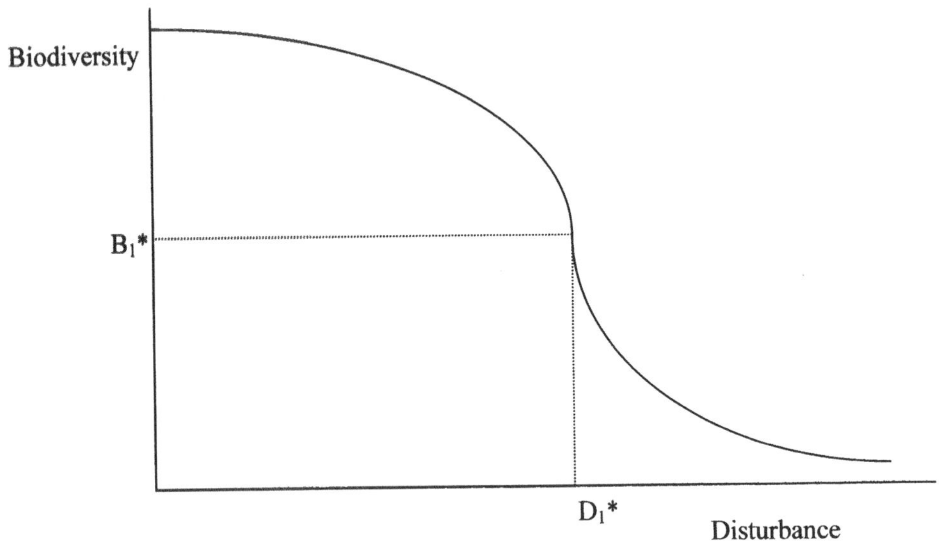

FIG. 7.1. Continuous relationship between biodiversity and disturbance with an ecological threshold.

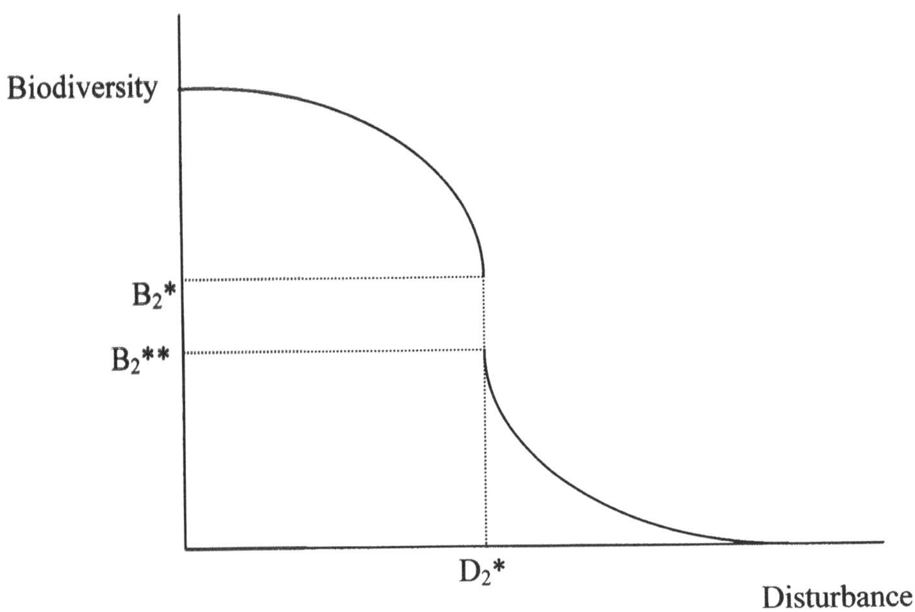

FIG. 7.2. Discontinuous relationship between biodiversity and disturbance with an ecological threshold.

bance subsides. For example, overgrazing, fire suppression, and increases in CO_2 concentrations have caused woody plant invasions into temperate savannas in various parts of the world. These invasions have caused grass-dominated communities to change to woody-plants-dominated communities (Fuhlendorf et al. 1996). Alleviating the human disturbance over a period of many years is often insufficient to restore the grass-dominated community (Walker et al. 1981).

Ecosystem Dynamics and Complexity

Ecosystem dynamics refers to ecosystem changes associated with ecological succession and human/natural disturbance (Franklin 1997). An example of ecosystem dynamics is the spruce-budworm cycle in boreal forests (Ludwig et al. 1978; Rinaldi and Muratori 1992). In this cycle, as tree foliage becomes dense, it produces an abundance of food sources, which contributes to increased populations of spruce budworms. The resulting infestation rapidly defoliates the trees. It takes several years for the trees to become reestablished, after which the cycle repeats itself. Human disturbance in the form of insecticide spraying to reduce spruce budworm infestation and natural disturbance in the form of fire influence the cycle. Other examples of cycles in ecosystems include the variability in prey and predator populations in the prey-predator cycle and periodic occurrences of forest fires. Economic systems exhibit dynamics as well. For example, the cobweb theorem explains year-to-year cycles in the agricultural markets whereby surpluses and low prices in one year lead to cutbacks in production in the following year resulting in deficits and high prices. Ecosystem complexity arises because ecosystems "are characterized by strong (usually nonlinear) interactions between the parts, complex feedback loops that make it difficult to distinguish cause from effect, and significant time and space lags, discontinuities, thresholds, and limits" (Costanza et al. 1993). Dynamics and complexity of ecosystems are highly interactive. Another example of ecosystem dynamics and complexity is the relationship between kokanee salmon and freshwater shrimp in the Flathead System (see sidebar).

At the core of ecosystem dynamics is the concept of stability, which refers to variability over time and space in ecological and economic variables. There are two interpretations of stability, engineering resilience and ecological resilience. Engineering resilience focuses on stability around a single equilibrium steady state and measures stability in terms of an ecosystem's resistance to disturbance and its capacity to return to that equilibrium following disturbance (Pimm 1984; Tilman and Downing 1994). Ecological resilience concentrates on instabilities that cause an ecosystem to deviate from an equilibrium steady state; it refers to the capacity of an ecosystem to absorb disturbance without changing its structure or patterns of behavior (Holling 1973, 1986). The two interpretations of stability have quite different implications for managing protected area

Kokanee Salmon and Freshwater Shrimp in the Flathead System

Human and natural disturbances in and outside of protected areas are dynamic and reversible. A classic example of the dynamic, reversible nature of ecological impacts originating outside of protected areas is the relationship between kokanee salmon (*Oncorhynchus nerka kennerlyi*) and freshwater shrimp (*Mysis relicta*) in the Flathead System in northwestern Montana (see map and photos). Kokanee salmon are a landlocked form of Pacific sockeye salmon that migrate from freshwater lakes to freshwater streams to spawn. They were introduced into Lake Mary Ronan (west of Flathead Lake) in 1918, apparently via a shipment of chinook salmon eggs from Oregon, and were found in Flathead Lake in 1919. The first fishing season for kokanee occurred in 1933 and resulted in the landing of 200,000 kokanee. During their heyday, kokanee made up more than 90 percent of the game fish caught in the Flathead System. The Montana Department of Fish, Wildlife, and Parks estimated that anglers spent more than $3 million on fishing tackle for kokanee in Flathead County (the southern boundary of Flathead County bisects Flathead Lake). Along with expenditures on fuel, food, lodging, and other items, kokanee-related expenditures amounted to $10 million per year, a significant economic value for the area.

After spending three to four growing seasons in Flathead Lake, where they feed almost exclusively on tiny crustaceans and other zooplankton, mature kokanee return to their natal waters to spawn, usually from late September to mid-November. McDonald Creek below Lake McDonald in Glacier National Park was a major spawning ground for kokanee during the 1960s and 1970s. It supported 74 percent of the Flathead kokanee spawning population. Large concentrations of kokanee in McDonald Creek attracted thousands of bald and golden eagles, numerous grizzly bears, and other wildlife that fed on kokanee (adult female kokanee die after spawning). During the kokanee spawning, McDonald Creek had the highest seasonal concentrations of bald eagles in the lower 48 states. The abundance of wildlife in McDonald Creek during the spawning season attracted thousands of bird and wildlife watchers. Despite the fact that the McDonald Creek kokanee run attracted a wide variety of wildlife, Glacier National Park did not view it with favor because kokanee were not native to the waters of the park.

In an effort to enhance native production of kokanee in Flathead Lake, the Montana Department of Fish, Wildlife, and Parks transplanted mysis shrimp from Canada's Waterton Lakes National Park, adjacent to Glacier National Park, to lowland lakes upstream of Flathead Lake. The replanting scheme was based on the premise that freshwater shrimp would increase production of kokanee; a premise based on faulty information on trial stockings in Kootenay Lake, British Columbia. Mysids were discovered in Flathead Lake in the early 1980s and their population grew exponentially. During the day, they avoid visual feeders, such as kokanee and cutthroat trout, by staying in cold, dark areas at the bottom of the lake. At night, mysids migrate from the bottom to the top of the lake to feed. This feeding-rest cycle makes mysids available for bottom-feeding juvenile lake trout and lake whitefish, but unavailable to open water fishes like kokanee.

The introduction of mysids hurt the kokanee population three ways. First, since the feeding patterns of mysids and kokanee are very different, mysids were not a food source for kokanee. Second, mysids competed with kokanee for zooplankton, which reduced food supplies for kokanee. Third, the population of lake trout and whitefish grew because these species fed on mysids and kokanee predators (lake trout and whitefish) increased due to the increased food base provide by the shrimp. Prior to the introduction of mysids to Flathead Lake, mountain lake fish, bull trout, and kokanee were caught in abundance. After the introduction of mysids, kokanee populations dropped precipitously. Today, the most commonly caught fish in Flathead Lake are lake trout and lake whitefish. Catching kokanee is a rarity in Flathead Lake.

As a result of the introduction of mysis shrimp to the Flathead System, kokanee salmon no longer spawn in McDonald Creek and this once nonnative species is no longer present in Glacier National Park. Human action brought kokanee to the Flathead and human action caused their demise.

Sources: Bonneville Power Administration and the Montana Department of Fish, Wildlife, and Parks (1985); Gildart (1986); Stanford and Ellis (2002).

ecosystems. Engineering resilience supports a command-and-control approach that emphasizes efficiency, control, constancy, and predictability, and management actions designed to achieve maximum production. In contrast, a view of ecosystem stability based on ecological resilience dwells on ecosystem properties like persistence, adaptiveness, variability, and unpredictability that support an adaptive management approach. Management approaches based on engineering resilience are likely to work for ecosystems exhibiting low uncertainty, but are prone to fail for "dynamic, evolving systems where variability and novelty results in high uncertainty" (Holling and Gunderson 2002). Such failures or resource management pathologies are well documented (Gunderson et al. 1995).

Five worldviews have been used to explain ecosystem stability: nature flat, nature balanced, nature archaic, nature resilient, and nature evolving (Holling et al. 2002b). *Nature flat* contends that the processes that affect nature, and hence ecosystem stability, are random and stochastic. This viewpoint implies that nature is "infinitely malleable and amenable to human control." Under the nature flat worldview, human ingenuity and knowledge make all things possible, including exponential growth. In this worldview, technology is expected to increase resource productivity over time and provide sufficient goods and services to meet people's needs even with exponential growth in population.

Nature balanced views nature as operating at or near a static or dynamic equilibrium, which is maintained through negative feedback. Negative feedback is self-reinforcing in the sense that changes in other variables induce movement toward equilibrium, which results in stability. Nature balanced supports views of sustainable development, maximum sustainable yield, and fixed carrying capacities. In the nature balanced worldview, scarce supplies and high prices of certain goods, such as wood, increase the prices of those goods thereby reducing their consumption (conserving wood) and stimulating the production of substitute goods such as plastic.

Nature archaic views nature as fundamentally unstable due to processes of growth and collapse. It is dominated by positive feedbacks. Positive feedback is self-weakening in the sense that changes in other variables induce movement away from equilibrium, which causes instability. For example, beyond some point, climate change could result in higher temperatures and more extreme weather events even though greenhouse gas emissions are being reduced. Nature archaic supports policy prescriptions that favor low-impact technologies and the "small is beautiful" paradigm (Schumacher 1999), maintenance of the status quo, and use of the precautionary principle.

Nature resilient is a worldview that envisions multiple stable ecosystem states punctuated by periods of exponential growth, brittleness, readjustment or collapse, and renewal through reorganization. In a nature resilient worldview, human disturbances are not of concern because nature can adjust to those disturbances without becoming unstable. In other words, nature is very forgiving.

A fifth worldview called *nature evolving* sees ecosystems as undergoing abrupt, unpredictable, and transforming change that is best managed by flexible policies and actions (Holling et al. 2002a). For example, uncertainty regarding how protected area ecosystems are going to respond to reintroduction of an extirpated species, such as the wolf or grizzly bear, warrants developing actions and policies for reintroduction in an adaptive management framework (Holling 1986). Chapter 8 provides more details about adaptive management.

The adaptive cycle depicted in figure 7.3 and the panarchy depicted in figure 7.4 are useful constructs for explaining the nature evolving worldview and ecosystem dynamics. The four-phase adaptive cycle diagram measures increases in potential inherent in the accumulated resources of biomass and nutrients (capital) on the vertical axis and the degree of connectedness among controlling variables on the horizontal axis. High connectedness means the elements and processes of the ecosystem are dominated by internal relationships that control or moderate the influence of external variability. Low connectedness means the elements and processes of the ecosystem are loosely connected causing their behavior to be more sensitive to external variability.

The exploitation (r) phase of the adaptive cycle is characterized by colonization of disturbed areas by species that are r-strategists, meaning they have high dispersal ability (they spread easily) and grow quickly. The arrows in the r-phase have short tails and are closely spaced indicating relatively rapid change. Potential and connectedness are both low in the r-phase.

In the conservation (K) phase, accumulation and storage of energy and material occur slowly, hence the arrows have longer tails and are farther apart. This phase is dominated by species that are K-strategists, which have slower growth rates and more specialized resources. Potential and connectedness increase during the K-phase.

In the release (Ω) phase, the biomass and nutrients that have accumulated during the conservation phase and are tightly bound together are released from the system as a result of disturbance. This process is known as creative destruction because biomass and nutrients are released for new development and growth processes. Potential and connectedness decrease in the release phase.

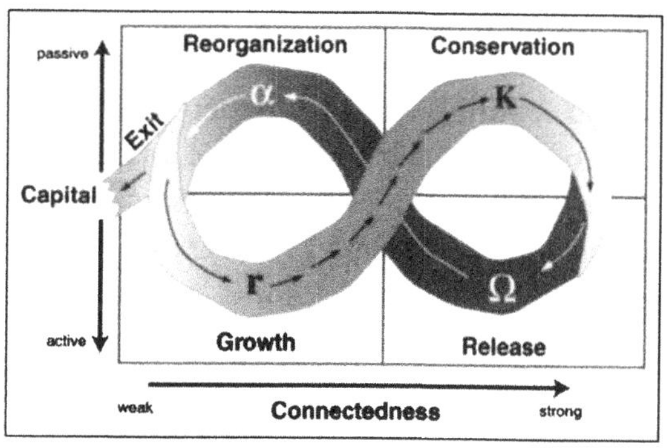

FIG. 7.3. Example of adaptive cycle. (From Holling and Gunderson 2002.)

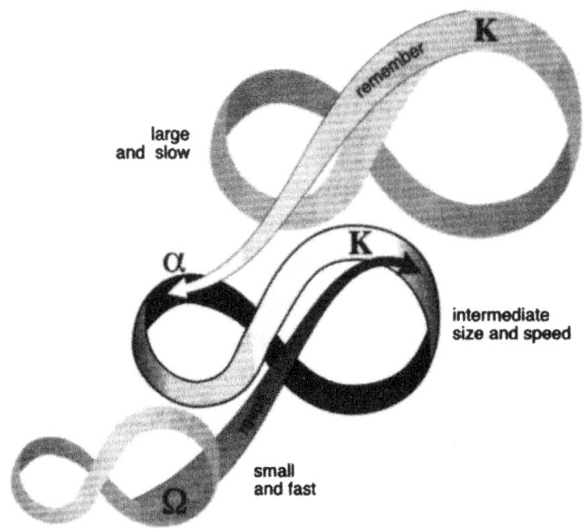

FIG. 7.4. Example of a three-nested panarchy. (From Holling et al. 2002b. Copyright © 2002 Island Press. Reproduced by permission of Island Press, Washington, DC.)

In the reorganization (α) phase, soil processes act to minimize further losses in nutrients and available nutrients are reorganized and supplied to pioneer species that operate in the exploitation phase. Processes that occur between conservation and release, release and reorganization, and reorganization and exploitation occur rather quickly. During the reorganization phase, the ecosystem could shift to a less productive but more organized system. A third dimension not shown on figure 7.3 is ecosystem resilience. Ecological resilience is high in the exploitation and reorganization phases and low in the conservation and release phases. The spruce budworm example of ecosystem dynamics given earlier illustrates the adaptive cycle.

The dynamics of economic systems can be explained in terms of adaptive cycles as well. Consider the evolution of Internet-based businesses. Rapid growth in purchases of PCs and demand for Internet access created market conditions that favored investment and rapid growth in Internet-based businesses (r-phase). As business growth overtook customer growth, the market became saturated (K phase). In an effort to survive, companies increase competition for customers, which caused some businesses to lose market share and become less profitable. Increased competition and resulting lower profits caused some companies to leave the industry (release phase). This released human and reproducible capital for other forms of investment and business enterprise (reorganization phase).

A panarchy is a set of adaptive cycles that are nested one within the other across spatial and temporal scales (Holling et al. 2002b). Consider an adaptive cycle operating in a small space at fast speed in the Ω (creative destruction) phase. The resulting collapse can cascade up to the next level in the panarchy, which is an adaptive cycle operating in an intermediate space and speed (see fig. 7.4). If the cascade occurs when the intermediate cycle is in the K phase where resilience is low, then a crisis can occur. In this case, a revolt is said to occur because a fast and small-scale event overwhelms a slow and intermediate-scale event. In turn, the crisis at the intermediate scale can cascade to the next higher level in the panarchy. An ecological example of a panarchy would be the spread of a fire from a campfire, which operates at ground level, to the crown of a tree, then to a forest patch, and finally to a whole stand of trees. In this case, the ground fire (disturbance) cascades from processes operating at fast speed in a small space (campfire) all the way up to processes operating at slow speed and in a large space (tree stand). Similar cascades occur in social panarchies when, for example, a local political movement parlays into a national political movement, which results in a change in public policy.

Natural Regulation

The natural regulation approach to managing protected areas surfaced in 1963 with publication of the Leopold Report on Wildlife Management in National Parks (Leopold et al. 1963). The report recommended that natural processes in national parks be allowed to operate and dominate because of their importance for ecological succession and landscape change. It did not, as some believe, advocate passive management. Rather it urged "active management of park resources using the most advanced techniques" (Craighead 1991). Natural regulation attempts to sustain or restore ecological processes, such as fire and trophic-level interactions between plants and herbivores. In natural regulation, fire is viewed as an essential and dynamic driver of ecological processes.

Natural regulation for wildlife is grounded in the principle that as a wildlife population increases, density-dependent changes (declines in birthrates and survivorship) cause the population to achieve an equilibrium density (Boyce 1991). Prior to 1967, the National Park Service killed elk in order to reduce what was perceived as overgrazing of Yellowstone's northern range. In 1967, after intense negative public reaction to this practice, the National Park Service initiated a natural regulation policy in Yellowstone National Park by terminating the culling of elk. Between 1967 when natural regulation was implemented and 1979, Yellowstone's elk herd increased from 5,000 to about 12,000. It has since increased to about 20,000.

Four criteria have been proposed for judging the success of natural regulation in Yellowstone National Park (Houston 1976):

1. High elk numbers should have no detrimental effects on other wildlife species, particularly other ungulates, within the park.

2. Elk should not cause major changes in vegetation to exotic or less palatable plants.
3. Soil erosion should not increase in the absence of elk culling.
4. Elk populations should stabilize at some equilibrium density.

Despite the fact that all of these conditions were violated in Yellowstone, Boyce argues it is "inappropriate or misleading" to judge natural regulation based on these conditions because they do not constitute a reasonable or convincing case for the success of natural regulation. With respect to the first criterion, Schullery (1997) notes, "Species had come and gone in Yellowstone for thousands of years before whites arrived; how would the disappearance of a plant species, much less a shift in its abundance, prove that something had gone wrong? And if we couldn't prove that something had gone wrong, how would we know if the 'great experiment' [of natural regulation] was really succeeding?"

In 1982, research based on ten years of monitoring of national regulation in Yellowstone National Park was published (Houston 1982). This and other research (Despain et al. 1986) concluded that culling of elk was not justified. Schullery (1997) comments that Houston's study showed that "rather than being an example of an unhealthy landscape, [Yellowstone] was instead an example of a normal wild landscape; the problem was that such landscapes had become so rare that they seemed to our eyes somehow unnatural or wrong or in need of repair." What is natural to our eyes in terms of range condition is that which occurs when livestock stocking densities on managed range are kept within economic carrying capacity, which is substantially less than a range's ecological carrying capacity. The latter occurs in unmanaged natural conditions, such as exist in Yellowstone.

In 1986, Congress authorized funding to study whether overgrazing was occurring in Yellowstone's northern range. Study results provided overwhelming evidence that "vegetation of the northern range was not overgrazed... the range's ecological processes seemed to be working and...traditional definitions of overgrazing did not seem applicable" (Schullery 1997; see chapter 10 for a more extended treatment of this issue). Implementation of natural regulation in Yellowstone National Park eventually led to protection of all wildlife species in the park, discontinuation of the practice of feeding human refuse to grizzly bears, introduction of wolves, and implementation of a fire management policy that allowed wildfires to burn as long as they did not threaten life and property.

The application of natural regulation is not without challenges and detractors. For example, prior to European settlement, it was common for Native Americans to use hunting to control wildlife populations and setting fires to enhance rangeland productivity (Wagner 1969; Chase 1987). Are hunting and fire setting by Native Americans natural processes that should be maintained under natural regulation? These activities have been disallowed in most national parks, including

Yellowstone, that claim to practice natural regulation. In terms of detractors, Craighead (1991) argues that natural regulation is not appropriate for Yellowstone National Park because its boundaries are artificial, the park is too small to encompass an intact natural system, and the number of visitors and development on adjacent land is intrusive. He blames implementation of natural regulation in Yellowstone for overpopulation of elk and bison and associated declines in white-tail and mule deer, beaver, and other species, and greatly reduced populations of grizzly and black bear.

There are disagreements about the operation of self-regulating mechanisms in an ecosystem. At one time, self-regulation of wildlife populations was thought to be the product of density dependence. Both density-dependent and density-independent factors are now considered determinants of wildlife populations (Horn 1968). Plant-herbivore interactions, which have a major influence on population dynamics for herbivores, result in complex ecosystem dynamics that undermine the idea that wildlife populations converge to some equilibrium density (Caughley and Lawton 1981; Schaffer 1988). Rather, the biophysical processes that determine elk populations in Yellowstone National Park are nonlinear and stochastic resulting in unpredictable variability in elk populations.

The dynamic nature of elk and other ungulate populations in the park flies in the face of the Leopold Report's recommendation that parks be managed as if they were a "vignette [idealized view] of primitive America" (Leopold et al. 1963). Fire is another dynamic ecological process that, if allowed to function, results in unpredictable, but nonetheless useful, landscape mosaics. One concept that incorporates the nonlinear and stochastic nature of ungulate populations is dynamic equilibrium (Houston 1982). A dynamic equilibrium implies that elk populations in Yellowstone National Park are not driven to a static equilibrium. Rather, equilibrium density fluctuates over time in response to variability in climate and other environmental agents. One model of dynamic equilibrium for elk in Yellowstone National Park predicted an equilibrium population of 19,000 in the winter of 1987–88 following five relatively mild winters, and an elk population of 15,000 in an average winter (Merrill and Boyce 1991). Another factor that appears to have reduced equilibrium elk population is the introduction of wolves into the park in 1995. While natural variability in wildlife populations is a natural element of park ecosystems, it could reduce visitation rates and/or visitor satisfaction in national parks acclaimed for their wildlife populations, such as Yellowstone.

In order to reduce misunderstandings associated with the term "natural regulation," a new term has been proposed, namely, "ecological-process management" (EPM) (Boyce 1991). The objective of EPM is to "allow ecological processes of nutrient cycling, plant succession, fire, decomposition, competition, predation, symbiosis, birth, and death to operate unimpeded by human intervention." EPM does not prohibit human intervention in an ecosystem. Rather, it allows interventions designed to restore ecological integrity. Examples of this in Yellowstone National Park include discontinuing the feeding of human refuse to grizzly bears, introduction of wolves, and only suppressing fires that threaten life and property. Boyce (1991) concludes that it is better to implement EPM in Yellowstone National Park through "carefully reasoned intervention with a directed goal," and monitor the resulting ecosystem changes than to arbitrarily intervene. In other words, EPM is not a totally hands-off approach. It requires active management of parks, especially in places like Yellowstone National Park where what goes on outside the park is sometimes more important in terms of preserving ecological integrity than what goes on inside the park. Examples of this phenomenon in the Yellowstone area include the loss and fragmentation of habitat in areas adjacent to the park and the killing of bison as they leave the park during the winter. In other words, natural regulation can be implemented within the context of ecosystem management and adaptive management. Although EPM is more descriptive of what natural regulation means, the term is not widely used by protected area managers.

Carrying Capacity

What is considered the first documented concern about carrying capacity in national parks occurred in the mid-30s when the National Park Service posed the question: "How large a crowd can be turned loose in a wilderness area without destroying its essential qualities?" and the retort that recreation use be kept "within the carrying capacity" (Sumner 1936). The 1978 National Parks and Recreation Act (P.L. 95-625) requires carrying capacities to be determined for each park as part of the process of developing a general management plan. Specifically, amendments to Public Law 91-383 (84 State. 824, 1970) require general management plans developed for national park units to include "identification of and implementation commitments for visitor carrying capacities for all areas of the unit" and determination of whether park visitation patterns are consistent with ecological and social carrying capacities. The National Park Service defines carrying capacity as "the type and level of visitor use that can be accommodated while sustaining desired resources and social conditions that complement the purpose of a park unit and its management objective" (National Park Service 1997). Amendments to the National Trails System Act (Public Law 90-543, 1968) mandate "an identified carrying capacity of the trail and a plan for its implementation" be developed in comprehensive trail planning. Regulations implementing the National Forest Management Act of 1976 dictate that, in wilderness management planning, provision be made "for limiting and distributing visitor use of specific areas in accord with periodic estimates of the maximum levels of use that allow natural processes to operate freely and that do not impair the values for which wilderness areas were created." Similarly, the National Outdoor Recreation Plan requires "each federal recreation land managing agency [to] determine the carrying capacity of its recreation lands" (Bureau of Outdoor Recreation 1973).

Two forms of carrying capacity are relevant to protected areas: those that relate to the human experience and biological or ecological degradation (Seidl and Tisdell 1999). Thomas Malthus gave, perhaps, the earliest analysis of human carrying capacity. He postulated that human population growth would outstrip the land's capacity to produce food resulting in food shortages (Malthus 1798). Biological or ecological carrying capacity is fundamental in range and wildlife management. It is defined as the maximum population of a particular species a habitat area can support in a given period of time without damaging the area or reducing the future ability of the area to support the species (Miller 1990; Hawden and Palmer 1994; Hanley et al. 1999). Leopold defined it as the maximum density a range is capable of supporting (Dhondt 1988). Exceeding an area's ecological carrying capacity increases the risk of irreversible ecosystem change, such as declines in plant community structure or species diversity (Caughley and Lawton 1981; Wallace 1999). Other ecological effects include loss of soil and vegetation, damage to trees, and wildlife disturbance (Manning 1998; Leung and Marion 2000). However, these definitions oversimplify the dynamic interactions between animal populations and landscapes, which are characterized by nonlinear dynamics and population thresholds (Seidl and Tisdell 1999).

In the mid-1960s, the carrying capacity concept for protected areas was expanded beyond just environmental effects to include social or experiential effects of visitation (Wagar 1964). Examples of such effects include crowding, use conflicts, and excess resource degradation (Manning 1998; Leung and Marion 2000). Visitor carrying capacity for protected areas is defined as the maximum number and type of visitors an area can sustain without causing irreversible deterioration of the physical environment and appreciable loss of visitor satisfaction (Seidl and Tisdell 1999; Shelby and Heberlein 1986). Since there are likely to be different considerations for the human, ecological, and economic components of visitor carrying capacity, it is difficult to define. Biophysical characteristics of an area (vegetation type, topography, and climate), human factors (location and mode of travel, season of use, group size, and behavior of other visitors), and management policies (use limitations) are more important determinants of ecological and social (visitor) carrying capacities than simply the size of the population or number of visitors. Accordingly, contemporary definitions of carrying capacity are cast in terms of the acceptability of human, ecological and economic impacts of visitation. In addition to these impacts, increased use of a protected area influences its management. Specifically, increased use is likely to result in more intensive management practices, such as periodic rest and rotation of degraded areas, construction of new roads and trails, and others (Manning et al. 1996b). In general, carrying capacity depends on value judgments, institutional arrangements, technologies, consumption patterns, and human goals (Seidl and Tisdell 1999).

Since protected area decisions that affect visitation are easier to defend when based on quantitative measures of ecological and social carrying capacities (Peterson 1996), several quantitative measures of carrying capacity have been developed. The three most common ones are Limits of Acceptable Change (LAC), Visitor Impact Management (VIM), and Visitor Experience and Resource Protection (VERP). Rather than defining carrying capacity as the maximum number of visitors allowed in an area, the LAC method evaluates the acceptability of visitor impacts on key biophysical and social processes (Stankey et al. 1985; McCool and Cole 1997). Impact acceptability is judged by comparing a set of indicators of biophysical and social processes to standards of quality that "define the minimum acceptable condition of indicator variables" or limits of acceptable change (Newman et al. 2001). The latter define the desired future conditions for resource, social, and managerial settings (Merigliano 1990; Manning 1999; Newman et al. 2001). In essence, limits of acceptable change articulate the management objectives for an area (Frissell and Stankey 1972; Manning et al. 1996a; Manning 1999). For example, managers of the wilderness areas in Yosemite National Park established indicators and standards for ecological, social, and managerial aspects of wilderness experiences. Ecological aspects included signs of human use at campsites as measured by the extent of root exposure, and signs of stock or stock use as measured by trail impacts, tree scars, and manure. Social aspects included encounters with other groups along the trail and encounters with other groups at campsites. Managerial aspects included camping regulation (designated campsites vs. freedom to camp anywhere) and availability of permits (Newman et al. 2001). If indicators exceed established standards, then a management action is taken to bring indicators into conformance with standards.

The LAC method, developed by the U.S. Forest Service, is an extension of the Recreation Opportunity Spectrum (ROS) Framework (Clark and Stankey 1979, 1990). ROS divides the landscape into zones or settings that supply a spectrum of recreational opportunities from primitive to developed areas. Managers match the supply of recreational opportunities with visitors' demands for those opportunities. The LAC method complements the ROS Framework by identifying in cooperation with stakeholders opportunities for recreational experiences, specifying indicators and standards for resource and social conditions in each zone, and developing appropriate management actions for cases where monitoring indicates unacceptable changes relative to the indicators and standards (Stankey et al. 1985; Knopf 1990; McCool 1990a and 1990b). The LAC method has been applied in Hells Canyon National Park in the United States, Yoho National Park in Canada, and elsewhere.

The VIM method, developed by researchers and the National Parks and Conservation Association, is very similar to the LAC method. It evaluates visitor impacts by comparing standards for key indicators of natural resources, cultural resources, and visitor experiences to monitored values of those indicators, and identifies and implements appropriate management action when standards are violated (Graefe et al. 1990). LAC and VIM have been applied to backcountry management planning in Shenandoah National Park in the United States, the Columbia Icefields in Jasper National Park in Canada, and others (Marion et al. 1985; Vaske 1994).

In 1992, the National Park Service (NPS) established the VERP method to evaluate carrying capacity in developing general management plans for park units (National Park Service 1997). The VERP method was first implemented in Arches National Park (Hof et al. 1994; Manning 2001) and has since been used in a number of other national parks in the United States including Isle Royale and Acadia (Vande Camp et al. 2001). Like the LAC and VIM methods, the VERP method determines the amounts and kinds of visitor use a management zone can sustain without causing unacceptable resource and social impacts (Shelby and Heberlein 1986; Manning et al. 1996a). Resource impacts include loss in vegetation, tree damage, soil erosion and compaction, and wildlife disturbance; social impacts encompass crowding, use conflicts (snowmobiling vs. cross-country skiing), reduced quality of visitor experiences due to excessive resource degradation, and other factors that diminish visitor satisfaction (Leung et al. 2002).

Implementation of the VERP method requires managers to (1) select appropriate management objectives for different zones within a protected area; (2) translate the objectives for each zone into indicators and standards of quality for resource and social impacts; (3) implement a monitoring program to measure indicators; (4) design and implement a new management action when the standards are violated; and (5) monitor the new management action for compliance with the standards (Manning 2001; Leung et al. 2002). The LAC, VIM, and VERP methods have several elements in common, namely, (1) determining the types of recreation opportunities to be provided in different zones, (2) defining opportunities in terms of specific indicators and standards of quality, (3) monitoring indicators for compliance with standards, and (4) implementation of appropriate management actions when standards are violated (Manning 1999). Other visitor management frameworks include Visitor Activity Management Planning (Nilsen and Grant 1998) and Tourism Optimization Management Model (Manidis Roberts Consultants 1997).

Ecological footprint is a sustainability metric related to carrying capacity (Wackernagel et al. 2002). It measures the extent to which human demands for land, in terms of the amount a country needs for production of energy, food, and timber and waste disposition, can be satisfied from that country's land area. If the amount of land needed exceeds the amount available, then the country has a positive ecological footprint (or ecological deficit), which means the country is depleting its natural capital and/or importing energy, food, and timber from other countries. This relates to protected area management because countries with positive ecological footprints have a harder time supporting the creation and maintenance of parks and protected areas. If all countries in the world have a positive ecological footprint, then the world system is unsustainable. A recent study indicated that the ecological footprint was about 70 percent of the Earth's regenerative capacity in 1961 and at least 120 percent in 1999 (Wackernagel et al. 2002). This result indicates the Earth's carrying capacity is being exceeded by 20 percent, which is unsustainable in the long run.

A new method for evaluating carrying capacity is called the Multiple Attribute Scoring Test for Capacity (MASTEC) (Prato 2001). MASTEC integrates elements of the LAC, VIM, and VERP methods. It allows managers to assess whether the current state of a protected area ecosystem is in compliance with established standards for carrying capacities when the manager is uncertain about the state of a protected area ecosystem before and after new management actions are implemented. In cases where the standards are violated, the MASTEC method allows the manager to identify the best management action for complying with the standards.

The MASTEC method judges carrying capacity using a multiple attribute evaluation procedure. Multiple attribute evaluation has been used in a variety of decision-making contexts, including water resources management (Haimes and Hall 1974), environmental management (Backus et al. 1982; Janssen 1992), food security (Haettenschwiler 1994), forest management (Kangas and Kuusipalo 1993; Kangas 1994; Penttinen 1994), agricultural-environmental management (Xu et al. 1995), protection of natural areas (Anselin et al. 1989; Gehlbach 1975; Sargent and Brande 1976; Smith and Theberge 1986; Smith and Theberge 1987), regional water quality analysis (Makowski et al. 1995), management of agroecosystems (Prato et al. 1996a), wildlife management (Prato et al. 1996b), and soil and water management (Prato 1998b). MASTEC is well suited for assessing carrying capacities for protected areas because it allows consideration of the multiple dimensions of carrying capacity, supports transactive planning (stakeholders and planning team work cooperatively; Laven et al. 2001), and provides quantitative support for decisions (Schmoldt et al. 1994; Peterson 1996).

The MASTEC method allows a manager to rank feasible management actions for complying with social and ecological carrying capacities (see fig. 7.5). Feasible management actions satisfy carrying capacity constraints that limit the frequency (probability) with which social and ecological capacity standards are exceeded. A carrying capacity constraint requires an attribute of carrying capacity to be equal to or greater than the established standard for that attribute with a reliability level chosen by the manager, for example, 95 percent. In essence, each attribute standard specifies the limits of acceptable change for that attribute. Standards are determined based on resource and social surveys, and statutory requirements.

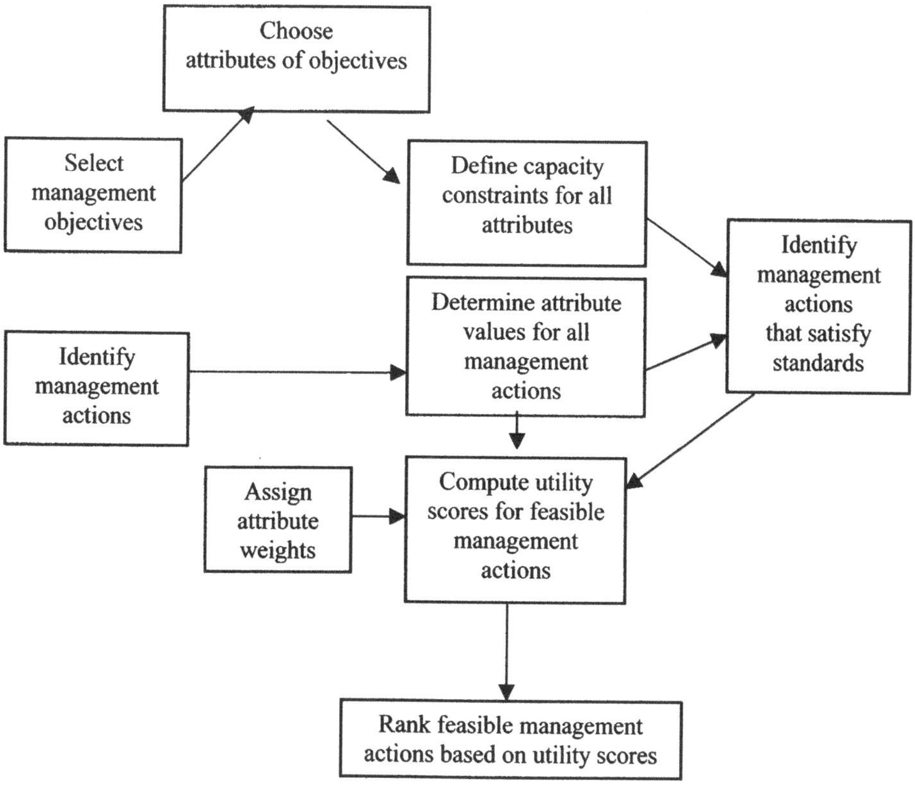

FIG. 7.5. Schematic of MASTEC method.

Ranking of feasible management actions is based on utility scores for management actions. A utility score for a management action is the weighted sum of the attribute values for that action. Attribute values for each management action are estimated based on expert opinion, and simulation models and attribute weights are determined by the manager. The highest ranked management action is considered the best management action for complying with carrying capacities.

In order to apply MASTEC, the manager must determine (1) management actions, (2) management objectives, (3) attributes of objectives, (4) values of attributes for each management alternative, (5) standards for attributes, (6) reliability levels for achieving standards, and (7) weights for attributes. Technical details of the MASTEC method are provided by Prato (2001).

Consider how the MASTEC method could be used to determine the best transportation mode for a national park. Suppose feasible transportation modes (management actions) identified by the manager include conventional bus; articulated bus, monorail, self-propelled rail transit, and cog railway. Objectives for transportation modes include accommodating expected visitation or ridership levels; reducing private vehicle traffic; lowering vehicle-related air pollution; decreasing energy consumed by transportation; employing proven, available transportation systems; complying with Americans with Disabilities Act; allowing unrestricted movement of wildlife across and along transportation systems; controlling visitor encroachment into sensitive natural areas; and others (U.S. Department of the Interior 1994). Possible attributes for these objectives are visitor delays in using transportation modes, maintaining an acceptable level of comfort and viewing, providing flexibility and minimizing delays in disembarking and embarking on to transportation vehicles, complying with air quality standards, minimizing interference with wildlife migration and degradation of sensitive nature areas, and others.

The carrying capacity methods described above involve complex interaction between environmental, social, and management considerations, specification of indicators of biophysical/social processes and management objectives, determination of standards of quality, and, in the case of MASTEC, specification of reliability levels for achieving standards and weights for attributes of carrying capacity. For this reason, use of these methods, even within the same protected area zone, is not likely to result in a single carrying capacity. Instead, the methods provide a range of carrying capacities. Lack of definitiveness regarding carrying capacity and other criticisms of the concept have led to other approaches, including:

1. Placing more emphasis on management objectives and less emphasis on use limits (Becker and Jubenville 1982; Jubenville and Becker 1983; Stankey et al. 1984)
2. Utilizing design capacity rather than carrying capacity
3. Deemphasizing use limitations (Washburn 1982; Burch 1984; Stankey et al. 1984)
4. Recognizing that not all areas require use limits and not all user-caused change is undesirable
5. Judging the undesirability of change relative to management objectives (Stankey 1974)

VERP will continue to receive more use than MASTEC because it is better known. However, MASTEC offers several advantages relative to VERP that make it well suited to assessing carrying capacity. While there have been no applications of MASTEC to protected areas, it is likely to receive more attention and use as it becomes better known.

Assessing Ecosystem Capacity

This section describes an analytical framework for assessing the productive capacity of a protected area ecosystem based on the strong sustainability criterion and identifying the best management action to take when capacity is decreasing. Strong sustainability is typically measured using either a stock attribute, such as area of a particular habitat or the value of buildings, machinery, and equipment, or a flow attribute, such as net primary productivity or gross domestic product. Consider a situation in which productive capacity of an ecosystem is evaluated by a set of stock and/or flow attributes. Strong sustainability of an ecosystem requires maintaining minimum values for all attributes of productive capacity. It requires a manager to implement a best management action when one or more attributes fall below their respective threshold levels as described below. A biological attribute of sustainability is species biodiversity. Thresholds for biodiversity can be defined in terms of minimally viable populations and/or minimum suitable habitat areas for threatened and endangered species. Environmental attributes could be established for water quality and air quality. Socioeconomic attributes could include income and employment levels generated by tourism and recreation.

Suppose the protected area manager chooses n attributes to evaluate the sustainability of an ecosystem, namely $z_1,\ldots, z_n$, and n corresponding threshold levels, namely, $v_1,\ldots, v_n$. Strong sustainability of the ecosystem requires the following conditions to be satisfied:

$$z_i \geq v_i \text{ for all } i = 1,\ldots, n$$

These conditions are quite restrictive because they allow no flexibility in terms of allowing the attributes to be below their corresponding threshold values. A less restrictive formulation of the strong sustainability conditions allows the conditions to be met a large percentage of the time. The next section, which deals with adaptive management, describes a decision rule for strong sustainability that allows for more flexibility in determining when strong sustainability is achieved by accounting for the random variability in attributes.

If any of these conditions are violated, then the protected area manager must take action to bring the ecosystem to a sustainable state. In other words, the manager must select the best management action for achieving sustainability. Selection of the best management action involves two kinds of decisions: namely, identifying the sustainable and efficient combinations of attributes and associated management actions, and choosing the most desirable combination of attributes from those that are sustainable and efficient.

Consider a situation in which the manager has identified six technically, administratively, and politically feasible combinations of two attributes, biodiversity (z_1) and regional income (z_2). Figure 7.6 illustrates the six combinations of the two attributes, designated as $m_1, \ldots, m_6$. Each combination corresponds to a unique management action. For example, m_5 provides z_1 units of biodiversity and z_2 units of regional income. Suppose values of biodiversity above the horizontal line labeled v_1 ($z_1 \geq v_1$) are sustainable in terms of biodiversity and values of biodiversity to the right of the vertical line labeled v_2 ($z_2 \geq v_2$) are sustainable in terms of regional income. Only m_5 and m_6 are sustainable because they are the only combinations that satisfy both the biodiversity and regional income conditions for sustainability, that is, m_5 and m_6 satisfy $z_1 \geq v_1$ and $z_2 \geq v_2$. If current management of the ecosystem results in the combination of attributes given by m_1, then the ecosystem is not sustainable because regional income is below the threshold level of v_2.

The manager must select a management action that provides a sustainable and efficient combination of biodiversity and regional income. It has already been determined that m_5 and m_6 are sustainable. But, are they efficient? Determination of the efficient combinations of attributes is illustrated in figure 7.7. Action m_2 is inefficient because other management actions, namely, m_1, m_4, m_5, and m_6, provide more biodiversity and regional income than m_2. Action m_1 is inefficient because it provides less regional income than m_4. In general, all combinations of attributes on segment AB (except for B) are inefficient because they provide less of both attributes than B. Similarly, m_3 is inefficient because it provides less biodiversity and regional income than m_5. In general, all combinations of attributes on segment CD (except for C) are inefficient because they provide less amounts of both attributes than C. Therefore, segment BC contains the efficient combinations of attributes. Of the six combinations of attributes, only m_4, m_5, and m_6 are efficient. The efficient combinations of metrics define the efficiency frontier. In this case, the efficiency frontier contains three combinations of attributes and underlying management actions, namely m_4, m_5, and m_6. When there are a large

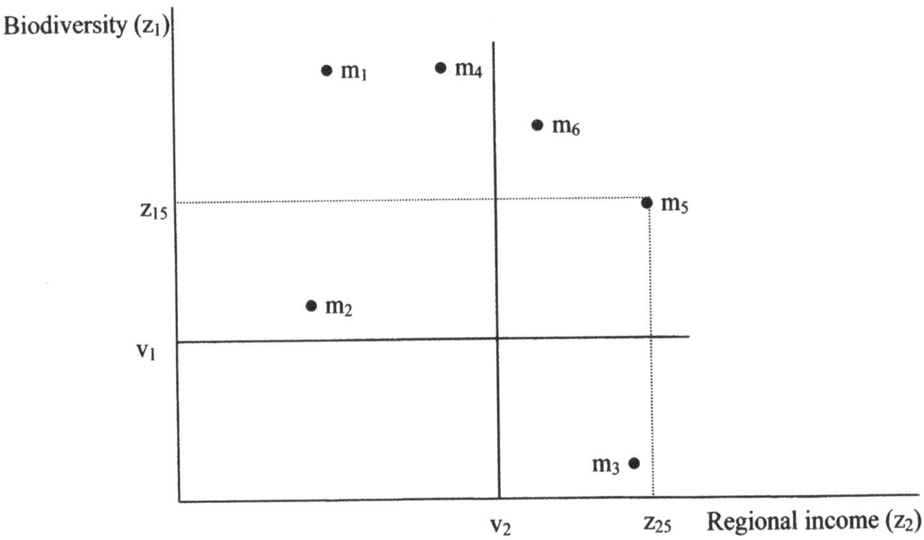

FIG. 7.6. Sustainable combinations of biodiversity and regional income.

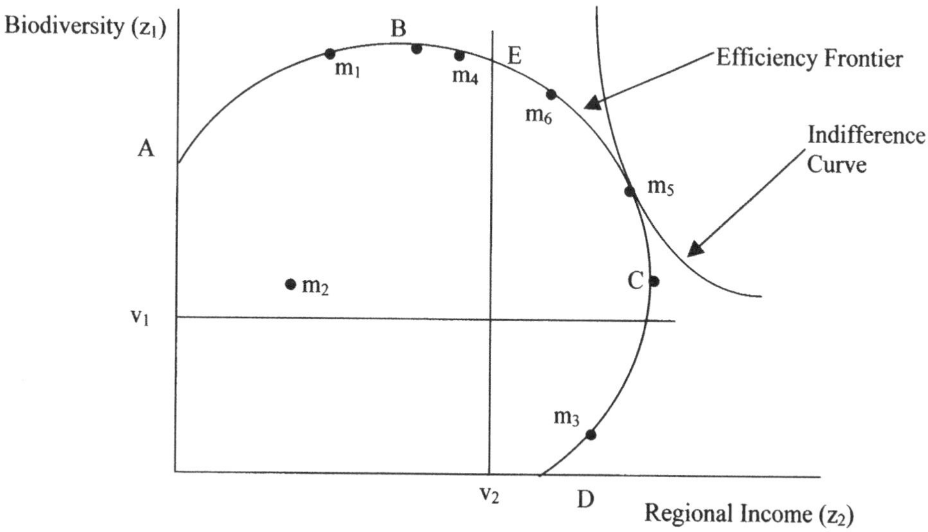

FIG. 7.7. Best management action for a manager.

number of efficient management actions, the efficiency frontier can be approximated by the continuous line labeled BC.

Delineation of the efficiency frontier is important because it separates efficient management actions from inefficient management actions. Combinations of attributes below the efficiency frontier are inefficient, combinations on the efficiency frontier are equally efficient, and combinations above the frontier are not achievable given current knowledge and technology. To ensure that limited human, financial, and natural resources are being used efficiently, the best management action for a protected area must provide a combination of attributes on the efficiency frontier. Moving up or down the efficiency frontier implies trade-offs between biodiversity and regional income. Actions that provide more regional income result in less biodiversity, and actions that provide more biodiversity result in less regional income. Trade-offs between attributes along the efficiency frontier cause the frontier to be concave (bowed outward). Combining the sustainability conditions and efficiency frontier indicates that management actions m_5 and m_6 are both sustainable and efficient. Therefore, the best management action for bringing the ecosystem to a sustainable state is m_5 or m_6.

The question then becomes, is m_5 or m_6 the best management action? If the only factor influencing the selection of m_5 or m_6 is the manager's preferences for the two attributes, then the decision only requires consideration of the manager's preferences for attributes. Based on chapter 4 (see fig. 4.10a), the manager's preferences for attributes can be represented by an indifference map composed of indifference curves. Each curve contains all combinations of attributes that provide the same level of satisfaction to the manager. The manager's objective is to reach the highest indifference curve possible without exceeding the efficiency frontier as shown in figure 7.7. The tangency of the efficiency frontier and indifference curve at m_5 signifies it is the most desirable combination of attributes. Hence, the best management action is m_5.

Selection of the best management action is more complex when considering the preferences of diverse stakeholders for the protected area. This is the case in countries where protected area managers are required to consider the preferences of stakeholders in selecting a management action. For example, the National Environmental Policy Act in the United States requires preparation of an environmental impact statement and solicitation of public comments on all actions expected to have significant environmental impacts. Selecting a best management action for two stakeholders (I and II) having different preferences for biodiversity and regional income is illustrated in figure 7.8. Stakeholder I favors regional income relatively more than biodiversity and stakeholder II favors biodiversity relatively more than regional income. For stakeholder II, m_5 is the most preferred management action, and for stakeholder I, m_6 is the most preferred management action. Hence, there is a trade-off or conflict between satisfying stakeholder I and stakeholder II. The situation becomes more complicated when there are many stakeholders.

Conflicts among stakeholders can be resolved several ways. For example, if it is possible to develop a management action that provides a combination of attributes on the efficiency frontier between m_5 and m_6, then that action should be implemented. If this is not possible, then the manager must choose m_5 or m_6. Typically, the choice of management action depends on whether m_5 and m_6 are closer to the manager's preferred management action, the political and economic power held by different stakeholders, and the success of competing stakeholders in courts of law. Conflicts are even more pronounced when stakeholders have not only different preferences for attributes, but also different threshold values for attributes. It is possible under these conditions for the best management action for one stakeholder to provide levels of one or more attributes that are below the threshold values for another stakeholder. This possibility is illustrated in figure 7.8, which shows that stakeholder II's threshold value for regional income is v'_2 instead of v_2. Selecting m_6 as the best management action still maximizes stakeholder I's satisfaction, but provides a level of income that is considered unsustainable by stakeholder II.

Regional Scale Conservation

Planning and management for protected areas occurs at several scales. The preceding sections dealt with ecosystem management at the scale of an individual

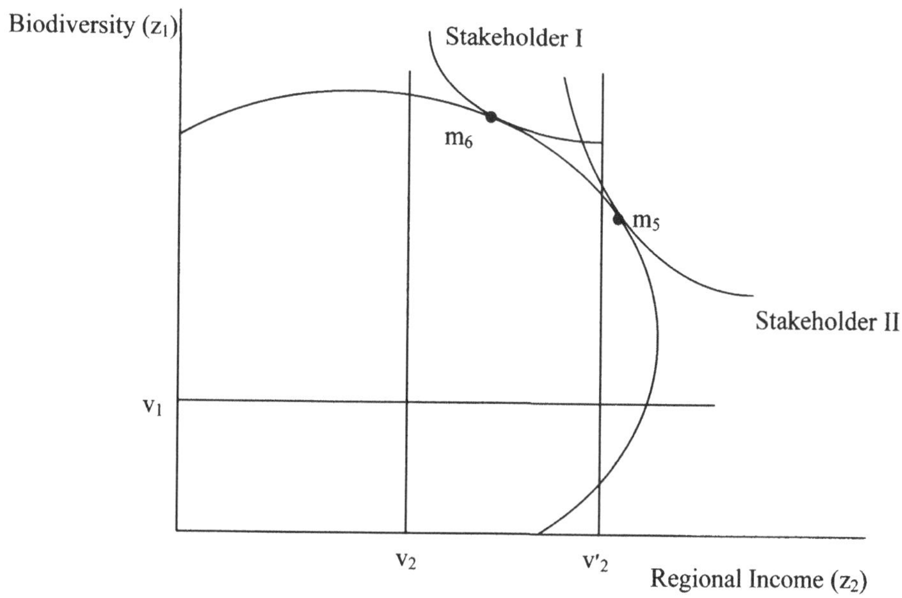

FIG. 7.8. Best management actions with two stakeholders.

protected area, which is the scale at which protected area managers are best able to manage for ecological integrity. The next scale up from the protected area is the region or greater ecosystem. Planning and management at the greater ecosystem scale is justified by new ecological ideas like island biogeography, landscape ecology, conservation biology, and empirical studies (Noss and Cooperrider 1994) that substantiate the complex and dynamic interactions occurring between protected areas and surrounding private and public lands. Other approaches argue for connectivity among parks and protected areas and surrounding lands and waters in order to reduce isolation, fragmentation, and other processes that impair biodiversity. Theberge and Theberge (2002) conclude that "because parks and reserves are not ecologically self-sufficient, regional environmental management is absolutely necessary." Hence, there is ample evidence that most protected areas are not large enough to provide sufficient habitat for a full array of megafauna, especially grizzly bear, elk, wolf, bison, and others. Regional scale planning and management of natural and human resources is especially important in greater ecosystems experiencing rapid growth, such as the Greater Yellowstone Ecosystem (GYE) (Keiter 1991; Franklin 1997).

The GYE is one of the last large, nearly intact ecosystems in the northern temperate zone of the Earth (Reese 1984; Keiter and Boyce 1991). The GYE includes 7.3 million ha (18 million acres) of public and private lands in Wyoming, Montana, and Idaho (see fig. 7.9). Significant public lands in the GYE include two national parks (Yellowstone and Grand Teton), portions of seven national forests (Bridger-Teton, Shoshone, Caribou, Targhee, Gallatin, Custer, and Beaverhead), three national wildlife refuges (National Elk Refuge, Red Rocks Lakes, and Gray Lake), lands managed by the Bureau of Land Management and states, and private lands. About 2.4 million ha (6 million acres) of the GYE is in national parks managed by the National Park Service and wilderness areas managed by the Forest Service. Nonwilderness national forest lands, which are managed for multiple uses like logging, oil and gas development, hard-rock mining, grazing, recreational activities, and wildlife/watershed protection, account for another 2.4 million ha (6 million acres). The remaining 2.4 million ha (6 million acres) are state and privately owned. Other significant ecosystems that are large and relatively intact include the Sierra Nevada Ecosystem, with national parks and national forests stretching along the crest of the mountains in California, and Cascadia, a mountainous region in southern British Columbia, Canada, and northern Washington, U.S.A.

The GYE is considered ecologically significant because it (Keiter 1991):

- Includes a park-wilderness complex that constitutes one of the largest wildland areas in the United States and is the core of a regional ecosystem characterized by rich wildlife diversity and natural features, such as geysers, mud pots, hot springs, deep canyons, and tall waterfalls
- Contains one of the world's few remaining free-roaming bison herds
- Provides critical habitat for rare and endangered species including grizzly bear, whooping crane, bald eagle, peregrine falcon, and trumpeter swan
- Includes the world's largest, diverse and intact collection of geysers and hot springs
- Contains the headwaters for three major river systems: Snake-Columbia, Green-Colorado, and Yellowstone-Missouri
- Contains vast forests of lodgepole pine and spruce.

The existence of such ecologically significant resources in a region experiencing rapid population and economic growth causes conflict between preservation of natural values and enhancement of economic values. In particular (Keiter and Boyce 1991):

- Intensive livestock grazing on public lands causes degradation of riparian areas, streambank erosion, and sedimentation of rivers.
- Since nearly 40 percent of occupied grizzly bear habitat on public lands is leased for livestock grazing, these areas have greater mortality for bears.
- Wildlife and cattle compete for scarce forage on critical winter rangelands.
- Thousands of bison that have left the park to access grazing areas in winter have been shot to reduce the risk of brucellosis in cattle.

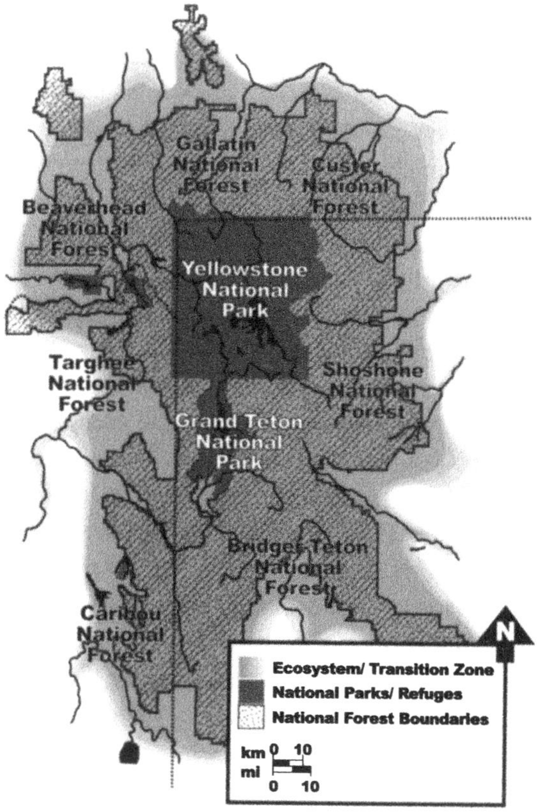

FIG. 7.9. Greater Yellowstone Ecosystem. (From Greater Yellowstone Coalition 2002.)

- Conflicts between recreational activities and the presence of livestock are growing in backcountry areas.
- Snowmobile use in and around Yellowstone National Park generates significant income and employment benefits for West Yellowstone and surrounding areas, but has negative effects on wildlife, air quality, and tranquillity.

Several nonprofit organizations and groups advocate for ecosystem management and resource protection, lobby for policies and management actions to reduce threats to ecological integrity, and conduct research and educational programs for the GYE. For example, the Greater Yellowstone Coalition has 10,000 members and one hundred conservation and sportsmen's organizations dedicated to protecting the GYE and its full range of life. The Greater Yellowstone Coordinating Committee is made up of representatives from the National Park Service and Forest Service that coordinate land management activities related to ecosystem management. Increasingly, more federal employees are dedicated to monitoring and interfacing with other entities beyond the boundaries of their park or forest to ensure that policies and activities are not in conflict and are mutually supportive. The Interagency Grizzly Bear Committee coordinates across numerous jurisdictions to ensure effective bear management at the greater ecosystem scale. The U.S. Geological Survey's Northern Rocky Mountain Science Center is dedicated to providing research that addresses complex natural resource management issues. The Big Sky Institute for Science and Natural History at Montana State University promotes the distribution of science-based knowledge related to natural ecosystems and the human communities that depend on them. The Yellowstone Center for Resources is an arm of Yellowstone National Park that works to understand and preserve Yellowstone's cultural and natural resources. The Sonoran Institute works collaboratively with local people and interests to conserve and restore important natural landscapes in western North America.

Legislation is another way to protect parks from external threats. Canada's National Parks Act was amended to require that "maintenance of ecological integrity through the protection of natural resources shall be the first priority when considering park zoning and visitor use in a management plan" (McNamee 2002). The focus on ecological integrity was critical because according to the Panel of Ecological Integrity of Canada's National Parks, "ecological integrity in Canada's national parks is under threat from many sources and for many reasons" (Parks Canada Agency 2000). Managing for ecological integrity "compels the government to act against threats to park resources that emanate from areas outside the parks...[and requires parks to] take action to define and eliminate the range of internal and external threats to park resources" (McNamee 2002).

Continental Scale Conservation

There is growing recognition that most protected areas and greater ecosystems are too small and disconnected from one another to provide sufficient habitat for wide-ranging large animals, particularly top carnivores (grizzly bears, wolves, and cougars). Achieving viable populations of these species "require[s] extensive, connected, relatively unaltered, heterogeneous habitat" (Soule and Noss 1998). The need for habitat connectivity and buffer zones that soften external threats to parks have been recognized as high priorities by park managers at the regional scale. However, the landscape fragmentation issue is perhaps most critical when considered beyond an individual park and its surrounding ecosystem. This perspective has led to the concept of continental scale conservation, or rewilding, which attempts to restore complete, self-regulating, and fully functioning ecosystem networks.

A primary objective of continental scale conservation is to provide more favorable habitat for keystone species, primarily top carnivores. Island biogeography provides the scientific foundation for this objective (Newmark 1987, 1995). Research indicates that the size of an area is positively correlated with biodiversity and negatively correlated with extinction rates; the smaller the area, the higher the extinction rates. In addition, species extirpation and ecosystem disintegration are more likely in small areas because of greater vulnerability to edge effects, and invasions of exotic plants, animals, and pathogens. Key arguments for rewilding are as follows (Soule and Noss 1998):

- Structure, resilience, and diversity (integrity) of ecosystems are often maintained by top-down interactions that are initiated by top predators (see also Terborgh et al. 1999).
- Wide-ranging predators usually require large cores of protected landscape for secure foraging, seasonal movement, and other needs, which justifies protected area largeness.
- Connectivity is required because core reserves are typically not large enough in most regions; hence, they must be linked to ensure long-term viability of wide-ranging species.

Attaining continental scale conservation requires designing and establishing reserves that have: large, strictly protected core reserves, buffer areas, and corridors connecting core areas. Core areas are "areas where conservation of biodiversity, ecological integrity, wilderness, or similar values take precedence over other values and uses" (Noss et al. 1999). They (1) provide places where nature can operate with little or no human disturbance, (2) allow humans to experience wild nature, (3) permit collection of baseline data for judging ecosystem health (Arcese and Sinclair 1997), (4) serve as control areas for ecosystem management (Agee and Johnson 1988; Grumbine 1994; Christiansen et al. 1996) and setting standards of biotic integrity (Karr 1991), (5) maintain populations of space-demanding and disturbance sensitive species, and (6) reduce the spread of invasive species. Since very few core areas are undisturbed by human activities, a primary role of core areas is to restore a natural process regime. This requires active management, such as reestablishing fire regimes and river flows; reintroducing top carnivores, large ungulates, and other extirpated species; and controlling invasive species (Noss et al. 1999). Restoring core areas could require the removal of dams and roads, elimination or cessation of fire suppression, and regulation, and possibly elimination, of hunting, grazing, logging, mining, and motorized sports.

Core areas are selected and designed to maintain or restore the natural qualities of an area. National parks, wilderness areas, public and private nature preserves, and certain state and provincial parks are examples of core areas. There are three primary design features for core areas, namely, (1) limiting access by humans, which typically means no roads or low road density; (2) maintaining natural conditions; and (3) incorporating entire environmental gradients and including viable examples of all habitat types (Noss and Cooperrider 1994).

A buffer area is an "area that permits a greater range of human uses than core reserves, but still is managed with native biodiversity as the preeminent concern" (Noss and Cooperrider 1994). Buffers are often multiple-use lands that function as core habitat for some species, such as plants, insects, amphibians, and animals with small and restricted ranges, and, if properly managed and connected to core areas, as supplemental habitat for other species (Groom et al. 1999). The effectiveness of buffers in providing habitat depends on the species in question, land uses within the buffer, and change in land use over time. Residential and commercial development, resource extraction (logging, grazing, and mining), agricultural production, and motorized recreation impair natural processes within buffers, such as nutrient cycling, reproduction, and fire. While any one of these activities may not cause significant degradation of buffers, the cumulative impacts of these activities can have profound effects over time (Knight et al. 1995). Buffers are a critical element in the protection of Biosphere Reserves (UNESCO 1974) as explained in chapter 2.

Landscape fragmentation around core areas causes them to become "islands of natural habitat in a sea of human development" (Janzen 1986, 1989). In particular, landscape fragmentation impedes the natural movement of animals, seeds, spores and pollen, and nutrient/energy flows, and in general contributes to local population and species extinction and decreases in biodiversity (Laurance and Bierregaard 1997; Heywood 1995; Dobson et al. 1997). One way to alleviate the island effect of landscape fragmentation is to establish natural corridors between core areas. Corridors are defined as "large, regional connections that are meant to facilitate animal movements and other essential flows between different sections of the landscape" and serve as habitat for animal feeding and breeding (Bennett 1990; Dobson et al. 1999). There are two primary benefits of corridors. First, they allow daily or seasonal movements of animals and increase access to the required resources and breeding success. Second, they facilitate the dispersal of animals from their birthplace to adult home range where breeding occurs (Dobson et al. 1999). The second benefit is usually more important than the first. Because species have different behaviors and habitat requirements, no one approach to improving connectivity of core areas is likely to yield equal benefits for all species and processes. The important point here is that it is not just the *amount* of regional habitat that is a determinant of animal or plant populations but the *access* to key resources that is a limiting factor. Otherwise high-quality habitat that is missing one critical resource will not support a viable population unless a habitat corridor can provide access to that resource in another core area. Thus, corridors can allow somewhat degraded landscapes to be nearly as functional as an intact landscapes.

Perhaps the most important benefit of connecting core areas is to expand the effective habitat and viability of top predators because they often are affected first during landscape fragmentation because they are wide-ranging. The latter is key to continental scale conservation because losses in top predators have several effects. This situation

1. often leads to biotic simplification and loss of habitat and species (e.g., extirpation of the wolf in Yellowstone National Park led to increases in elk populations in the Northern Range causing decreases in aspen regeneration and associated declines in beaver populations, beaver ponds, and riparian habitat);
2. causes major increases in smaller predators (mesopredators), such as house cats, opossums, and foxes (top predators prey on mesopredators), resulting in decreases in songbirds and other small prey animals;
3. increases competition among former prey species for food and space causing elimination of many species; and
4. poses a greater risk for population viability because the reproductive rate of top predators is much lower than for smaller mammals (carnivores produce fewer young and nurture them over a longer time period, which limits their capacity to recover from overhunting and extirpation campaigns) (Soule and Noss 1998).

To illustrate the significance of linking isolated protected areas for top predators, consider the case of grizzly bears in the Greater Yellowstone Ecosystem (GYE). Large mammals require a minimum population size of 2,000 to ensure sufficient genetic diversity to rebound from the challenges of weather, disease, and other threats to viability. Based on the large size of a grizzly bear's home range, up to 1,364 km^2 (900 mi^2), the GYE can probably support 500 bears (Schullery 1997). Hence, the Greater Yellowstone Ecosystem is too small to ensure population viability for grizzly bears. Continental scale conservation is likely to benefit grizzly bears to the extent that it connects protected areas and greater ecosystems.

The Wildlands Project is a primary proponent of continental scale conservation. The project is "dedicated to the restoration and protection of wilderness and biodiversity [and] is working to design and implement systems of protected areas—wildlands networks across the continent" (*Wild Earth*, various issues). The Wildlands Project's mission statement recognizes that national parks, wilderness areas, and wildlife refuges "are too small, too isolated, and represent too few types of ecosystems to perpetuate biodiversity of the continent" (Soule and Terborgh 1999). For this reason, the mission statement identifies the need for "vast landscapes without roads, dams, motorized vehicles...where evolutionary and ecological processes...can continue" (Foreman et al. 1992). The Wildlands Project is involved in several networks set up to foster continental scale conservation. Wildlands networks have been established for (1) Sky Islands, an archipelago of 40 mountain ranges rising from desert and grassland areas in southeast Arizona, southwest New Mexico, and northern Mexico; (2) Cebadillas, in

Mexico's northern Sierra Madre; (3) Yellowstone-to-Yukon (discussed below); and (4) Maine, which covers the 10-million-acre legendary Maine Woods (The Wildlands Project 2002).

Yellowstone-to-Yukon Conservation Initiative

The goals and challenges of continental scale conservation are examined for the Yellowstone-to-Yukon Ecoregion (Y2YE). The Y2YE extends from near Riverton, Wyoming, in the United States to the Peel River at the northern end of the Mackenzie Mountains in Canada's Yukon Territory (see fig. 7.10). Y2YE is 3,030 km (2,000 mi) long and up to 455 km (300 mi) wide, encompassing approximately 758,000 km^2 (500,000 mi^2). Sixty-two percent of the Y2YE's land area is in Canada and the balance is in the United States. The Y2YE consists of rolling grasslands in the eastern foothills, ice-covered alpine peaks, spectacular valleys of the Rocky Mountain Trench, and lush cedar/hemlock forests and salmon-filled rivers of the western watersheds. The Y2YE encompasses 10 major river systems, including the Columbia, Fraser, Mackenzie, and Missouri Rivers that drain into the Pacific, Arctic, and Atlantic Oceans. About 60 percent of the region is forested, 20 percent is bare rock and tundra, and less than 3 percent is agricultural land. The Y2YE contains 11 national parks and numerous wilderness areas, state and provincial parks, other nature reserves, and traditional territories of 31 native North American tribes (Chadwick 2000; Canadian Parks and Wilderness Society 2002). "With its unsurpassed collection of reserves and the potential for ensuring habitat linkages and movement corridors between them, Y2Y is fundamentally about the vitality of nature and our connection to it over the centuries to come" (Chadwick 2000).

The Y2Y Conservation Initiative is an international cooperative network of over 320 organizations and individuals dedicated to ensuring that the Y2YE's rich and diverse wilderness, wildlife, native plants, and natural processes continue to function as an interconnected web of life, capable of supporting all of the natural and human communities that reside within it, for present and future generations. The initiative recognizes that protected areas in the Y2YE are not large enough to support healthy populations of animals having large home ranges like wolves, grizzly bears, wolverine, cougar, elk, and caribou. The primary focus is on identifying ways to expand habitat for far-ranging wildlife and manage critical wildlife travel corridors in a manner that allows wildlife to navigate around and through human communities, highways, and industrial complexes, and the developments associated with forestry and oil and gas exploration. The primary goal of the initiative is to establish an interconnected wildlife/wildlands system of large core areas and corridors for terrestrial, aquatic, and avian species (Canadian Parks and Wilderness Society 2002). Bart Robinson, coordinator of the Y2Y Conservation Initiative, pointed out that "Y2Y represents the best chance anywhere on

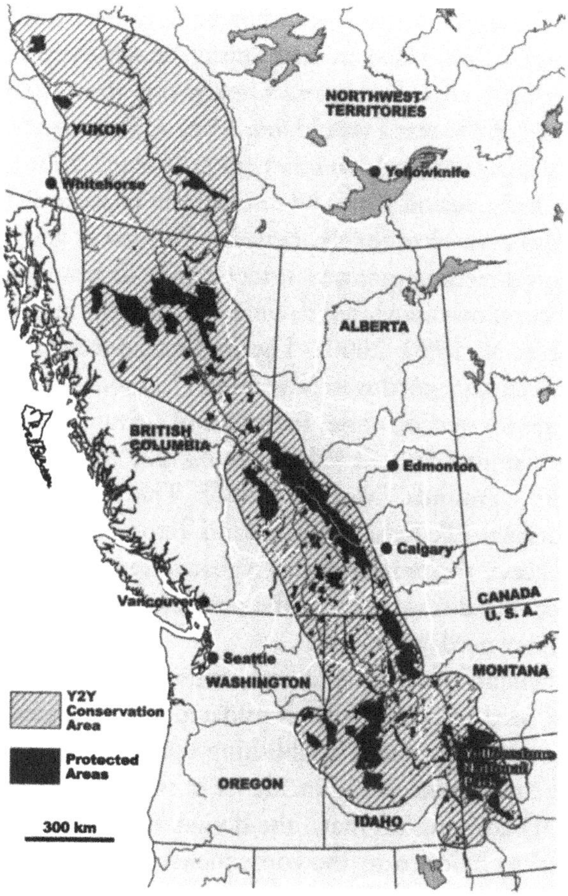

Fig. 7.10. Yellowstone-to-Yukon Ecoregion. (From Canadian Parks and Wilderness Society 2002.)

Earth to ensure the survival of a functioning mountain ecosystem" (Chadwick 2000).

Achieving the goal of the Y2Y Conservation Initiative is challenging because the Y2YE contains private ranch land (Bienen 2002). Conflicts have and will continue to occur between ranching and protected areas. Consider for example the controversy surrounding the killing of bison that leave Yellowstone National Park in the winter to seek grazing lands north of the park. In the severe winter of 1996–97, 1,100 bison were killed by the Montana Department of Livestock sparking one of the worst political controversies in the history of Yellowstone bison management (Schullery 1997). The agency justified the killings on the grounds that the livestock industry would incur large economic losses if brucellosis were transmitted from bison to cattle, even though there is no documented case of this occurring. Cattle that contract brucellosis abort their fetuses. Should this occur, the state of Montana would lose its brucellosis-free status and would not be allowed to ship beef cattle to other states (Bienen 2002). The state spent over $30 million in the period 1985–89 to eradicate brucellosis (Yannone 2003).

Brucellosis is also a problem for elk. Between 30 percent and 45 percent of the elk fed hay on winter-feeding grounds (under a government-sponsored program) test positive for brucellosis compared to only 1–2 percent for elk outside feeding grounds (Thorne et al. 1996, 2000). The highest concentration of winter elk feeding grounds occurs in northwest Wyoming (Jackson-Pinedale region) where 23 feedgrounds are located on state, federal, and private land. Each year, from three hundred to five hundred elk calves are lost due to brucellosis on the state and federal feeding grounds (Mockler 2002). The most famous elk feeding ground is the National Elk Refuge near Grand Teton National Park in Jackson, Wyoming. In this case, the protected area is inadvertently harboring and increasing the prevalence of a disease because the protected area is not large enough to accommodate winter needs of the elk.

Although the benefits of continental scale conservation are clear, there are negative aspects that need to be carefully considered and balanced against benefits. One concern about Y2YE is that establishing corridors to link protected areas would increase the transmission and incidence of wildlife diseases (Bienen 2002). Tabor et al. (2001) admonishes that, "the design and management of conservation reserves need to be open to the consequences of disease threats—not as a marginal issue of concern, which is the case at this moment in time, but as one more central in achieving long-run conservation success."

Ecological Economics

Ecological economics is a transboundary discipline that addresses the complex interrelationships between ecological and economic components of ecosystems (Costanza et al. 1991). It views the ecological component of ecosystems, namely, the Earth and all its resources, as being finite and nonexpanding. In contrast, neoclassical economics, which is at the core of mainstream economics, does not consider natural capital as a constraint or limitation on economic growth for two reasons. First, continued improvements in technology are expected to increase productivity in terms of output of goods and services per unit of natural resources, such as food output per unit of land. Second, reproducible (human knowledge and human-made) capital and natural capital are viewed as substitutes. The latter implies that as *nonrenewable resources*, such as oil and natural gas, become scarcer due to depletion, and or *renewable resources*, such as forests and fisheries, decline due to overharvesting (harvest rates in excess of regeneration), goods and services produced using natural capital do not decrease because reproducible capital is substituted for natural capital. For example, as agricultural and forest land become scarcer due to conversion to residential and commercial development, neoclassical economics says that land does not become a constraint on food and wood production because reproducible capital is substituted for land. As a result, the productivity of the remaining agricultural and forest land increases. In other words, substitution of reproducible capital for natural capital allows for greater production of agricultural and forest products. According to neoclassical economics, improvements in technology and substitution of reproducible capital for natural capital allow economic growth to continue indefinitely, unconstrained by natural capital.

Ecological economics has a different view of the relationship between reproducible capital and natural capital than neoclassical economics. The former maintains that technological innovation often increases rather than decreases the use of natural capital and/or results in environmental degradation. For example, adoption of more advanced timber and fish harvesting technologies requires more not less natural capital. Building and operating larger and more powerful timber harvesting equipment and fishing vessels requires more natural capital in the form of energy minerals, such as oil, natural gas, and coal, nonrenewable, nonenergy minerals, such as aluminum and iron ore, and water. Moreover, more advanced technologies often lead to unsustainable harvesting practices, such as deforestation and overexploitation of forests and fisheries, and environmental degradation in the form of global warming, habitat loss and fragmentation, and/or species extinction. Ecological economics maintains that reproducible capital and natural capital are complements, not substitutes. When reproducible capital and natural capital are complements, they cannot be substituted for one another. In other words, to increase the production of goods and services requires greater use of both forms of capital. The complementary relationship between reproducible capital and natural capital is compatible with the strong sustainability criterion mentioned earlier.

A major goal of ecological economics is to determine how best to maintain long-term sustainability of ecosystems. Achieving ecosystem sustainability requires finding the optimum scale of the economic system in terms of output of goods and services relative to the ecological system and implementing policies and institutions to maintain that scale. The notion of an optimum scale is important because if the scale of the economy is too large relative to the ecosystem, then ecological degradation occurs. The latter occurs in many parts of the developing world where population is so high relative to natural capital that there are inadequate supplies of consumer and ecological goods and services. As a result of this phenomenon, many developing countries have high rates of starvation and poverty, and soil, air, and water are polluted. Ecological economics is concerned with achieving and maintaining the optimum scale for the economy. The latter requires maintaining the capacity of natural capital to assimilate the residuals (undesirable substances) resulting from economic activities, ethical concerns for future generations (intergenerational equity), and the survival of nonhuman species (biodiversity).

When the optimal scale for the economy is exceeded, which occurs when economic growth is too high, economic activities have undesirable consequences, such as soil, air, and water pollution; global climate change; loss of biodiversity; and other forms of resource degradation. Achieving the optimal scale of the economy is not automatic. As Daly (1991) points out "there [is not] anything that allows the price system to determine the best scale of throughput according to ecological criteria of sustainability." The optimum scale for the economy is likely to be exceeded with the neoclassical economics paradigm whose objective is to maximize short-term economic output (gross domestic product). One risk of exceeding the optimum scale for the economy is degradation of the ecological integrity of protected areas. For example, air emissions from power plants upwind of Smoky Mountain National Park is causing significant impairment of visibility (see chapter 9 for a more detailed discussion of air quality in national parks).

The optimal scale for the economy has been defined in terms of the physical size of the total capital stock, which includes people (human capital), reproducible capital, and natural capital (Daly 1991). Total capital stock changes when inflow is different from outflow. Inflow includes things like population growth and accumulation of reproducible and natural capital. Outflow includes things like human death, depreciation of reproducible capital, and depletion of natural capital. If inflow exceeds outflow, then the scale increases. Conversely, if outflow exceeds inflow, then the scale decreases. When the scale is less than optimum, inflow can exceed outflow until the optimum is reached, which necessitates growth. On the other hand, when the scale exceeds the optimum, then outflow needs to exceed inflow until the optimum is reached, which necessitates negative growth. Therefore, maintaining the optimum scale for the economy requires balancing inflow and outflow.

While the above explanation indicates how to achieve and maintain an optimum scale, it says nothing about how to determine the optimum scale. In ecological economics, optimum scale is defined as a scale that keeps the throughput flow (production) of goods and services below ecological limits, and controls the degree of inequality in the distribution of goods and services among people (Daly 1991). The notion here is that ecological degradation occurs when production of goods and services requires that natural resources be used at rates that result in overexploitation of natural capital and emissions of pollutants that exceed the capacity of the ecosystem to assimilate them. Social injustice occurs when a disproportionately small percentage of the population consumes a disproportionately high percentage of the goods and services.

Achieving an optimum scale for the economy is relevant to protected areas. While true protected areas are insulated from certain forms of economic and resource development, their ecological integrity is nonetheless threatened by high levels of visitation, which is directly related to population and wealth, and resource extraction and residential development on adjacent lands, which is driven by economic growth. In other words, when the optimal scale for the economy is exceeded, there is a greater risk that the ecological integrity of protected areas will be compromised. Expanding the definition of ecosystems to include interactions between humans and the natural environment as is done in ecological economics is particularly relevant to protected areas because it focuses attention on how protected areas support economic values and how economic values support protected areas. In this regard, Ehrlich and Ehrlich (1981) identify several ecosystem benefits, namely:

> 1) maintaining atmospheric quality by regulating gas ratios and filtering dust and pollutants; 2) controlling and ameliorating climate through the carbon cycle and effects of vegetation in stimulating local and regional rainfall; 3) regulating freshwater supplies and controlling flooding; 4) generating and maintaining soils through the deposition of organic matter and the relationships between plant roots and mycorrhizal fungi; 5) disposing of residuals, including domestic sewage and by-products from industrial and agricultural production, and cycling of nutrients; 6) controlling pests and diseases; and 7) pollinating crops and useful wild plant species by insects, bats, hummingbirds, and other pollinators.

Protected areas provide many of these ecological benefits. On the other hand, economic development often makes it easier to preserve protected area ecosystems. In particular, increased wealth and the heightened awareness of environmental issues that often accompany economic development increase human and financial investment in environmental protection and restoration, including the creation of protected areas. For example, the United States has spent more than US$541 billion to reduce water pollution under the aegis of the 1972 Clean Water Act (Akobundu and Riggs 2000).

The ecological economics literature is replete with theoretical and empirical studies about the interrelationships between the economy and ecosystems at various spatial scales, including watersheds, large ecosystems, entire countries, continents, and the globe. In terms of temporal scales, the underlying goal of ecological economics is to maintain the long-term productivity of ecosystems (at whatever spatial scale) to ensure that natural capital and the ecological goods and services that they provide are available for future generations (intergenerational equity). Stated another way, the overall goal of ecological economics is to achieve sustainable resource use and management.

Summary

Traditional approaches to resource management have yielded to ecosystem management, an approach that attempts to manage people and natural resources at larger spatial scales, for longer time periods and with more variables than commodity-based resource management. Ecosystem management strives for sustainable productivity of an entire ecosystem, including its socioeconomic and ecological elements. An ecosystem's productive capacity can be evaluated based on a weak or strong sustainability criterion. Weak sustainability requires nondecreasing amounts of total capital, which consists of reproducible (human knowledge and human-made) capital and natural capital. Strong sustainability requires both nondecreasing amounts of reproducible capital and nondecreasing amounts of natural capital.

Newer approaches to evaluate losses in biodiversity emphasize its role in maintaining the functioning and resilience of ecosystems. Ecological functioning refers to the biophysical processes of ecosystems and ecological resilience is the capacity of an ecosystem to maintain its structure and patterns of behavior after disturbance. Ecological thresholds are useful in evaluating ecological resilience.

Ecosystems change over time and space in response to natural and human disturbances. Dynamic behavior of ecosystems can be explained in terms of an adaptive cycle that has phases for exploitation, conservation, creative destruction and reorganization, and a panarchy. The latter is a nested set of adaptive cycles having different space and time dimensions.

Natural regulation or ecological-process management is an approach to protected area management that attempts to sustain and restore natural processes, such as fire and trophic-level interactions between plants and herbivores, in an effort to enhance ecological integrity. It requires active management of ecosystems. Carrying capacity for protected areas is based on the acceptability of ecological and human impacts of visitation and is assessed using various methods, including Limits of Acceptable Change, Visitor Impact Management, Visitor Experience and Resource Protection, and Multiple Attribute Scoring Test for Capacity.

Regional scale conservation applies ecosystem management to areas larger than a single protected area, such as the Greater Yellowstone Ecosystem. This form of conservation is important because landscape fragmentation and poor management practices on adjacent private and public lands threaten the ecological integrity of protected areas. Most protected areas and greater ecosystems are deemed too small and disconnected from one another to provide sufficient habitat for wide-ranging, large keystone species, notably top carnivores (grizzly bears, wolves, and cougars). The goal of continental scale conservation or rewilding is to restore complete, self-regulating, and fully functioning ecosystems that support viable populations of top carnivores and other species. Key elements of continental scale conservation are core areas, such as national parks and wilderness areas, natural corridors that provide safe passage of species between core areas, and buffers that protect core areas from human disturbance. A prime example of continental scale conservation is the Yellowstone-to-Yukon Conservation Initiative. Rewilding is the primary goal of the Wildlands Project.

Ecological economics is a transboundary discipline that views the economy as a component of the world ecosystem, which is finite, technological improvements as increasing the use of natural capital, and reproducible capital and natural capital as complements. A primary goal of ecological economics is to determine how best to maintain the long-term sustainability of an ecosystem. Achieving this goal requires maintaining the optimum scale of the economy relative to the ecological system by implementing effective policies and institutions. The ecological integrity of protected areas depends on whether or not the optimal scale for the economy is achieved.

References

Agee, J. K., and D. Johnson (Eds.). 1988. *Ecosystem management for parks and wilderness*. Seattle, WA: University of Washington Press.

Akobundu, E., and D. W. Riggs. 2000. Pervasive permitting: The EPA's proposed TMDL rules. *Water Resources Impact* 2: 4–6.

Anselin, A., P. M. Meire, and L. Anselin. 1989. Multicriteria techniques in ecological evaluation: An example using the analytic hierarchy process. *Biological Conservation* 49:215–229.

Arcese, P., and A. R. E. Sinclair. 1997. The role of protected areas as ecological baselines. *Journal of Wildlife Management* 61: 587–602.

Backus, G. J., W. G. Stillwell, S. M. Latter, and M. C. Wallerstein. 1982. Decision making with applications for environmental management. *Environmental Management* 6: 493–504.

Barbier, E. B., J. C. Burgess, and C. Folke. 1995. *Paradise lost? The ecological economics of biodiversity*. London, England: Earthscan Publications, Ltd.

Becker, R., and A. Jubenville. 1982. Forest recreation management. *Forest Science* 28: 335–355.

Bennett, A. F. 1990. Habitat corridors: Their role in wildlife management and conservation. Department of Conservation and Management, Melbourne, Australia.

Bienen, L. 2002. Informed decisions: Conservation corridors and the spread of infectious diseases. *Conservation in Practice* 3: 10–17.

Bonneville Power Administration and the Montana Department of Fish, Wildlife, and Parks. 1985. *Project brief: Kokanee salmon in the Flathead system*. http://www.efw.bpa.gov/Environment/EW/EWP/DOCS/REPORTS/GENERAL/I510.pdf (accessed April 22, 2003).

Boyce, M. S. 1991. Natural regulation or the control of nature? In *The Greater Yellowstone Ecosystem: Redefining America's wilderness heritage*, ed. R. B. Keiter and M. S. Boyce, 183–208. New Haven, CT: Yale University Press.

Burch, W., Jr. 1984. Much ado about nothing—Some reflections on the wider and wilder implications of social carrying capacity. *Leisure Sciences* 6: 487–486.

Bureau of Outdoor Recreation. 1973. *Outdoor recreation: A legacy for America*. Washington, DC: U.S. Government Printing Office.

Canadian Parks and Wilderness Society. 2002. http://www.castlewilderness.ca/y2y.html (accessed May 2002).

Caughley, G., and J. H. Lawton. 1981. Plant-herbivore systems. In *Theoretical ecology: Principles and applications*, ed. M. S. Boyce and L. D. Hayden-Wing, 132–166. Oxford, England: Blackwell Science.

Chadwick, D. H. 2000. *Yellowstone to Yukon*. Washington, DC: National Geographic Society.

Chase, A. 1987. *Playing God in Yellowstone: The destruction of America's first national park*. Boston, MA: Atlantic Monthly Press.

Christiansen, N. L., A. M. Bartuska, J. H. Brown, S. Carpenter, C. D'Antonio, R. Francis, J. F. Franklin, J. A. MacMahon, R. F. Noss, D. J. Parsons, C. H. Peterson, M. H. Turner, and R. G. Woodmansee. 1996. The report of the Ecological Society of America Committee on the Scientific Basis for Ecosystem Management. *Ecological Applications* 6: 665–691.

Clark, R. N., and G. H. Stankey. 1979. The recreation opportunity spectrum: A framework for planning, management and research. General Technical Report PNW-98, Pacific Northwest Research Station, USDA Forest Service, Seattle, WA.

———. 1990. The recreation opportunity spectrum: A framework for planning, management and research. In *Toward serving visitors and managing our resources*, ed. R. Graham and R. Lawrence, 127–158. Tourism Recreation and Education Center, University of Waterloo, Ontario, and Canadian Parks Service, Environment Canada.

Costanza, R., et al. 1997. The value of the world's ecosystem services and natural capital. *Nature* 387: 253–260.

Costanza R., H. E. Daly, and J. A. Bartholomew. 1991. Goals, agenda, and policy recommendations for ecological economics. In *Ecological economics: The science and management of sustainability*, ed. R. Costanza. New York: Columbia University Press.

Costanza, R., L. Wainger, C. Folke, and K-G. Maler. 1993. Modeling complex ecological economic systems: Toward and evolutionary, dynamic understanding of people and nature. *BioScience* 43: 545–555.

Council for Agricultural Science and Technology (CAST). 1999. *Benefits of biodiversity*. Task Force Report No. 133, Ames, IA.

Craighead, J. J. 1991. Yellowstone in transition. In: *The Greater Yellowstone ecosystem: Redefining America's wilderness heritage*, ed. R. B. Keiter and M. S. Boyce, 27–39. New Haven, CT: Yale University Press,.

Daily, G. C. 1997. Introduction: what are ecosystem services? In *Nature's services: Societal dependence on natural ecosystems*, ed. G. C. Daily, 1–10. Washington, DC: Island Press.

Daly, H. 1991. *Steady-state economics* 2nd ed. Washington, DC: Island Press.

Despain, D., D. Houston, M. Meagher, and P. Schullery. 1986. *Wildlife in transition: Man and nature on Yellowstone's northern range*. Boulder, CO: Roberts Renehart.

Dhondt, A. A. 1988. Carrying capacity: A confusing concept. *Acta Oecol.* 9: 337–346.

Diaz, N. M., and S. Bell. 1997. Landscape analysis and design. In *Creating forestry for the 21st century: The science of ecosystem management*, ed. K. A. Kohm and J. F. Franklin, 255–270. Washington, DC: Island Press.

Digital Studios. 1997. Task force environmental investigations kit: Salmon vs. Hydropower, Aptos, CA.

Dobson, A., K. Ralls, M. Foster, M. E. Soule, D. Simberloff, D. Doak, J. A. Estes, L. S. Mills, D. Mattson, R. Dirzo, H. Arita, S. Ryan, E. A. Norse, R. F. Noss, and D. Jones. 1999. In *Continental scale conservation: Scientific foundations of regional reserve networks*, ed. M. E. Soule and J. Terborgh, 129–197. Washington, DC: Island Press.

Dobson, A. P., J. P. Rodriguez, W. M. Roberts, and D. S. Wilcove. 1997. Geographic distribution of endangered species in the United States. *Science* 275: 550–553.

Ehrlich, P. R., and A. H. Ehrlich. 1981. *Extinction: The causes and consequences of the disappearance of species*. New York: Random House.

Foreman, D., J. Davis, D. Johns, R. Noss, and M. E. Soule. 1992. The Wildlands Project mission statement. *Wild Earth* (special issue): 2–3.

Franklin, J. F. 1993. The fundamentals of ecosystem management with applications in the Pacific Northwest. In *Defining sustainable forestry,* ed. G. H. Aplet, N. Johnson, J. T. Olson, and V. A. Sample, 127–144. Washington, DC: Island Press.

———. 1997. Ecosystem management: An overview. In *Ecosystem management: Applications for sustainable forest and wildlife resources,* ed. M. S. Boyce and A. Haney, 21–53. New Haven, CT: Yale University Press.

Frissell, S., and G. Stankey. 1972. Wilderness environmental quality: Search for social and economic harmony. Proceedings of the Society of American Foresters Annual Conference. Society of American Foresters, Hot Springs, AR..

Fuhlendorf, S., F. Smeins, and W. Grant. 1996. Simulation of a fire-sensitive ecological threshold: A case study of Ashe juniper on the Edwards Plateau of Texas, USA. *Ecological Modelling* 90: 245–255.

Gehlbach, F. R. 1975. Investigation, evaluation, and priority ranking of natural areas. *Biological Conservation* 8: 79–88.

Gildart, R. C. 1986. *Montana's Flathead County.* No. 14, *Montana Magazine,* Inc. Helena, MT.

Graefe, A. R., F. R. Druss, and J. Vaske. 1990. Visitor impact management: The planning framework. Washington, DC: National Parks and Conservation Association.

Greater Yellowstone Coalition. 2002. http://www.greateryellowstone.org/home_gye_map.html (accessed May 24, 2002).

Groom, M., D. B. Jensen, F. L. Knight, S. Gatewood, L. Mills, D. Boyd-Heger, L. S. Mills, and M. E. Soule. 1999. Buffer zones: Benefits and dangers of compatible stewardship. In *Continental scale conservation: Scientific foundations of regional reserve networks,* ed. M. E. Soule and J. Terborgh, 171–197. Washington, DC: Island Press.

Grumbine, R. E. 1994. What is ecosystem management? *Conservation Biology* 8: 27–38.

Gunderson, L. H., C. S. Holling, and S. S. Light (Eds.). 1995. *Barriers and bridges to the renewal of ecosystems and institutions.* New York: Columbia University Press.

Haettenschwiler, P. 1994. *Decision support systems applied to Swiss federal security policy and food supply.* International Institute of Applied Systems Analysis Workshop, Laxenburg, Austria.

Haeuber, R., and J. Franklin. 1996. Forum: Perspectives on ecosystem management. *Ecological Applications* 6: 692–693.

Haimes, Y. Y., and W. A. Hall. 1974. Multiobjectives in water resource systems: The surrogate trade-off method. *Water Resources Research* 10: 615–624.

Hanley, N., I. Moffatt, R. Faichney, and M. Wilson. 1999. Measuring sustainability: A time series of alternative indicators for Scotland. *Ecological Economics* 28: 55–73.

Hawden, S., and L. J. Palmer. 1994. Reindeer in Alaska. *U.S. Department of Agriculture Bulletin* 1089: 1–70.

Heywood, V. H. (Ed.). 1995. *Global biodiversity assessment.* Cambridge, England: Cambridge University Press.

Hof, M., J. Hammett, M. Rees, J. Belnap, N. Poe, D. W. Lime, and R. E. Manning. 1994. Getting a handle on visitor carrying capacity: A pilot project at Arches National Park. *Park Science* 14: 11–13.

Holling, C. S. 1973. Resilience and stability of ecological systems. *Annual Review of Ecological Systems* 4: 1–23.

———. 1986. The resilience of terrestrial ecosystems: Local surprise and global change. In *Sustainable development of the biosphere,* ed. W. Clark and R. Mund. UK: Cambridge University Press.

Holling, C. S., and L. H. Gunderson. 2002. Resilience and adaptive cycles (chap. 2). In *Panarchy: Understanding transformations in human and natural systems,* 25–62. L. H. Gunderson and C. S. Holling. Washington, D.C.: Island Press.

Holling, C. S., L. H. Gunderson, and D. Ludwig. 2002a. In quest of a theory of Adaptive change (chap. 1). In *Panarchy: Understanding transformations in human and natural systems,* ed. L. H. Gunderson and C. S. Holling, 3–22. Washington, DC: Island Press.

Holling, C. S., L. H. Gunderson, and G. D. Peterson. 2002b. Sustainability and panarchies. In (Editors), *Panarchy: Understanding Transformations in Human and Natural Systems,* ed. L. H. Gunderson and C. S. Holling, 63–102. Washington, DC: Island Press.

Horn, H. S. 1968. Regulation of animal numbers: A model counter-example. *Ecology* 49: 776–778.

Houston, D. B. 1976. Research in ungulates in northern Yellowstone National Park. *Research in the parks: Transactions of the National Park Centennial Symposium,* National Park Service, Washington, DC, pp. 11–25.

———. 1982. *The northern Yellowstone elk: Ecology and management.* New York: Macmillan.

International Union for the Conservation of Nature and Natural Resources (IUCN). 1980. World Conservation Strategy, Gland, Switzerland.

IUCN/UNEP/WWF. 1991. *Caring for the Earth: A strategy for sustainable living.* Gland, Switzerland: World Conservation Union.

Janssen, R. 1992. *Multiobjective decision making for environmental management.* The Netherlands: Kluwer Academic Publishers.

Janzen, D. H. 1986. The eternal external threat. In *Conservation biology: The science of scarcity and diversity,* ed. M. E. Soule. Northampton, MA: Sinauer Associates, Inc.

——— 1989. The evolutionary biology of national parks. *Conservation Biology* 3: 109–112.

Johnson, K. N. 1997. Science-based assessments of the forests of the Pacific northwest. In *Creating forestry for the 21st century: The science of ecosystem management,* ed. K. A. Kohm and J. F. Franklin, 397–409. Washington, DC: Island Press.

Jubenville, A., and R. Becker. 1983. Outdoor recreation management planning: Contemporary schools of thought. *Recreation Planning and Management* 10: 303–319.

Kangas, J. 1994. An approach to public participation in strategic forest management planning. *Forest Ecology and Management* 70: 75–88.

Kangas, J., and J. Kuusipalo. 1993. Integrating biodiversity into forest management planning and decision-making. *Forest Ecology and Management* 61:1–15.

Karr, J. R. 1991. Biological integrity: A long-neglected aspect of water resource management. *Ecological Applications* 1: 66–84.

Keiter, R. B. 1991. An introduction to the ecosystem management debate. In *The Greater Yellowstone ecosystem: Redefining America's wilderness heritage,* ed. R. B. Keiter and M. S. Boyce, 3–18. New Haven, CT: Yale University Press.

Keiter, R. B., and M. S. Boyce (Eds.). 1991. *The Greater Yellowstone Ecosystem: Redefining America's wilderness heritage.* New Haven, CT: Yale University Press.

Knight, R. L., G. N. Wallace, and W. E. Riebsame. 1995. Ranching the view: Subdivisions versus agriculture. *Conservation Biology* 9: 459–461.

Knopf, R. C. 1990. The limits of acceptable change (LAC) planning process: Potentials and limitations. In *Toward serving visitors and managing our resources,* ed. R. Graham and R. Lawrence, 201–211. Tourism Recreation and Education Center, University of Waterloo, Ontario, and Canadian Parks Service, Environment Canada.

Laurance, W. F., and R. P. Bierregaard (Eds.). 1997. *Tropical forest remnants: Ecology, management, and conservation of fragmented communities.* Chicago, IL: University of Chicago Press.

Laven, D., R. Manning, D. Johnson, and M. Vande Kamp. 2001. Integrated subsistence use and users into park and wilderness management. *George Wright Forum* 18: 52–61.

Lee, K. N. 1993. *Compass and gyroscope: Integrating science and politics for the environment.* Washington, DC: Island Press.

Leopold, A. S., S. A. Cain, C. M. Cottam, I. N. Garbielsonm, and T. L Campbell. 1963. Wildlife management in the national parks. *Transactions of the North American Wildlife Resources Conference* 28: 29–44.

Leung, Y.-F., and J. L. Marion. 2000. Recreation impacts and management in wilderness: A state-of-knowledge review. In *Wilderness science in a time of change conference—Vol. 5: Wilderness ecosystems, threats, and management,* ed. D. N. Cole, S. F. McCool, W. T. Borrie, and J. O'Loughlin. RMRS-P-

15-VOL-5, U.S. Department of Agriculture-Forest Service, Intermountain Research Station, Ogden, UT.

Leung, Y.-F., N. Shaw, K. Johnson, and R. Duhaime. 2002. More than a database: Integrating GIS data with the Boston Harbor Islands visitor carrying capacity study. *George Wright Forum* 19: 69–78.

Likens, G. 1992. *The ecosystem approach: Its use and abuse.* Oldendorf/Luhe, Germany: Ecology Institute.

Ludwig, D., D. D. Jones, and C. S. Holling. 1978. Quantitative analysis of insect outbreak systems: The spruce budworm and forest. *Journal of Animal Ecology* 47: 315–322.

MacKenzie, S. H. 1996. *Integrated resource planning and management: The ecosystem approach in the Great Lakes Basin.* Washington, DC: Island Press.

Makowski, M., L. Somlyody, and D. Watkins. 1995. *Multiple criteria analysis for regional water quality management: The Nitra River case.* Working Paper WP-95-022. Laxenburg, Austria: International Institute of Applied Systems Analysis.

Malthus, T. R. 1798. *An essay on the principle of population* [reprinted in 1986]. London, England: Pickering.

Manidis Roberts Consultants. 1997. Developing a tourism optimization management model (TOMM), A model to monitor and manage tourism on Kangaroo Island, South Australia. Surry Hills, New South Wales.

Manning, R., N. Ballinger, J. Marion, and J. Roggenbuck. 1996b. Recreation management in natural areas: Problems and practices, status and trends. *Natural Areas Journal* 16:142–146.

Manning, R. E. 1998. To provide for significant enjoyment: recreation management in national parks. *The George Wright Forum* 15: 6–20.

———. 1999. *Studies in outdoor recreation: Search and research for satisfaction* 2nd ed. Corvallis: Oregon State University Press.

———. 2001. Visitor experience and resource protection: A framework for managing the carrying capacity of national parks. *Journal of Park and Recreation Administration* 19: 93–108.

Manning, R. E., D. W. Lime, and M. Hof. 1996a. Social carrying capacity of natural areas: Theory and application in the U.S. national parks. *Natural Areas Journal* 16: 118–127.

Marion, J., D. Cole, and D. Reynolds. 1985. Limits of acceptable change: A framework for assessing carrying capacity. *Park Science* 6: 9–11.

McCool, S., and D. Cole (Eds.). 1997. *Proceedings—Limits of acceptable change and related planning processes: Progress and future directions.* General Technical Report INT-371, U.S. Department of Agriculture-Forest Service.

McCool, S. F. 1990a. Limits of acceptable change: Evolution and future. In *Toward serving visitors and managing our resources,* eds. R. Graham and R. Lawrence, 185–193. Tourism Recreation and Education Center, University of Waterloo, Ontario, and Canadian Parks Service, Environment Canada.

———. 1990b. Limits of acceptable change: Some principles. In *Toward serving visitors and managing our resources,* ed. R. Graham and R. Lawrence, 195–200. Tourism Recreation and Education Center, University of Waterloo, Ontario, and Canadian Parks Service, Environment Canada.

McNamee, K. 2002. From wild places to endangered spaces: A history of Canada's national parks. In *Parks and protected areas in Canada: Planning and management,* D. Dearden and R. Rollins, 21–50. 2nd ed. Ontario, Canada: Oxford University Press.

Merigliano, L. 1990. *Indicators to monitor the wilderness recreation experience.* St. Paul, MN: University of Minnesota.

Merrill, E. H., and Boyce, M. S. 1991. Summer range and elk population dynamics in Yellowstone National Park. In *The Greater Yellowstone ecosystem: Redefining America's wilderness heritage,* ed. R. B. Keiter and M. S. Boyce, 263–273. New Haven, CT: Yale University Press.

Miller, G. T., Jr., 1990. *Resource conservation and management.* Belmont, CA: Wadsworth Publishing Co.

Mockler, K. 2002. Are Wyoming's feedgrounds a hotbed of disease? *High Country News* 34: 4–5, April 29.

Muradian, R. 2001. Ecological thresholds: A survey. *Ecological Economics* 38: 7–24.

National Park Service. 1997. *VERP: A summary of the visitor experience and resource protection (VERP) framework. A handbook for planners and managers.* Denver, CO: Denver Service Center, United States Department of the Interior,

National Research Council. 1992. *Restoration of aquatic ecosystems: Science, technology, and public policy.* Washington, DC: National Academy Press.

Newman, P., J. L. Marion, and K. Cahill. 2001. Integrating resource, social, and management indicators of quality into carrying capacity decision-making. *George Wright Forum* 18: 28–40.

Newmark, W. D. 1987. Mammalian extinctions in western North American parks: A landbridge perspective. *Nature* 325: 430–432.

——— 1995. Extinction of mammal populations in western North American national parks. *Conservation Biology* 9: 512–526.

Nilsen, P., and T. Grant. 1998. A comparative analysis of protected area planning and management frameworks. In *Proceedings—Limits of acceptable change and related planning processes: Progress and further direction,* S. F. McCool and D. N. Cole, 49–57. Ogden, UT: USDA Forest Service Rocky Mountain Research Station.

Northwest Power Planning Council. 1991. Columbia River Basin Fish and Wildlife Program. Northwest Power Planning Council, Portland, OR.

———. 2003. *Fish and Wildlife Program. Introduction.* http://www.nwcouncil.org/library/2000/2000-19/intro.htm (accessed April 2, 2003).

Noss, R. F., and A. Y. Cooperrider. 1994. *Saving nature's legacy: Protecting and restoring biodiversity.* Washington, DC: Island Press.

Noss, R. F., E. Dinerstein, B. Gilbert, M. Gilpin, B. J. Miller, J. Terborgh, and S. Trombulak. 1999. Core areas: Where nature reigns. In *Continental scale conservation: Scientific foundations of regional reserve networks,* M. E. Soule and J. Terborgh, 99–128. Washington, DC: Island Press.

Parks Canada Agency. 2000. *Unimpaired for future generations? Protecting ecological integrity with Canada's national parks. Vol. 2: Setting a new direction for Canada's national parks.* Report of the Panel on the Ecological Integrity of Canada's National Parks. Ottawa, Canada.

Pearce, D., E. Barbier, and A. Markandya. 1990. Economic appraisal and the natural environment. *Sustainable development: Economics and environment in the Third World.* London, England: Earthscan Publications, pp. 57–66.

Penttinen, M. 1994. *Forest owner's decision support systems—a management solution for nonindustrial private forest owners.* Laxenburg, Austria: International Institute of Applied Systems Analysis Workshop,.

Peterson, D. L. 1996. Research in parks and protected areas: Forging the link between science and management. In *National parks and protected areas: Their role in environmental protection,* ed. R. G. Wright. Cambridge, MA: Blackwell Science, Inc.

Pimm, S. L. 1984. The complexity and stability of ecosystems. *Nature* 307: 321–216.

Prato, T., 1998a. *Natural resource and environmental economics.* Ames: Iowa State University Press.

———. 1998b. Protecting soil and water resources through multi-objective decision making. In *Multiple objective decision making for land, water and environmental management,* ed. S. A. El-Swaify and D. S. Yakowitz, 385–394. Delray Beach, FL: St. Lucie Press.

———. 1999. Multiple attribute decision analysis for ecosystem management. *Ecological Economics* 30: 207–222.

———. 2001. Modeling carrying capacity for national parks. *Ecological Economics* 39: 321–331.

Prato, T., C. Fulcher, and Y. Zhou. 1996b. Integrated resource management using a decision support system. Southern African Wildlife Management Association Conference, Sustainable Use of Wildlife, University of Cape Town, South Africa, April 9–11.

Prato, T., C. Fulcher, S. Wu, and J. Ma. 1996a. Multiple-objective decision making for agroecosystem management. *Agricultural and Resource Economics Review* 25: 200–212.

Reese, R. 1984. Greater Yellowstone, the national park and adjacent wildlands. Helena, MT: Montana Geographic.

Rinaldi, S., and S. Muratori. 1992. Limit cycles in slow-fast forest-pest models. *Theoretical Population Biology* 41: 26–43.

Sargent, F. O., and J. H. Brande. 1976. Classifying and evaluating unique natural areas for planning purposes. *Journal of Soil and Water Conservation* May-June: 113–116.

Schaffer, W. M. 1988. Predicting order in the chaos of nature. In *Evolution of life histories of animals,* ed. M. S. Boyce, 313–350. New Haven, CT: Yale University Press..

Schmoldt, D. L., D. L. Peterson, and D. G. Silsbee. 1994. Developing inventory and monitoring programs based on multiple objectives. *Environmental Management* 18: 707–727.

Schowalter, T., E. Hansen, R. Molina, and Y. Zhang. 1997. Integrating the ecological roles of phytophagous insects, plant pathogens, and mycorrihzae in managed forests. In *Creating forestry for the 21st century: The science of ecosystem management,* ed. K. A. Kohm and J. F. Franklin, 171–189. Washington, DC: Island Press.

Schullery, P. 1997. *Searching for Yellowstone: Ecology and wonder in the last wilderness.* New York: Houghton Mifflin Co.

Schumacher, E. F. 1999. *Small is beautiful: Economics as if people mattered: 25 years later...with commentaries.* Vancouver, British Columbia: Hartley and Marks Publishers, Inc.

Seidl, J., and C. A. Tisdell. 1999. Carrying capacity reconsidered: From Malthus' population theory to cultural carrying capacity. *Ecological Economics* 31: 395–408.

Shelby, B., and T. A. Heberlein. 1986. *Carrying capacity in recreation settings.* Corvallis, OR: Oregon State University Press.

Smith, F. 1996. Biological diversity, ecosystem stability and economic development. *Ecological Economics* 16: 191–203.

Smith, P. G. R., and J. B. Theberge. 1986. A review of criteria for evaluating natural areas. *Environmental Management* 10: 715–734.

———. 1987. Evaluating natural areas using multiple criteria: Theory and practice. *Environmental Management* 11: 447–460.

Soule, M., and R. Noss. 1998. Rewilding and biodiversity: Complementary goals for continental conservation. *Wild Earth,* Fall, pp. 19–28.

Soule, M. E., and J. Terborgh. 1999. The policy and science of regional scale conservation. In *Continental scale conservation: Scientific foundations of regional reserve networks,* ed. M. E. Soule and J. Terborgh, 1–17. Washington, DC: Island Press.

Stanford, J. A., and B. K. Ellis. 2002. Natural and cultural influences on ecosystem processes in the Flathead River Basin (Montana and British Columbia). In *Rocky Mountain futures: An ecological perspective,* ed. J. S. Baron, 269–284. Washington, DC: Island Press.

Stankey, G. 1974. Criteria for the determination of recreational carrying capacity in the Colorado River Basin. *Environmental Management in the Colorado River Basin.* Logan, UT: Utah State University Press.

Stankey, G., D. Cole, R. Lucas, M. Peterson, S. Frissell, and R. Washburne. 1985. The limits of acceptable change (LAC) system for wilderness planning. General Technical Report INT-176, USDA Forest Service.

Stankey, G., S. McCool, and G. Stokes. 1984. Limits of acceptable change: A new framework for managing the Bob Marshall Wilderness Complex. *Western Wildlands* 10: 33–37.

Sumner, E. 1936. Special report on a wildlife study in the High Sierra in Sequoia and Yosemite National Parks and adjacent territory. U.S. National Park Service Records, National Archives, Washington, DC.

Tabor, G. M., et al. 2001. Conservation biology and the health sciences: defining the research priorities of conservation medicine. In *Research priorities in conservation biology,* M. E. Soule and G. H. Orians. 2nd ed. Washington, DC: Island Press.

Terborgh, J., J. A. Estes, P. Paquet. K. Ralls, D. Boyd-Heger, B. J. Miller, and R. F. Noss. 1999. The role of top carnivores in regulating terrestrial ecosystems. In *Continental scale conservation: Scientific foundations of regional reserve networks,* ed. M. E. Soule and J. Terborgh, 39–64. Covelo, CA: Island Press.

Theberge, J. C., and J. B. Theberge. 2002. Application of ecological concepts to the management of protected areas. In *Parks and Protected Areas in Canada: Planning and Management,* ed. D. Dearden and R. Rollins, 70–96. 2nd ed. Ontario, Canada: Oxford University Press.

Thomas, J. W. 1997. Foreword. In *Creating forestry for the 21st century: The science of ecosystem management,* ed. K. A. Kohm and J. F. Franklin, ix–xii. Washington, DC: Island Press.

Thorne, T., et al. 1996. Beating brucellosis. *Wyoming Wildlife* 60: 17–24.

———. 2000. United States Animal Health Association, 2000 Committee Reports. Report of the Committee on Brucellosis, Wyoming Game and Fish Department.

Tillman, D., and J. A. Downing. 1994. Biodiversity and stability in grasslands. *Nature* 367: 363–365.

U.S. Department of the Interior. 1994. Alternative transportation modes feasibility study volume III: Yellowstone National Park, Wyoming. National Park Service, Denver Service Center.

UNESCO. 1974. *Task Force on Criteria and Guidelines for the Choice and Establishment of Biosphere Reserves.* Man and the Biosphere report no. 22. Bonn, Germany: UNESCO.

Vande Camp, M., D. Johnson, and R. Manning. 2001. Application of visitor experience and resource protection (VERP) to Alaskan national park wilderness. Technical report NPS/CCSOUW/NRTR-2001-01. Cascadia Field Station, University of Washington, Seattle, WA.

Vaske, J. J. 1994. Social carrying capacity at the Columbia icefield: Applying the visitor impact management framework. Parks Canada, Department of Canadian Heritage, Ottawa, Ontario.

Wackernagel, M., N. B. Schulz, D. Deumling, A. C. Linares, M. Jenkins, V. Kapos, C. Monfreda, J. Loh|, N. Myers, R. Norgaard, and J. Randers. 2002. Tracking the ecological overshoot of the human economy. *Proceedings National Academy of Sciences* 99: 9266–9271.

Wagar, J. A. 1964. The carrying capacity of wild lands for recreation. *Forest Science Monograph 7,* Society of American Foresters, Washington, D.C.

Wagner, F. H. 1969. Ecosystem concepts in fish and game management. In *The ecosystem concept in natural resource management,* ed. G. M. Van Dyne et al. New York: Academic Press.

Walker, B., D. Ludwig, C. S. Holling, and R. M. Peterman. 1981. Stability of semi-arid savanna grazing systems. *Journal of Ecology* 69: 473–498.

Wallace, L. L. 1999. Evaluation and wildlife management. *The George Wright Forum* 16: 27–39.

Washburn, R. 1982. Wilderness recreational carrying capacity: Are numbers necessary? *Journal of Forestry* 80: 726–728.

Wild Earth. Various issues. Statement of purpose included in issues of *Wild Earth.*

Wildlands Project. 2002. Official website. http://www.twp.org/ (accessed July 17, 2002).

Williams, J. E., C. A. Wood, and M. P. Dombeck (Eds.). 1997. *Watershed restoration: Principles and practices.* Bethesda, MD: American Fisheries Society.

World Commission on Environment and Development. 1987. *Our common future.* New York: Oxford University Press.

World Resources Institute. 2000. What is an ecosystem approach? Fundamentals of an ecosystem approach. *World Resources 2000–2001—People and ecosystems: The fraying web of life.* Washington, DC. http://www.wri.org/wr2000/ecosys_approach.html (accessed April 2, 2003).

Xu, F., T. Prato, and J. C. Ma. 1995. A farm-level case study of sustainable agricultural production. *Journal of Soil and Water Conservation* 50: 39–44.

Yaffee, S. L., A. F. Phillips, I. C. Frenz, P. W. Hardy, S. M. Maleki, and B. E. Thorpe. 1996. *Ecosystem management in the United States: An assessment of current experience.* Washington, DC: Island Press.

Yannone, V. 2003. Feature creature: Bison. *Montana Magazine* January/February: 58–61.

CHAPTER

8

Approaches to Ecosystem Management

Ecosystem management is managing ecosystems so as to ensure their sustainability.

—Jerry F. Franklin

Several approaches are useful in implementing ecosystem management for protected areas. This chapter describes the following principles and methods: benefit-cost analysis, cost effectiveness analysis, economic impact analysis, landscape change and ecological impact analysis, multiple attribute decision making, adaptive management, optimization and simulation, integrated modeling, and spatial analysis. Important techniques in spatial analysis include geographic information systems (GIS), GIS-based environmental modeling, remote sensing, global positioning systems, interactive mapping, and computerized landscape visualization. Knowledge of these principles and methods is useful in evaluating a wide range of protected area planning and management issues.

A case study is used to facilitate understanding of the principles and methods described in this chapter. It centers on the hypothetical Greater Cimarron Ecosystem (GCE), which contains Cimarron National Park, a national forest, and several gateway communities. This example embodies attributes and issues from several real national parks but combines them here for ease of illustration. There is a symbiotic relationship between the park and outlying communities. Park visitors spend money on lodging, food, meals, gifts, entertainment, and many other services in gateway communities making those communities economically dependent on the park. Conversely, the park depends on the communities to provide adequate support facilities and services for park visitors. The GCE is experiencing rapid development because the environmental amenities provided by the park, and national forest in the ecosystem attract visitors and permanent and seasonal residents. The symbiotic relationship between the park and communities is being strained by landscape fragmentation caused by growth in gateway communities.

The park has been designated a recovery area for the threatened grizzly bear whose home range extends beyond the boundaries of the park. Landscape fragmentation not only threatens the recovery program for grizzly bears by reducing effective habitat, but also increases the number of encounters between humans and bears. Such encounters are potentially harmful to people and often lead to the killing of the offending bear. In fact, this is a major source of bear mortality and jeopardizes the bear recovery program. Ironically, while the presence of grizzly bears contributes to the wilderness character and attractiveness of the park as a place to visit and live near, it places the grizzly bear at greater risk of regional decline and eventual extirpation.

In an effort to resolve this dilemma, the planning staffs for the park and gateway communities, developers, and environmental groups in the GCE form the Cimarron Landscape Analysis Group (CLAG). The mission of CLAG is to evaluate the economic and ecological impacts of alternative futures specified by the group and to identify alternative futures that sustain economic activity without impairing grizzly bear recovery in the GCE. Since the grizzly bear is a keystone species, protecting its habitat would also protect many other species.

Benefit-Cost Analysis

CLAG can use benefit-cost analysis (BCA) to evaluate the economic efficiency of alternative futures. Four different although related efficiency-based criteria are used in BCA: net present value, annual net benefit, benefit-cost ratio, and internal rate of return (Prato 1998). Net present value of an alternative future is discounted total benefits minus discounted total costs:

$$PV = \sum_{t=0}^{T} (B_t - C_t)(1+r)^{-t}$$

B_t is nominal (undiscounted) benefit and C_t is nominal cost of an alternative future in year t. $(B_t - C_t)$ is net benefit, r is the discount rate, $(1+r)^{-t}$ is the discount factor, $(B_t - C_t)(1 + r)^{-t}$ is the present value of the net benefit occurring in year t. T + 1 is the number of years over which alternative futures are evaluated. Discounting causes the present value of a dollar of net benefit to decrease exponentially over time.

To illustrate the calculation of present value, suppose an alternative future is expected to provide the following benefits (B) and costs (C) over a five-year period:

	Year (t)				
	0	1	2	3	4
B_t	80	85	100	130	160
C_t	60	70	80	90	100

For an annual discount rate of 8 percent (1 + r = 1.08) and treating benefits and costs as though they occurred at the end of the year, present value is:

$$PV = (80 - 60)(1.08)^{-0} + (85 - 70)(1.08)^{-1} + (100 - 80)(1.08)^{-2} + (130 - 90)(1.08)^{-3} + (160 - 90)(1.08)^{-4} = \$134.26$$

This calculation is repeated for each alternative future. If an alternative future has a positive net present value, then it is economically efficient. Conversely, if an alternative future has a negative net present value, then it is not economically efficient. An efficient future increases and an inefficient future decreases economic welfare. The most efficient alternative future is the one with the highest positive net present value.

Annual net benefit (ANB) is the uniform annual payment over the evaluation period that has a present value just equal to the present value of the alternative future. ANB is the solution to the following equation:

$$\sum_{t=0}^{T} ANB\ (1+r)^{-t} = PV$$

namely:

$$ANB = PV\ r(1 + r)^{T}/[(1 + r)^{T} - 1]$$

For the benefits and costs given above, ANB equals \$40.60. An alternative future is economically efficient when ANB is positive and economically inefficient when ANB is zero or negative.

The benefit-cost ratio (BCR) is the ratio of the present value of benefits to the present value of costs, namely:

$$BCR = \frac{\sum_{t=0}^{T} B_t\ (1+r)^{-t}}{\sum_{t=0}^{T} C_t\ (1+r)^{-t}}$$

An alternative future is economically efficient when BCR is greater than or equal to one and economically inefficient when it is less than one. For the benefits and costs given above, BCR equals 1.4.

The internal rate of return (IRR) is the discount rate that makes the present value and ANB of an alternative future equal to zero and BCR equal to one. IRR is determined by solving the following equation:

$$\sum_{t=0}^{T} (B_t - C_t)\ (1+IRR)^{-t} = 0$$

Efficiency is evaluated by comparing the IRR to the interest rate on borrowed capital. An alternative future is economically efficient (inefficient) when IRR is greater than or equal to (less than) the market interest rate.

The four efficiency criteria (present value, annual net benefit, benefit-cost ratio, and internal rate of return) are generally consistent with one another. If one criterion indicates that an alternative future is economically efficient, then so do the others. Application of the four criteria requires all benefits and costs to be expressed in monetary terms. The benefits of future development are measured in terms of ecosystem-wide total economic output or income levels, which can be forecasted using an input-output model like the one discussed in the next section. Similarly, certain costs of future development, notably expenditures on roads, buildings, water, light, power, and other infrastructure, can be estimated in monetary terms because these goods and services have well-established markets. However, it is difficult to derive monetary values for the impacts of development on grizzly bear habitat because there are no markets or prices for grizzly bear habitat. In other words, BCA does not account for negative impacts of an alternative future on grizzly bear habitat and other ecological services provided by the GCE. Nonetheless, BCA is useful in evaluating the monetary benefits and costs of alternative futures for the GCE.

Benefit-cost analysis has other limitations. First, since it is an economic efficiency criterion, it does not consider the intragenerational and intergenerational fairness of the distribution of benefits and costs of development (Bishop 1993; Perrings 1994; Goulder and Kennedy 1997). Second, efficient alternative futures do not necessarily sustain the ecological services essential to humankind. Third, discounting cash flows for management actions designed to protect and/or restore ecosystem services that have long-term, uncertain benefits and high, short-term costs reduces the present value of those actions and could result in underinvestment in ecological protection or restoration. This deficiency arises, for example, in evaluating policies to reduce greenhouse gas emissions and their potentially adverse effects on global climate.

BCA has another deficiency that limits its usefulness in evaluating management actions that can cause an ecosystem to exceed some unknown ecological threshold. An example of this would be the decision of whether or not to expand the road system in a national park when the threshold beyond which increases in road density negatively impact large carnivores is unknown. In particular, exceeding the threshold for road density can result in extinction of certain large carnivores. Since the value of large carnivores is not determined in any market, there are no prices to signal when their availability is approaching a threshold (Bishop 1993; Folke 1999). This deficiency of BCA applies to all unpriced ecological goods and services. It implies that BCA does not account for the impacts of protected area management actions on unpriced ecological goods and services. This is significant because a majority of the ecological goods and services provided by protected areas is unpriced. Therefore, BCA is inappropriate for comparing protected area management actions that have irreversible consequences.

Cost-effectiveness Analysis

Cost-effectiveness analysis is a variant of BCA that compares the costs of alternative management actions for achieving a particular objective, such as conservation

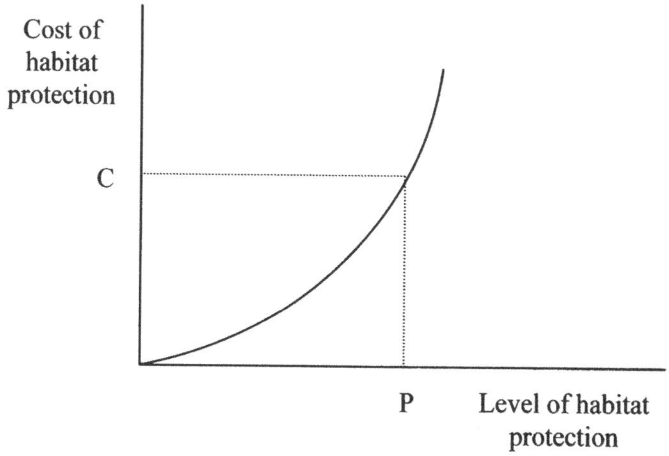

FIG. 8.1. Cost of increasing levels of habitat protection.

of grizzly bear habitat. It does not require placing monetary values on ecological services. The economic cost of management actions for improving grizzly bear habitat is given by the cost-effectiveness curve illustrated in figure 8.1. The curve is convex indicating that the incremental cost of habitat protection increases with the level of protection. If the protected area manager decides that the level of habitat protection needed to achieve the recovery goal is P, then C is the minimum cost of achieving P. The most cost-effective management action for achieving P is the one corresponding to C. Note that cost-effectiveness analysis does not allow determination of the most efficient level of habitat protection.

Economic Impact Analysis

Economic impact analysis is used to estimate the economic impacts of alternative futures. One way for CLAG to specify alternative futures is to apply various annual rates of growth to final expenditures on goods and services currently produced in all economic sectors in the GCE. Impacts of alternative futures on total output, household income, and employment can be estimated using the Impact Modeling for PLANning (IMPLAN) model. IMPLAN is a menu-driven computer software program developed by the U.S. Forest Service that permits non-survey regional input-output analysis of any county or combination of counties in the United States (see sidebar). IMPLAN predicts changes in total output, household income, and employment for final expenditure changes in up to 528 economic sectors (Lindall and Olson 1993). The basic geographic unit of analysis in IMPLAN is the county because intersectoral money flows in the model are calculated on a county basis.

IMPLAN Assessment for Big Muddy National Fish and Wildlife Refuge

Prior to settlement, the natural hydrograph for the Missouri River indicated much greater seasonal variation in flows, and the river coursed through a braided channel that varied from 1,285 m (1,500 feet) to 1.61 km (1 mile) in width and supported diverse riverine and floodplain habitat. Construction and operation of dams on the upper Missouri River and channelization of the lower river dramatically altered the hydrograph (figure A). These modifications dramatically reduced the size of the meander belt and eliminated vast areas of bottomland forests, sandbars, wetlands, and wet prairies in the lower river. Channelization reduced the length of the river by 107 km (72 mi) and diminished riverine shoreline habitat by 190 km (127 mi). Urban and agricultural developments in the floodplain displaced almost 143,320 ha (354,000 acres) of aquatic habitat that once existed in the river's meander belt. Levees constructed for flood protection of agricultural and urban areas disconnected riverine off-channel habitats and wetlands from the river. Biological consequences of these modifications include reductions in the populations of many fish and bird species and lower river biodiversity. Currently, the least tern and pallid sturgeon are endangered species, and the piping plover is a threatened species in the Missouri River.

The U.S. Fish and Wildlife Service created the Big Muddy National Fish and Wildlife Refuge in 1994 with the goal of restoring some of the biodiversity and providing more habitat for waterfowl and migratory birds. Specific objectives for the refuge are to (1) restore portions of the Missouri River floodplain, (2) improve and restore wetland habitat, (3) improve fishery and wildlife resources, and (4) provide public opportunities for outdoor recreation (hunting and fishing, wildlife observation, photography, and hiking) and environmental education. Initial size of the refuge was 4,211 ha (10,400 acres). In 1996, the refuge received authorization to expand to 24,292 ha (60,000 acres) in 25 to 30 units in the Missouri River floodplain from Kansas City to St. Louis, Missouri, and the lower 10 miles of major tributaries located in any of the 20 counties that lie along the Missouri River (figure B). Refuge land is floodplain cropland purchased from willing landowners. Following the extensive flooding in the lower Missouri River in 1993, many landowners expressed interest in selling their flooded cropland to the service. Money to expand the refuge comes from the Land and Water Conservation Fund. Operation of the refuge is done by a six-person staff with a FY 2002 budget of $556,000. The refuge has about 5,000 visitors per year.

Since refuge expansion constitutes a major federal action, the service was required to prepare an environmental impact statement (EIS) for the pro-

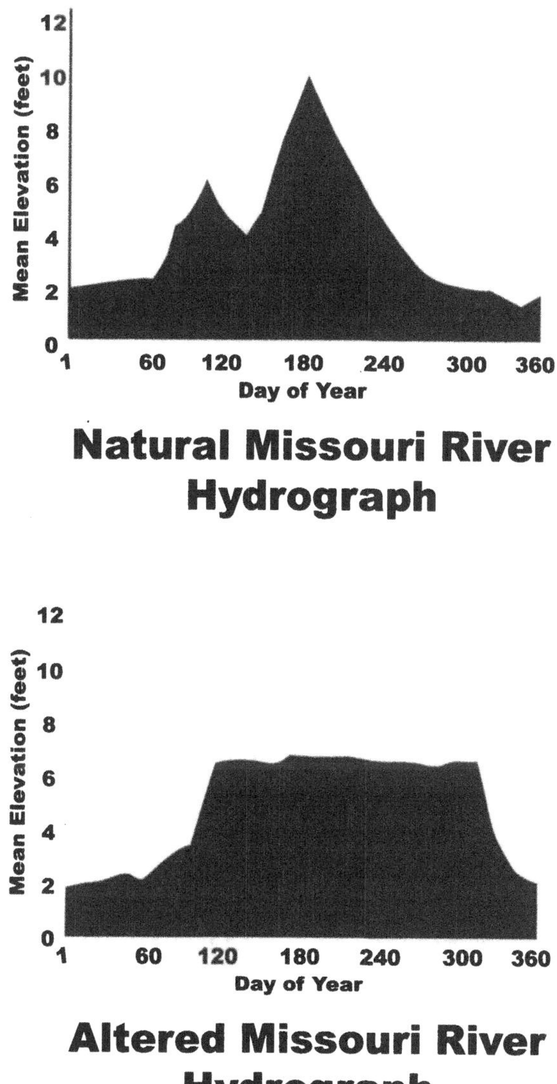

A. Hydrographs of Missouri River.

posed expansion. One of the chapters in the EIS used the IMPLAN input-output model to estimate the economic impacts of two alternatives: (1) acquiring 6,732 ha (16,628 acres) in seven units that had already been approved for purchase by the service, referred to as the no-action alternative; and (2) expanding the refuge by an additional 17,560 ha (43,372 acres) to 24,292 ha (60,000 acres), referred to as the preferred alternative. The location of the

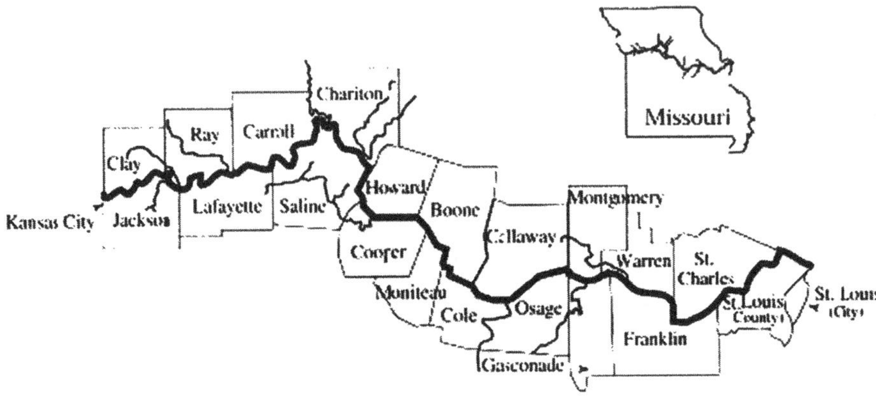

B. Missouri counties adjacent to Missouri River.

acreage for the no-action alternative is known. This is not the case for the additional 17,560 ha (43,372 acres) that would be purchased under the preferred alternative. Therefore, it was assumed that the additional acreage would be allocated to three geographic regions along the Missouri River between Kansas City and St. Louis, Missouri, based on three hypothetical acreage allocation options. Economic impacts of three categories of benefits and two categories of costs were estimated using IMPLAN. Benefit categories were: (1) public recreation, (2) refuge revenue sharing payments, and (3) income and employment from refuge construction and operation. Cost categories were (1) losses in agricultural production from converting cropland to refuge land and (2) loss in local property taxes. Private land purchased by the federal government is no longer subject to personal property tax, which results in a loss in tax revenue to the community. In an effort to compensate for such losses, federal revenue sharing payments are made to communities impacted in this way. IMPLAN estimates economic impacts in terms of changes in local, regional, and/or state economic output, household income, and employment.

Economic impacts of agricultural income losses associated with the preferred alternative were evaluated for three economic impact assessment scenarios, which specify:

1. Whether cropland purchased by the service is protected from floods (levees in place)
2. Whether cropland purchased by the service was enrolled in the regular Wetland Reserve Program (the WRP provides federal payments to private landowners who convert farmland to wetland) prior to purchase

3. The extent to which purchasing cropland for the refuge generates agricultural losses. Agricultural losses refer to the decline in the value of agricultural production that occurs when cropland is purchased for the Refuge.

Results of the IMPLAN analysis indicated that the bulk of the economic impacts of the no-action and preferred alternatives comes from the economic impacts from expected loss in agricultural production. Such impacts are substantially lower when acreage purchased for the refuge is not levee protected and enrolled in the Emergency WRP or regular WRP. Specifically, compared to the case where acreage purchased for the refuge is levee protected and not enrolled in the WRP, expected impacts of losses in agricultural production for the preferred alternative are (1) 38 percent lower when acquired acreage is protected by levees and is not enrolled in the WRP, and (2) 63 percent lower when acquired acreage is not protected by levees and is enrolled in the WRP.

The IMPLAN analysis showed that expected agricultural economic losses and land acquisition cost for the preferred alternative can be minimized by purchasing agricultural land that is enrolled in the WRP and not protected by levees. If land is acquired in levee-protected areas, then land acquisitions should be targeted to cropland enrolled in the Emergency WRP or regular WRP.

Sources: Zuerlein et al. (1995); Pfleiger and Grace (1997); Prato and Hamed (1999); U.S. Fish and Wildlife Service (2003a, b).

Total output includes direct, indirect, and induced economic impacts of expenditure changes in a primary sector, such as agriculture, recreation, and construction. Direct impacts include changes in final expenditures in the primary sector. Indirect impacts include the value of additional expenditures in related sectors as a result of expenditure changes in the primary sector. Induced impacts include changes in total household income or wages and salaries paid to employees in all sectors caused by expenditure changes in the primary sector. In addition, changes in total output alter the number of persons employed in the regional economy for the GCE.

Economic linkages in IMPLAN are as follows. Final expenditures on goods and services are the sum of personal consumption expenditure, investment expenditure, government expenditure, and net exports (exports minus imports).

Changes in final expenditures cause changes in total output, household income, and employment. Increases in final expenditures are serviced two ways. First, goods and services flow into a county from other regions. Money flow generated in this manner constitutes the export sector. Second, increases in final expenditures are serviced by production of goods and services within a county for local consumption by individuals, households, businesses, and government. The flow of money generated by county-level economic activities denotes the secondary sector (Summers and Field 2000). Growth in export and secondary sectors stimulates residential and commercial development, food and fiber production, and construction of government facilities and services, transportation networks, and community infrastructure (water, sewer, light, schools, hospitals, etc.), all of which increase the demand for land. Since the quantity and quality of land are relatively constant, increases in, say, residential demand for land require conversion of land from agriculture and forest to residential. The next section describes how to evaluate ecological impacts of changes in land cover.

IMPLAN calculates gross expenditures for an economic sector, such as accommodations (hotels/motels) or food, as follows:

$$x_{rut} = c'_{ru}f_{ut}$$

where x_{rut} is gross expenditures in sector r in county u in year t, c'_{ru} are input-output coefficients that measure the total direct and indirect outputs required in sector r to produce one dollar of final expenditures in all sectors in the county, and f_{ut} is the final expenditure for all sectors in county u in year t. Similar equations are used to explain household income (y_{rut}) and employment (e_{rut}). IMPLAN is used to estimate economic impacts for all counties in the GCE.

County-level total output, household income, and employment are determined as follows. First, CLAG specifies the final expenditures in all sectors of a county in future year t*, namely, f_{ut^*}. The latter could be determined, for example, by applying an annual rate of growth to current year's expenditures. Future year's expenditures are substituted in the above equation and similar equations for household income and employment to estimate total output (x_{ut^*}), total household income (y_{ut^*}), and total employment (e_{ut^*}) in the county in future year t*. In summary, IMPLAN is a widely used and accepted model for assessing the economic impacts of growth in a regional economy.

Landscape Change and Ecological Economic Impact Analysis

Changes in land cover and use are major drivers of global environmental change. During the past several decades, changes in land cover/use have reduced ecological integrity (Turner and Meyer 1994; Solecki 2001). On a national basis, more than 90 percent of the land in the lower 48 states has been logged, plowed, mined, overgrazed, paved, or otherwise modified from presettlement conditions (Terborgh and Soule 1999). Between 1982 and 1997, 121,000 km^2 (46,718 mi^2) of nonfederal lands were converted to urban developments (NRCS 1999). These conversions have contributed to a dramatic 1,000-fold increase in species extinction over the past 400 years (Stoltzenburg 1996). During the last three centuries, nearly 1.2 million km^2 (0.46 million mi^2) of forest and woodland and 5.6 million km^2 (2.17 million mi^2) of grassland and pasture have been converted to other uses; cropland has increased by 12 million km^2 (4.6 million mi^2) on a global basis (Ramankutty and Foley 1999).

Landscape change has altered grizzly bear habitat. There was a 400 percent increase in rural residential development in the Montana and Wyoming portions of the Greater Yellowstone Ecosystem between 1970 and 2000 (Williams 2001). As a result, current and potential grizzly bear habitat on private lands in the Greater Yellowstone Ecosystem have been degraded and fragmented. Continuation of this trend could impede grizzly bear recovery in the region (Johnson 2001).

Future economic growth in the GCE is likely to alter landscapes in ways that reduce the quantity and quality of grizzly bear habitat. Ecological economic impacts of landscape change in the GCE can be modeled as two processes, namely, effects of future economic growth on landscape change and impacts of landscape change on habitat suitability. A modeling approach for both processes is described below.

The first process models landscape change in two stages. In the first stage, changes in final expenditures on goods and services specified by an alternative future or economic growth scenario for the GCE are translated into future changes in land cover/use. Since the total amount of land in the ecosystem is fixed, economic growth requires conversion of land from undeveloped land uses, such as agricultural and forested lands, to developed land uses, such as residential housing and commercial businesses. In the second stage, future land use requirements associated with an alternative future are allocated across the landscape. Both stages are described below.

The first stage derives land-use requirements for an alternative future as follows. First, IMPLAN is used to estimate the changes in total economic output (sales) by sector and employment for an alternative future. Changes in sales and employment are translated into changes in land-use requirements. For example, increases in the demand for land for the retail sector of the GCE can be determined by multiplying each \$1,000 increase in final sales in the retail sector by the amount of land required per \$1,000 increase in retail sales. A similar procedure can be used to determine changes in land use by the nonretail sector. Changes in employment associated with changes in demands for all economic sectors influence the demand for land for residential housing. For example, if employment in the GCE increases by 1,000 people, then additional acreage must be converted to housing for those 1,000 people.

The second stage allocates changes in future land cover/use determined in the first stage to land parcels for which development is possible. For example, the additional land needed for housing and retail/nonretail establishments involves converting agricultural, forested, and other undeveloped land to housing, and retail/nonretail establishments. The conversion process can be modeled using transition probabilities between undeveloped and developed land estimated using historical land-cover data (Hall et al. 1988; Baker 1989; Luque 2000). Transition probabilities indicate the probability that land converts from undeveloped land uses to developed land uses and are estimated using historical changes in land cover/use. For example, suppose the first stage indicates that an alternative future for the CGE requires additional housing units that would occupy 197 hectares (500 acres). Let the transition probabilities for conversion of undeveloped land to housing be 0.5 for the forested/mining land (FM), 0.7 for the agricultural land (G), and 0.9 for the other undeveloped land. A random number between 1 and 2.1 is selected (2.1 = 0.5 + 0.7 + 0.9). The 197 hectares (500 acres) is taken from forested/mining land if the number is in the interval [0, 0.5], from the agricultural land if the number is in the interval [0.6, 1.2], and from the other undeveloped land if the number is in the interval [1.3, 2.1]. Alternatively, the 197 hectares (500 acres) can be split among agricultural, forested/mining, and other undeveloped land in proportion to their transition probabilities. This procedure is repeated for all conversions from undeveloped to developed land until all county-level land requirements for an alternative future are achieved.

Transition probabilities do not need to be the same throughout the GCE. Several approaches can be used to make transition probabilities vary across a landscape. They include using regression models to link transition probabilities to underlying drivers and landscape attributes (Theobold and Hobbs 1998). Examples of this method include logistic regression (Landis 1994), hedonic price models (Alig 1986; Geoghegan et al. 1997), and artificial neural networks (Pijanowski et al. 2000).

The next step in the second stage is to determine the specific locations for conversions of land from undeveloped to developed uses in the GCE. This step addresses, for example, the issue of which undeveloped land parcels should be converted to residential developments to achieve the additional 197 hectares (500 acres) needed for housing. Several schemes are suitable for this purpose. The first scheme determines which parcels of undeveloped land should be converted to developed uses based on cellular automata techniques. With these techniques, each parcel- or cell-sized area acts independently based on rules that take spatial properties of neighboring locations into account (Deadman et al. 1993; Clarke et al. 1997; Agarwal et al. 2000).

The second scheme models changes in land cover/use over time at a particular (x, y) location as follows:

$$f_{xyt+1} = P_{xyt}f_{xyt}$$

where f_{xyt+1} and f_{xyt} indicate the fractions of location (x, y) in particular land-cover classes at time t + 1 and t, respectively, and P_{xyt} indicates the local transition probabilities for conversion between land cover/use in location (x,y) in year t. Each spatially dependent transition probability is estimated by examining a neighborhood around a particular cell or parcel. Local transition probabilities can be calculated using a moving window such as a 5 × 5 block of cells (Brown et al. 2000). P_{xyt} is calculated for each window.

The third scheme determines the specific locations of land-use changes using multiple attribute evaluation methods. These methods involve scoring parcels of land available for development based on their suitability for different kinds of developments. Attributes and attribute weights are specified by the user (Smith and Therberge 1987; Prato 2000; Compas and Sugumaran 2004). For example, conversion of agricultural land to residential development in the CGE can be based on criteria such as proximity to roads and market centers, land slope, soil types, and proximity to ecologically sensitive areas such as rivers and wetlands. Multiple attribute evaluation is discussed in more detail in the next section.

The second process (impacts of landscape change on habitat quantity and quality) can be evaluated using concepts from landscape ecology. A common way to evaluate the impacts of changes in land cover/use on landscape structure is to apply FRAGSTATS to land-cover maps for alternative futures (O'Neill et al. 1988; McGarigal and Marks 1995; Hansen et al. 2001). Landscape metrics employed by FRAGSTATS include patch number, size, shape and perimeter, patch size coefficient of variation, isolation, connectivity, relative richness, relative evenness, relative patchiness, matrix porosity, diversity, dominance, fractal dimension, nearest neighbor probability, contagion, edges, and vegetative cover (Forman and Godron 1986; Turner et al. 2001).

Changes in landscape structure metrics are used to assess changes in habitat quality for particular species. Several relationships have been established between landscape structure metrics and habitat suitability. For example, (1) increases in the number of small patches influence the movement and persistence of organisms and the redistribution of matter and nutrients (Turner 1989); (2) matrix porosity, which measures density of patches, indicates the degree of species isolation and potential genetic variability present within populations of animals and plants; (3) patch size appears to be positively correlated with species and/or habitat diversity (Burgass and Sharpe 1981); (4) patch area is a good predictor of number of tree and bird species (Rudis and Ek 1981); (5) the amount of edge and interior spaces affects the abundance and diversity of organisms and wetland functions (Harris 1988; Quinn and Harrison 1988; Robinson and Quinn 1988); (6) a lower edge-to-interior ratio is likely to favor rare species (Geoghegan et al. 1997); and (7) the contagion index is negatively correlated with water quality (Hunsaker et al. 1992).

In summary, CLAG can evaluate the economic ecological impacts of alternative futures for the GCE by using IMPLAN to determine sector-by-sector changes in final demands for goods and services corresponding to those futures and translating those changes to land-use requirements for different sectors. The latter are then allocated to different undeveloped land categories and then to specific parcels of undeveloped land in the GCE. Landscape structure metrics estimated for an alternative future are then used to interpret changes in habitat suitability for grizzly bears.

Multiple Attribute Decision Making

CLAG can use multiple attribute decision making (MADM) to rank alternative futures based on their economic and ecological attributes and members' preferences for attributes (see sidebar for an application of MADM to design of a marine reserve). MADM has several advantages relative to benefit-cost analysis (BCA) (Joubert et al. 1997), particularly when development is likely to impair grizzly bear habitat. First, MADM facilitates public participation, allows comparisons of alternative futures based on member's preferences for attributes, and is well suited for collaborative decision making (Yaffee and Wondolleck 1997) and scientific assessments (Johnson 1997). These features are especially important in the case study because the members of CLAG have diverse interests. Second, MADM avoids some of the ethical, theoretical, and practical shortcomings of BCA. Third, MADM does not require ecological costs, such as degradation of grizzly bear habitat, to be expressed in monetary terms. Fourth, whereas BCA reduces the multiplicity of criteria and objectives underlying most decisions to a single monetary criterion (e.g., present value), MADM does not require attributes to be expressed in a common unit, such as dollars. This feature is advantageous particularly when determining whether an ecosystem is strongly sustainable. Fifth, MADM is easier to apply in developing countries because it does not require asking people their willingness to pay for the ecological and economic attributes of alternative futures. This line of questioning is not appropriate in developing countries with low incomes and high rates of illiteracy.

Overview of MADM

For simplicity, it is assumed that CLAG has excluded from further consideration alternative futures that are not technically and institutionally feasible. For example, an alternative future that attempts to confine bears to Cimarron National Park, as is done for big game species in private game reserves in Africa, is not technically feasible because grizzly bear habitat extends beyond the borders of the park. It is not feasible because fencing of national parks is not considered appropriate.

Multiple Attribute Evaluation for Asinara Island National Marine Reserve

The Istituto Centrale per la Ricerca Applicata al Mare (ICRAM) used multiple attribute evaluation (MAE) to develop a repeatable, rigorous approach for systematically "determining the suitability of various marine areas for different uses and levels of protection" in the Asinara Island National Marine Reserve. The reserve surrounds Asinara Island, a 50.9 km^2 (19.6 mi^2) island off the northwestern coast of Sardinia. Four levels of protection were evaluated: (1) an integral reserve consisting of either a no-entry, no-take zone that only allows authorized personnel to enter the reserve for monitoring, research, and maintenance; or an entry, no-take zone that allows recreational activities such as scuba diving and snorkeling in the reserve; (2) a general reserve that allows low-impact tourism like self-guided trails, swimming, and fishing with restrictions on number, size, and types of boats and fishing techniques; and (3) a partial reserve that serves as a buffer between areas outside the reserve and more restrictively protected zones containing administrative and educational facilities, recreational navigation, and certain types of sporting and commercial fishing.

ICRAM chose an MAE approach for the assessment because it provides a more objective and less arbitrary and unfair process for creating and evaluating zoning proposals for the reserve. In particular, spatialized MAE was used to create maps of concordance scores that indicate the agreement between measured attributes (or classification variables) for a particular area of the Asinara reserve and the four levels of protection indicated above. Concordance scores were determined by combining the measured attributes for the area with attribute weights that indicate the relative importance or priority of the attributes for each protection level. Attributes can be quantitative or semiquantitative.

ICRAM aggregated available data for Asinara Island into five higher-level attributes each of which was mapped: (1) natural value of the marine environment, (2) natural value of the coastal environment, (3) value of an area for recreational activities, (4) value of an area for commercial exploitation of resources, and (5) the degree of accessibility of an area. Attribute weights can be assigned using a consensus approach or estimated using a procedure such as the analytic hierarchy process. ICRAM assigned weights to the five higher-level attributes for each protection level by synthesizing information obtained from interviews with and feedback from stakeholders involved with or concerned about establishment and management of the reserve. Stakeholders were municipal representatives, local environmentalists, marine scientists, ornithologists, geologists, and economists.

Based on the results, ICRAM developed a preliminary zoning proposal (specific levels of protection for different areas of the reserve) that was presented to stakeholders. Feedback from stakeholders was used to develop a final zoning proposal for the reserve that indicated the size and location of areas receiving each of the four levels of protection. ICRAM concluded that, "the collaborative process, centered around easily understandable concordance maps, was a great asset in minimizing conflict and in keeping the group's attention focused on the core issues" and that spatialized MAE "holds promise as a user-friendly yet rigorous tool for articulating priorities and developing planning options...in reserve planning."

Sources: Villa et al. (1996, 2002).

Application of MADM involves six steps:

1. Selecting attributes of alternative futures,
2. Quantifying those attributes
3. Identifying alternative futures providing efficient combinations of attributes
4. Eliminating unsustainable alternative futures
5. Accounting for stakeholder preferences for attributes
6. Ranking alternative futures based on a multiple attribute utility function (Prato 1999)

While not strictly a part of MADM, another step is to resolve conflicts in members' rankings of alternative futures. Each of these steps is discussed below.

Selection of Attributes

The section on economic impact analysis explained how IMPLAN is used to estimate the impacts on total output, household income, and employment of future changes in final expenditures on goods and services. Similarly, the section on landscape change and ecological economic impact analysis explained how to assess the impacts of changes in final expenditures on land cover and the impacts of changes in land cover on grizzly bear habitat. To allow a graphical analysis of MADM, the simplifying assumption is made that CLAG evaluates alternative futures based on two attributes: total output (TO), which is an economic attribute, and habitat suitability (HS) for grizzly bear, which is an ecological attribute.

Quantifying Attributes

Suppose CLAG uses IMPLAN and habitat suitability impacts of landscape change described in the previous sections to derive point (single-valued) estimates

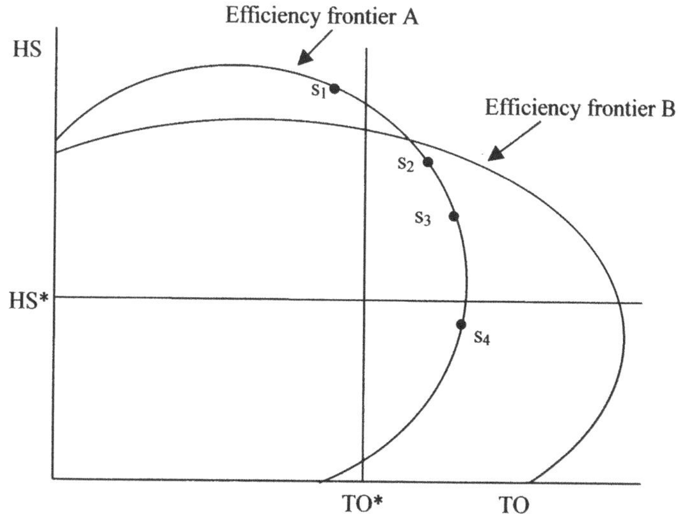

FIG. 8.2. Efficiency frontiers for habitat suitability (HS) and total output (TO).

of the combination of TO and HS provided by each alternative future. In the GCE case study, TO is measured in dollars (e.g., $200 million) and HS is an index that takes on values between 0 and 100 (e.g., 75). Alternatives to using point estimates are described in the section on handling uncertainty.

Efficient Combinations of Attributes

Not all combinations of TO and HS provided by alternative futures are efficient. Inefficient combinations and associated alternative futures are eliminated from further consideration because they provide less TO and/or HS than efficient combinations. The latter lie along the efficiency frontier labeled A in figure 8.2. This frontier contains four efficient combinations of TO and HS, namely s_1, s_2, s_3, and s_4. Note that the efficiency frontier has positively and negatively sloped segments. Combinations of TO and HS on the positively sloped segments are complementary because it is possible to have more of both TO and HS. Combinations on the negatively sloped segment are competitive because increases in TO occur at the expense of decreases in HS. Economically efficient combinations of TO and HS lie along the negatively sloped segment of the efficiency frontier.

Economic development typically involves increased human population and improved technology. For this reason, the shape and position of the efficiency frontier is likely to change over time, like the movement from frontier A to B in figure 8.2. This movement illustrates a case in which economic development favors TO more than HS. In particular, efficiency frontier B contains combinations of TO and HS that provide considerably more TO, but relatively few combinations with more HS.

Eliminating Unsustainable Alternative Futures

Requiring alternative futures to be technically and institutionally feasible and economically efficient does not mean they are sustainable. One way to ensure sustainable development is to eliminate from further consideration alternative futures that are not strongly sustainable as discussed in chapter 7. Strong sustainability places the following restrictions on TO and HS:

$$TO \geq TO^* \text{ and } HS \geq HS^*$$

TO* and HS* are the minimum acceptable levels of TO and HS, respectively, determined by CLAG. There are several ways CLAG can determine minimum acceptable levels of TO* and HS*, including consensus-based approaches in which members reach agreement on the levels, and majority-based decision making in which members vote on the minimum acceptable levels. An example of the consequences of imposing sustainability constraints is illustrated in figure 8.2. All alternative futures that provide levels of TO to the left of TO* and all levels of HS below HS* are not strongly sustainable and hence are ruled out. For the constraints illustrated in figure 8.2, the alternative future that provides combination s_1 is not economically sustainable because it provides a level of TO less than TO*. Similarly, the alternative future that provides combination s_4 is not ecologically sustainable because it provides a level of HS less than HS*. Therefore, the alternative futures that provide s_1 and s_4 are not sustainable. Only two alternative futures are efficient and sustainable, namely, those providing s_2 and s_3.

Accounting for Stakeholder Preferences

In order for CLAG to identify the most preferred alternative future, it must take account of members' preferences for HS and TO. If there is a small number of efficient and sustainable alternative futures as illustrated in figure 8.2, and members of the group have similar preferences for HS and TO, then it is relatively easy to reach agreement on the most preferred alternative future. However, if members of CLAG have dissimilar preferences, which is likely to be the case given the composition of CLAG, and a large number of efficient and sustainable alternative futures are being evaluated, then identifying the most preferred alternative future is more difficult. MADM is especially well suited to comparing a large number of efficient and sustainable alternative futures described by a large number of attributes. The remainder of this section explains the application of MADM under these conditions. Two issues are addressed. First, how does an individual member of CLAG rank alternative futures based on his or her preferences? Second, how might conflicts in rankings of alternative futures between group members be resolved? Graphical analysis becomes too complicated when there are multiple members and multiple attributes. Therefore, accounting for member preferences is explained mathematically.

Suppose CLAG decides to evaluate and rank alternative futures in terms of J economic attributes ($e_1,...,e_J$) and K ecological attributes ($g_1,...,g_K$). Each member of CLAG has unique preferences for these attributes that are represented by the following general utility function (see chapter 4 for a more detailed discussion of utility functions):

$$U_i(D_m) = U_k(e_{1m},...,e_{Jm}; g_{1m},...,g_{Km})$$

$U_i(D_m)$ is the level of satisfaction or utility that alternative future D_m provides to the ith member of the group, and $e_{1m},...,e_{Jm}$ are the amounts of economic attributes and $g_{1m},...,g_{Km}$ are the amounts of ecological attributes provided by D_m. For simplicity, all attributes are assumed to be positive, meaning that more of an attribute increases satisfaction or utility.

In order to rank the alternative futures, it is first necessary to specify a mathematical form for the utility function. While there is limited theoretical basis for selecting a particular mathematical form, the simple additive form is widely used due to its simplicity and relevance to real-world problems (Keeney and Raiffa 1976; Yakowitz et al. 1993; Foltz et al. 1995; Tecle et al. 1995; Prato and Hajkowicz 1999). The simple additive utility function for alternative futures is:

$$U_i(D_m) = \sum_{j=1}^{J} w_{ij}\, e^*_{jm} + \sum_{k=1}^{K} w_{ik}\, g^*_{km}$$

where e^*_{jm} is the value of the jth standardized economic attribute, and g^*_{km} is the value of the kth standardized ecological attribute for D_m. w_{ij} is the ith member's weight for the jth economic attribute and w_{ik} is the ith member's weight for the kth ecological attribute. Attribute values are standardized to avoid biases in the utility scores that could arise from differences in the units for measuring raw attributes. Since standardization converts each attribute to a [0,1] scale, it allows weighted summation of attributes expressed in different measurement units. Each weight is nonnegative ($w_{ij} \geq 0$ and $w_{ik} \geq 0$) and the weights sum to one, namely

$$\sum_{j=1}^{J} w_{ij} + \sum_{k=1}^{K} w_{ik} = 1$$

While the simple additive utility function is relatively easy to use, it is somewhat restrictive. It assumes that the person is risk neutral and attributes are utility independent. Risk neutrality implies that additional units of an attribute result in the same (constant) increment in utility. Utility independence of attributes means that the utility provided by an attribute depends only on the amount of that attribute, not on the amounts of other attributes.

Attribute weights for a member are estimated using fixed-point scoring, paired comparisons (Saaty 1987), or judgment analysis (Cooksey 1996). Fixed-point scoring requires a member to allocate 100 percentage points among the attributes. Attribute weights are set equal to the percentage points. Fixed-point scoring forces a member to consider trade-offs among attributes because it is not possible to assign a higher weight to one attribute without reducing the weight assigned to one or more of the remaining attributes. Paired comparisons are made using the Analytic Hierarchy Process, which derives quantitative weights for attributes by asking a member to evaluate the extent to which one attribute is more, less, or equally important relative to another attribute. In judgment analysis, a member is given the attribute values for all alternative futures and asked to score the alternative futures on a scale of 1 to 100. The weight for each attribute is estimated using a statistical method (multiple linear regression) that regresses the scores assigned to alternative futures on their corresponding attribute values. Other MADM methods that can be used to compare alternative futures include the surrogate worth trade-off method (Haimes and Hall 1974, 1977), free iterative search (Tecle et al. 1994), the Aspiration-Reservation Based Decision Support System (Makowski 1994; Fischer et al. 1996), and the Balancing and Ranking Method (Strassert and Prato 2002).

Ranking of Alternative Futures

The attribute values for the sustainable and efficient alternative futures and a member's attribute weights are substituted into the utility function described earlier to derive that member's utility scores for alternative futures. Scores are then used to rank alternative futures from highest to lowest. The most preferred alternative future is the one with the highest utility score. While attribute values for an alternative future are the same for all members of CLAG, attribute weights are likely to differ among members. This could result in different rankings for different members.

Differences in rankings of alternative futures do not necessarily imply conflicts among members. Consider two cases. In the first case, the members representing Cimarron National Park (A) and developers in gateway communities (B) rank alternative futures 1, 2, and 3 as follows:

A	B
2	1
3	3
1	2

Alternative future 2 is the most preferred by A and alternative future 1 is most preferred by B. There is a conflict. If A and B are equally powerful, then the conflict could be resolved by selecting alternative future 3. While alternative future 3 is not the first choice for either A or B, neither is it their last choice.

In the second case, park members (A) and environmental members (C) rank the three alternative futures as follows:

A	C
2	2
3	1
1	3

In this case, alternative future 2 is the first choice for A and C. There is no conflict because they can easily agree on alternative future 2. The difference in rankings of alternative futures 1 and 3 is of no consequence because both groups prefer alternative future 2 to alternative futures 1 and 3. Conflicts are likely to increase when there are more than two groups and three alternative futures.

Resolving Conflicts

When members' rankings result in conflict, CLAG needs to develop a conflict resolution procedure. Conflicts are likely to occur when members' rankings of alternative futures indicate that some members are willing to sacrifice ecological integrity, for example, impairment of grizzly bear habitat (lower HS), to achieve economic gains (higher TO) and others are not.

There are several ways to resolve conflicts in a MADM framework. First, consensus-based decision making could be used to reach agreement on a compromise set of weights for the economic and ecological attributes of alternative futures. Compromise is reached through discussions about the relative importance of weights. Utility scores for alternative futures are then calculated using the compromise set of weights. The preferred alternative future for CLAG is the one with the highest utility score. Second, consensus-based decision making can be applied to the alternative futures. In this approach, each member comes to the table with his or her own ranking of alternative futures and a consensus ranking of alternative futures is developed using a nominal group technique that involves facilitated responses, voting, and discussion (Meffe et al. 2002).

There is no guarantee that a consensus-based approach will resolve conflicts between members of CLAG. A consensus-based approach is more likely to succeed in smaller groups and more likely to fail in larger groups whose members have diverse preferences for attributes. While not based on MADM, a Citizen's Advisory Committee used a consensus approach to reach agreement on its preferred alternative for reconstruction of Going-to-the-Sun Road in Glacier National Park (National Park Service 2002).

Third, CLAG can vote on attribute weights or the alternative futures themselves. If voting is used, then the weights receiving the most votes are used to calculate utility scores and rank alternative futures. The most preferred alternative future is the one with the highest rank. A problem with voting on weights is determining which weights to include on the ballot. This problem could be resolved using a nomination process. Alternatively, members could simply vote for alternative futures. The alternative future receiving a majority of the votes is declared the most preferred. As with voting for weights, there is a problem of determining which alternative futures to place on the ballot.

Who gets to vote is another issue. If each member of CLAG has one vote and there is a disproportionate number of members from one interest group, then ballot results will favor that interest group. Voting has lower transaction cost than consensus approaches especially when there are many stakeholders and numerous alternative futures.

Fourth, CLAG could use the Analytical Hierarchy Process "to develop compromises between competing interests by pointing out areas of agreement, helping to isolate the areas of conflict, and illustrating the trade-offs between different options" (Kangas 1994).

Implications and Limitations

MADM has two general implications for managing protected areas. First, some alternative futures and management actions are likely to be inefficient in terms of providing combinations of attributes below the efficiency frontier. MADM provides a framework for identifying inefficient alternative futures, which should be eliminated from further consideration. Second, CLAG and stakeholders in general are likely to have different preferences for the economic and ecological attributes of alternative futures, which could result in different rankings of alternative futures. Different rankings can lead to conflicts regarding the preferred alternative future. When conflicts arise, it is necessary to devise a procedure for resolving them.

Like more conventional evaluation methods, MADM has certain limitations that are worth noting. First, MADM is static. In most applications of MADM, the ranking of alternatives is based on the current efficiency frontier, sustainability constraints, and preferences for attributes (utility function). Over time, the efficiency frontier changes in response to technological advancements and shifts in a region's or country's resource endowments. In addition as living standards increase through time, people are likely to become more concerned about the adverse ecological consequences of human activities and technology. Dynamics of this sort are likely to cause the efficient and sustainable alternative futures and an individual's ranking of those futures to vary over time. For these reasons, a MADM evaluation needs to be periodically updated.

Second, due to the complexity of economic and ecological systems, planning groups, such as CLAG, are not certain which combination of economic and ecological attributes is likely to result from a future alternative future. In other words, there is not a simple one-to-one correspondence between alternative futures and combinations of attributes as assumed in figure 8.2. An alternative future can result in more than one combination of attributes, which complicates the task of ranking alternative futures. Decision making under uncertainty is discussed in the next section.

Third, determining an individual's preferences for attributes, or attribute weights, is not a simple matter. Several kinds of biases can occur when individuals are asked questions about their preferences for attributes. Strategic bias occurs when an individual intentionally overstates or understates the weight for an attribute in order to achieve a more desirable outcome. For example, an individual who favors environmental protection may overstate the weight for ecological attributes relative to economic attributes in hopes of increasing the ranks of alternative futures that achieve greater environmental protection. Conversely, a development-oriented individual may overstate the weights for economic attributes relative to ecological attributes in hopes of stimulating development.

Adaptive Management (AM)

A common assumption made in evaluating management actions is that a protected area manager can determine how a protected area ecosystem responds to management actions before those actions are implemented. In other words, it assumes that the outcomes of different management actions are known. For several reasons, protected area managers do not know the impacts of management actions with certainty. First, variability in environmental, demographic, and other factors makes the response of ecosystems to management actions intrinsically uncertain. Second, sampling and measurement errors make it difficult to precisely measure the response of ecosystems to management actions. Third, even after accounting for the above sources of uncertainty, incomplete understanding of ecosystems prevents accurate prediction of their responses to management actions (Conroy 2000).

Many socioeconomic and environmental factors, some of which cannot be controlled by managers and are uncertain, influence the outcomes of management actions. Consider a manager faced with the decision of whether to reduce the number of ungulates (herd size) allowed to graze in a rangeland area of a national park. Overgrazing by ungulates and associated declines in the range productivity are more likely to occur when rainfall is low and the herd size is high. The manager can control herd size, but not rainfall. Hence, the effects of different herd sizes on range productivity are uncertain.

An AM approach can be used to handle the large uncertainties regarding ecosystem responses to management actions for protected areas (Woodley 2002). The concept of AM was developed in the mid-1970s at the International Institute for Applied Systems Analysis in Vienna, Austria, as a way to manage natural resource systems under uncertainty (Holling 1978; Irwin and Wigley 1993; Parma et al. 1998; Walters 1996; Walters and Holling 1990). The basic premise of AM is that "if human understanding of nature is imperfect, then human interactions with nature [e.g., management actions] should be experimental" (Lee 1993). Kohm and Franklin (1997) point out that "adaptive management is the only logical approach under the circumstances of uncertainty and the continued accumulation of knowledge." AM involves major investments in research, monitoring, and modeling to test alternative hypotheses about sustainable use and management of natural resources (Hilborn et al. 1995; Smith and Walters 1981; Walters and Green 1997). An AM approach is well suited for implementing ecosystem management. In adaptive ecosystem management, public land managers experimentally test management actions that maximize their capacity to learn about ecosystem responses to those actions while simultaneously attempting to satisfy the social and ecological objectives established for the ecosystem.

There are two forms of AM: passive and active. Passive AM formulates predictive models of ecosystem response to management actions, makes management decisions based on those models, and revises the parameters of the models based on monitoring data (Walters and Hilborn 1978; Hilborn 1992). Washington State's Timber, Fish, and Wildlife Agreement was implemented using a passive AM approach that involved "developing predictive tools that can be used for site-specific management" (Halbert 1993). Passive AM is relatively simple and inexpensive because monitoring is done without controls. Unfortunately, if replication and randomization, both of which are essential for statistically valid experimental design (Hulbert 1984), are lacking in passive AM, then it does not yield knowledge of cause-and-effect relationships between management actions (cause) and ecosystem responses (effect) (Wilhere 2002).

Active AM uses experiments to generate data for testing hypotheses about ecological and socioeconomic responses to alternative management actions. Each management action is viewed as a treatment in a statistically valid experimental design. Active AM consists of applying "experimentation to the design and implementation of natural-resource and environmental management policies" (Halbert 1993). Since the experiments incorporate replication and randomization of management actions (treatments), managers are able to test cause-and-effect relationships between management actions (cause) and associated ecological and socioeconomic responses (effect). Results of hypothesis testing reveal whether a particular management action achieves its intended objective, for example, movement of the ecosystem to a sustainable state. AM has evolved away from the passive mode toward the active mode (Wilhere 2002). Compared to the trial-and-error approach, which provides slow, random accumulation of information, AM provides reliable knowledge (Lee 1993).

AM utilizes concepts from adaptive control process theory, which attempts to design devices for making decisions that are capable of learning from experience, operations research, and management science (McClain and Lee 1996). The latter makes extensive use of science-based mathematical models to make decisions in settings characterized by complexity, changing conditions, and uncertainty regarding the relationships among key elements of a system (Thierauf and Klekamp 1975; Keen and Morton 1978; Haley 1990). Whereas operations research and management science assist decision makers in situations where it is too costly to conduct scientific experiments, active AM uses experimentation and mathematical modeling as a means of learning about the impacts of alternative management actions (Prato 2000). For example, the AM approach used to assess management policies for improving the salmon fishery in British Columbia employed models to "quantify and incorporate objectives from various interest groups...[develop] utility functions to help decision makers make trade-offs among competing objectives [and] identify misguided policies" (McClain and Lee 1996).

Several conditions favor the application of active AM (Lee 1993; Wilhere 2002). Following each condition (given in italic) is the reason for the condition and some obstacles to achieving it:

1. *There must be a mandate to take action in the face of uncertainty.* Protected area managers need to realize that the outcomes of most management actions are uncertain. Unfortunately, experimentation and learning in protected area ecosystems are at best secondary objectives that are likely to be dismissed or not even proposed if they conflict with primary objectives, such as recreation and resource protection.
2. *Preservation of pristine environments must no longer be an option.* While many protected areas have remarkable natural and cultural resources, very few are totally pristine. Even protected areas that are relatively pristine are subject to human disturbance both within and outside their boundaries and, hence, must be treated as managed ecosystems that are appropriate for AM.
3. *Human intervention must not be capable of producing desired outcomes predictably.* AM is not needed when ecosystem states and management outcomes are predictable. Unpredictability occurs because management actions have uncertain outcomes.
4. *There must be sufficient institutional stability to measure long-term outcomes.* AM experiments need to be long term in order to capture the relatively slow responses of ecosystems to human interventions. The institutional environment for many protected areas is stable because the management agency (e.g., National Park Service in the United States) is charged with managing the park in accordance with objectives established in an organic act. However, there are other sources of institutional instability, including turnover in upper management, declining budgets, and, in some countries, political instability and civil war. In any event, institutional stability is typically outside the control of experimenters and managers.
5. *It must be possible to formulate hypotheses.* Based on past experience and scientific knowledge, protected area managers should be able to design experiments to test hypotheses about ecosystem states and responses to management ac-

tions. However, it is generally not technically, financially, or politically feasible to test all relevant hypotheses.

6. *Theory, models, and field methods must be available to estimate and infer ecosystem-scale behavior.* While scientific knowledge about ecological and socioeconomic relationships for protected area ecosystems is incomplete, there is usually sufficient understanding of the ecosystem to design monitoring, research, and evaluation programs and sampling schemes to collect the data needed to test hypotheses regarding ecosystem states and responses to management actions. AM improves such understanding because it promotes learning. Preliminary results might indicate that the experimental design was faulty or certain management actions are not likely to achieve desirable ecosystem states or outcomes. The latter increases the pressure to change management actions before experiments are completed, which is disruptive.
7. *Decision makers need to view management actions as experiments and uncertainty about outcomes as potential hazards.* The idea of experimenting with protected area ecosystems, particularly highly visible ones, is not likely to be well received because it admits the possibility that management actions can fail to achieve desired social, economic, and ecological objectives. Protected area managers need to communicate these risks to the public and stakeholders and explain that experiments that do not achieve desirable outcomes are valuable because they improve knowledge of the ecosystem. Experiments and monitoring programs can be designed to minimize risk.

   AM for highly sensitive issues, like recovery of declining species, can be designed to minimize the risk that management actions do not recover the species. For example, risky experiments can be done in an area of a protected area ecosystem where human and environmental factors are easier to control. Minimizing the risk of adverse consequences needs to be balanced with maximizing the information value of experiments. Politicians seeking opportunities to increase their chances of reelection are likely to oppose experiments that could adversely affect the welfare of their constituents.
8. *Organizational culture must encourage learning from experience and the culture must value reliable information.* While most protected area managers are willing to learn from experience, they often evaluate their actions based on casual observation and unreplicated actions. AM requires setting in place monitoring, research, and evaluation programs that provide statistically reliable data for testing hypotheses about the effectiveness of management actions in achieving specific objectives. This approach is costly and requires institutional commitment.
9. *Resources must be sufficient to measure ecosystem-scale behavior.* Monitoring for AM is a long-term and expensive proposition, which makes AM vulnerable to insufficient budgets, changes in policy, and controversy. New managers and administrators may not understand or support an experimental approach.

10. *Decision makers must care about improving outcomes over biological time scales.* The cost of monitoring, controls, and replication is high at the outset relative to the costs of unmonitored trial and error. Protected area managers may not have the motivation, patience, and budget to implement long-term AM experiments especially if their term of office is significantly less than biological time scales, which is almost always the case. For example, a protected area manager under pressure to sustain and increase visitation may be less concerned with biological impacts of higher visitation especially when the research and monitoring programs needed to test hypotheses regarding those impacts require many years to complete.

    This situation may improve with the adoption of ecosystem management. For example, one of the guiding principles for ecosystem management in protected areas that is highly supportive of AM is to "acknowledge that ecosystem management should be viewed as a long-term experiment. Resource management decisions should be presented as hypotheses, with programs established to monitor their success in accomplishing desired objectives. Management must adapt in response to new information" (National Park Service 1993).

Of the 10 prerequisites for active AM, 1, 2, 3, 5, 6, and 7 are likely to be satisfied and 4, 8, 9, and 10 are unlikely to be satisfied.

There are two ways to improve the chances of successful AM. First, management actions should be substantial in order to ensure that natural variation in ecosystems does not overwhelm the effects of management actions. Second, a management action should be relatively simple to ensure detection of ecosystem responses to that action (Lee 1993). Due to ecological uncertainties, managers should expect to be surprised by certain management outcomes. Surprise should not be viewed as the failure to predict ecological responses, but rather as an opportunity to learn more about the ecosystem (Lee 1993).

Even if the conditions and method of implementing active AM are optimal, it has several drawbacks. First, it is more time-consuming, complex, and costly than other forms of management, such as passive AM, trial-and-error, and deferred action (Walters and Hilborn 1978; Walters and Holling 1990). Second, it can give faulty results when relevant variables are either ignored or not held constant (Smith 1997). Third, it has certain application pitfalls. For example, AM has been used in New Brunswick, Canada, to manage the spruce budworm problem and British Columbia, Canada, and the Columbia River Basin in the United States to manage conflicts between hydropower and fisheries. These applications have been criticized due to their heavy reliance on the use of linear systems models, discounting of nonscientific forms of knowledge, and giving insufficient attention to policy processes that promote the development of shared understandings among diverse stakeholders (McLain and Lee 1996). Fourth, AM runs the risk of implementing management actions that fail to achieve desired outcomes (Lee 1993). These limitations can be alleviated by incorporating knowledge from multiple sources, using several systems models, utilizing new forms of cooperative decision making (McLain and Lee 1996), and educating politicians and managers about the benefits and risks of AM.

Analytical Framework for Implementing Adaptive Management

While much has been written about AM, the analytical framework for implementing AM is poorly developed. This section describes an analytical procedure for implementing active AM and illustrates the procedure for a protected area manager who wants to determine whether a rangeland area in a national park is being overgrazed (Prato 2000). The remainder of this section utilizes probability theory. Readers unfamiliar with probability theory might want to proceed to the nontechnical summary given at the end of the section.

The procedure has two stages. In the first stage, the manager decides whether or not the current state of the rangeland area is sustainable. The current state, which is influenced by current and past uses, conditions, and management, is described by a set of multiple stochastic attributes chosen by the manager, possibly in collaboration with park stakeholders and scientists. If the first stage indicates the current state of the ecosystem is not sustainable, the manager proceeds to the second stage in which the best management action for achieving a sustainable ecosystem state is selected.

Use of this two-stage procedure requires defining sustainability. Suppose sustainability of the rangeland area is defined in terms of four mutually exclusive ecosystem states: M_1, M_2, M_3, and M_4. M_1 is highly unsustainable, M_2 is moderately unsustainable, M_3 is moderately sustainable, and M_4 is highly sustainable. Probabilities of the states before the experiments are implemented are called prior probabilities, namely $p(M_1)$, $p(M_2)$, $p(M_3)$, and $p(M_4)$. Prior probabilities sum to one. Suppose the manager selects two attributes to assess rangeland productivity, percentage of native species present (z_1) and percentage of forage production potential (z_2). Let there be four rangeland productivity conditions, R_1, R_2, R_3, and R_4, based on these attributes. R_1 is a poor condition signifying no more than 25 percent of native plant species and no more than 20 percent of forage production potential. R_2 is a fair condition denoting 26 percent to 49 percent of native plant species and 21 percent to 49 percent of forage production potential. R_3 is a good condition signifying 50 percent to 74 percent of native plant species and 50 percent to 80 percent of forage production potential. R_4 is an excellent condition denoting more than 75 percent of native plant species and more than 80 percent of forage production potential. These definitions of rangeland conditions were adapted from Miller (1990).

The first-stage decision is to determine whether the rangeland area is sustainable when current rangeland productivity is R_q where q takes on values between 1 and 4. This is not a trivial decision. For example, the manager would be relatively confident in viewing the rangeland area sustainable when the productivity condition is excellent (R_4) or unsustainable when the productivity condition is poor (R_1). However, the state of the ecosystem is not obvious when the current productivity condition is fair (R_2) or good (R_3).

Two kinds of decision errors can be made. First, the manager could decide the rangeland area is sustainable when it is actually unsustainable. This error causes the manager not to take action when in fact action should be taken. It causes the ecosystem to remain unsustainable and possibly deteriorate. Second, the manager could decide the rangeland area is unsustainable when it is actually sustainable. This error causes the manager to take action when none is needed. It leads to a change in management, when none is needed.

Ellison (1996) points out that "Bayesian inference and decision theory provide a quantitative framework and intelligible language in which to analyze and express adaptive management procedures." In the current example, Bayes' rule, which is central to Bayesian inference, is used to infer the current state of an ecosystem from a set of outcomes. In the rangeland example, there are 16 possible outcomes, namely (M_j R_q) for j, q = 1,..., 4. For example, the outcome of a highly sustainable ecosystem and excellent rangeland productivity is designated (M_4 R_4). Since the 16 outcomes are mutually exclusive (only one outcome is possible), the prior probability for productivity condition R_q is the sum of the joint probabilities:

$$p(R_q) = p(M_1\ R_q) + \ldots + p(M_J\ R_q)$$

where $p(M_j\ R_q)$ is the joint probability of the outcome (M_j R_q). Bayes' rule states that the posterior probability that the ecosystem is in state M_j given the productivity condition is R_q is:

$$p(M_j \mid R_q) = [p(R_q \mid M_j)\ p(M_j)]/[\sum_{j=1}^{J} p(R_q \mid M_j)\ p(M_j)]$$

where $p(R_q \mid M_j)$ is the likelihood function for M_j, namely, the probability of observing productivity condition R_q when the ecosystem state is M_j, and $p(R_q \mid M_j)$ $p(M_j)$ is the expected value of the likelihood function.

Determination of posterior probabilities requires knowledge of the prior probabilities, $p(M_1)$, $p(M_2)$, $p(M_3)$, and $p(M_4)$, and the likelihood functions, $p(R_q \mid M_j)$, for all states of the ecosystem. Bayes suggested making prior probabilities equal to one another, for example, $p(M_1) = p(M_2) = p(M_3) = p(M_4) = 1/4$. This assumption implies complete ignorance regarding the likelihood of ecosystem states. Ellison (1996) and Wolfson et al. (1996) argue that ecologists are not likely to be ignorant regarding possible ecosystem states. Mendenhall (1975) suggests that prior probabilities should be determined based on the decision maker's knowledge of the situation. For example, if the rangeland area were located in a relatively pristine national park, then the prior probabilities are likely to be higher for sustainable states than unsustainable states. The opposite would hold for rangeland areas in national parks heavily impacted by human presence. The likelihood functions are somewhat easier to specify than the prior probabilities because the probability of a productivity condition can be estimated based on knowledge of the ecosystem that generates that condition.

The above analytical procedure can be implemented as follows. First, a panel of experts composed of managers and scientists defines a finite number of states for the protected area ecosystem of interest. The example used above specifies four states. Second, the panel assigns prior probabilities to ecosystem states based on their knowledge of the region in which the ecosystem is located. Third, the most likely current state of the ecosystem is the one having the highest posterior probability. Table 8.1 gives an example of posterior probabilities for a hypothetical rangeland area whose current productivity condition is R_c (c stands for current condition). Since $p(M_1 \mid R_c) = 0.63$ is the highest posterior probability, the most likely state of the ecosystem is M_1. Since M_1 is an unsustainable state, the ecosystem state for the rangeland area is deemed to be unsustainable.

Suppose the protected area manager initiates experiments to determine which of several management actions is most likely to achieve a sustainable state for the rangeland area. Specifically, experiments are conducted to assess how the rangeland area responds to three technically, socially, and administratively feasible management actions (K = 3). The experiments provide measurements (data) on native species present and percentage of forage production potential. Other variables can also be measured. Suppose experiments lead to rejection of the hypothesis that the means of native species present are equal over the three management actions; likewise for forage production potential. The means of native species present and percentage of forage production potential for each management action

Table 8.1. Calculation of Posterior Probabilities for Four Hypothetical States of a Rangeland Area with Productivity Condition R_c

State	$p(M_j)$	$p(R_c \mid M_j)$	$p(M_j \mid R_c)$[a]
M_1[b]	0.4	0.5	<u>0.63</u>
M_2[b]	0.3	0.3	0.28
M_3[c]	0.2	0.1	0.13
M_4[c]	0.1	0.1	0.06

[a] $[p(R_c \mid M_j)\, p(M_j)]/[\Sigma_j\, p(R_c \mid M_j)\, p(M_j)]$ (j = 1,..., 4)
[b] Unsustainable states.
[c] Sustainable states.

are then used to determine the productivity condition for that management action. For example, a management action yielding a mean of 60 percent for native plant species and a mean of 75 percent for forage production potential implies a productivity condition of R_3. Suppose experiments show that each of the three management actions results in a different resource condition, namely, management action 1 results in R_4, management action 2 results in R_1, and management action 3 results in R_3. Bayes' rule is used to compute three posterior probabilities, $p(M_j \mid R_4)$, $p(M_j \mid R_1)$, and $p(M_j \mid R_3)$ for each ecosystem state. For a management action to be considered sustainable, the highest posterior probability for a given resource condition must be for M_3 or M_4. It is possible for more than one management action to result in a sustainable state. In this case, the manager would use other attributes to choose the best management action, such as cost effectiveness and public acceptability. Alternatively, states of nature can be defined in terms of all relevant attributes.

Table 8.2 contains a hypothetical example of posterior probabilities for the four rangeland states with the three management actions. M_2 has the highest posterior probability with management action 1 (0.46), M_3 with management action 2 (0.38), and M_4 with management action 3 (0.38). Therefore, the most likely state of the ecosystem is M_2 with management action 1, M_3 with management action 2, and M_4 with management action 3. Ecological sustainability is achieved with management actions 2 and 3, but not with management action 1.

Another example of the application of Bayes' rule in AM is given in the following sidebar. Further details about the application of Bayesian statistics in AM are given in Bergerud and Reed (1998).

In summary, application of Bayes' rule allows a protected area manager to determine whether the current state of an ecosystem is sustainable. In cases where

Table 8.2. Calculation of Posterior Probabilities for Four Hypothetical States of a Rangeland Area with Three Management Actions

		Management action[a]					
		1		2		3	
State	$p(M_j)$	$p(R_4 \mid M_j)$	$p(M_j \mid R_4)$[b]	$p(R_1 \mid M_j)$	$p(M_j \mid R_1)$[c]	$p(R_3 \mid M_j)$	$p(M_j \mid R_3)$[d]
M_1[e]	0.4	0.2	0.31	0.1	0.19	0.05	0.13
M_2[e]	0.3	0.4	<u>0.46</u>	0.2	0.29	0.1	0.19
M_3[f]	0.2	0.2	0.15	0.4	<u>0.38</u>	0.25	0.31
M_4[f]	0.1	0.2	0.08	0.3	0.14	0.6	<u>0.38</u>

Note: Highest posterior probabilities for each management action are underscored.

[a]Management action 1 results in R_4, management action 2 results in R_1, and management action 3 results in R_3.

[b]$[p(R_4 \mid M_j)\, p(M_j)]/[\Sigma_j\, p(R_4 \mid M_j)\, p(M_j)]$

[c]$[p(R_1 \mid M_j)\, p(M_j)]/[\Sigma_j\, p(R_1 \mid M_j)\, p(M_j)]$

[d]$[p(R_3 \mid M_j)\, p(M_j)]/[\Sigma_j\, p(R_3 \mid M_j)\, p(M_j)]$

[e]Unsustainable states.

[f]Sustainable states.

it is not, AM experiments are designed to test hypotheses about which state of sustainability the ecosystem is in. Resource conditions resulting from the management actions and prior probabilities determined by the manager are combined in Bayes' rule to calculate posterior probabilities for ecosystem states. The ecosystem state having the highest posterior probability for a given management action is the most likely state of the ecosystem.

Application of Bayes' Rule in Adaptive Management

Active adaptive management embraces uncertainties regarding how ecosystems respond to management actions by evaluating the validity of alternative hypotheses regarding those responses. Alternative hypotheses and underlying models regarding how landscape fragmentation influences wildlife habitat can be experimentally tested using frequentist or Bayesian statistical inference.

Frequentist statistics, which is the dominant method of statistical inference, would use experimental data to estimate a model that relates carnivore populations to landscape fragmentation. For example, suppose it was possible to design an experiment to measure how the population of a particular carnivore species responded to various treatments (degrees of landscape fragmentation). Randomly assigning treatments to different areas and replicating each treatment would provide a suitable data set for estimating the relationship between carnivore population and landscape fragmentation metrics and possibly other ecological variables. The estimated relationship could be used to test specific hypotheses about the effects of landscape fragmentation on carnivore population.

While frequentist statistics provides useful insights about the relationship between carnivore population and landscape fragmentation, it cannot be used to determine which of several hypotheses about that relationship has the highest probability of being true, or the probability an estimate of the relationship is reasonably close to some unknown true relationship. This is precisely the kind of information needed to implement active adaptive management and provided by Bayesian statistical inference.

Bayesian statistics allows a manager to directly quantify uncertainty. As explained in the section on adaptive management given in this chapter, Bayes' rule or theorem allows determination of a posterior probability by combining the prior probability of a particular hypothesis regarding the relationship between carnivore population and landscape fragmentation and the likelihood function or probability of observing the newly collected data, given that hypothesis holds. The prior probability is a subjective probability because it reflects the decision maker's personal knowledge and beliefs about the hypotheses being evaluated before new data are collected. The importance of the prior probability relative to the likelihood function in determining the posterior probability decreases (increases) as the amount of newly collected data increases (decreases).

The following hypothetical example illustrates how Bayes' theorem is used to make a decision about whether or not to remove roads from a large section of a national forest in order to improve habitat for grizzly bears. Suppose there are two possible states of nature for the section: θ_S meaning the section provides suitable habitat and θ_U meaning the section provide unsuitable habitat for grizzly bears. A random sample of 12 equally sized plots is drawn (n = 12) from the section and the number of plots out of 12 (X) containing insufficient habitat (IH) is determined. The posterior probabilities for θ_S and θ_U are as follows:

$$p(\theta = \theta_S \mid X = X_0) = [p(\theta = \theta_S) * p(X = X_0 \mid \theta = \theta_S)]/p(X = X_0)$$
$$p(\theta = \theta_U \mid X = X_0) = [p(\theta = \theta_U) * p(X = X_0 \mid \theta = \theta_U)]/p(X = X_0)$$

where $p(X = X_0) = [p(\theta = \theta_U)^* \, p(X = X_0 \mid \theta = \theta_U) + p(\theta = \theta_S)^* \, p(X = X_0 \mid \theta = \theta_S)]$ and X_0 is a value between 0 and 12. The state of nature with the highest posterior probability is selected.

The following table summarizes hypothetical likelihoods or probabilities that X plots out of 12 are IH given each value of θ, the posterior probability that the section is in state θ_U for all possible values of X when the $p(\theta_U) = 0.84$ (implying $p(\theta_S) = 0.16$), the most likely state of the section and the resulting management decision.

Number of plots with insufficient habitat(X)	Likelihood that plots are IH when $\theta = \theta_U$	Likelihood that plots are IH when $\theta = \theta_S$	Posterior probability for $\theta = \theta_U$ $p(\theta = \theta_U \mid X)$	Most likely state of the section	Management decision regarding roads
0	0.000	0.001	0.000	θ_S	No removal
1	0.000	0.008	0.000	θ_S	No removal
2	0.000	0.034	0.000	θ_S	No removal
3	0.000	0.092	0.000	θ_S	No removal
4	0.001	0.170	0.030	θ_S	No removal
5	0.003	0.222	0.066	Unclear	Ambiguous
6	0.016	0.212	0.284	Unclear	Ambiguous
7	0.053	0.149	0.651	Unclear	Ambiguous
8	0.133	0.076	0.902	Unclear	Ambiguous
9	0.236	0.028	0.978	θ_U	Remove
10	0.283	0.007	0.995	θ_U	Remove
11	0.206	0.001	0.999	θ_U	Remove
12	0.069	0.000	1.000	θ_U	Remove

Source: Adapted from Bergerud and Reed (1998).

For example, the probability that X = 7 is

$$p(X = 7) = 0.84 * 0.053 + 0.16 * 0.149 = 0.0684$$

The posterior probability for X = 7 is

$$p(\theta = \theta_U \mid X = 7) = [0.84 * 0.053] / 0.0684 = 0.651$$

For the hypothetical data given in the table, roads are not removed from a section if four or fewer plots have insufficient habitat because the posterior probability for these values of X is less than 0.05, which is the most commonly specified type I error (probability of deciding not to remove roads when the true state of the section is unsuitable habitat). Roads are removed from a section if X is between 9 and 12 because the posterior probability for these values of X is greater than 0.95. The latter is the probability of deciding to remove roads from the section when the true state of the section is unsuitable (power of test). If X is between 5 and 8, then the decision is ambiguous because the posterior probabilities for these values of X are between 0.05 and 0.95.

Source: Bergerud and Reed (1998).

Other Approaches for Handling Uncertainty

The MADM framework discussed earlier assumes that CLAG knows with certainty the economic and ecological attributes provided by alternative futures. This is not the case due to uncertainty. On the landscape and ecological impact side, there is uncertainty regarding (1) location and extent of landscape fragmentation that occurs under different alternative futures, (2) impacts of landscape fragmentation on grizzly bear habitat, and (3) extent to which habitat degradation causes grizzly bear population to decline. On the economic side, there is uncertainty about economic impacts of development due to spatial and temporal variation in input-output coefficients. The economic impact analysis described earlier assumes the input-output coefficients in IMPLAN remain constant over time. This assumption is likely to be violated the more distant the future years. Uncertainty regarding economic and ecological impacts of alternative futures can lead to incorrect decisions. For example, if uncertainty leads to the selection of an alternative future that is not expected to impair grizzly bear habitat but in fact it does, then the alternative future would be deemed sustainable when it is not.

The previous section described how AM and Bayes' rule are used to evaluate the sustainability of an ecosystem under alternative management actions when there is uncertainty. Is the AM framework suitable for evaluating the ecological and economic impacts of alternative futures? It is not for three reasons. First, management actions are fundamentally different from alternative futures. Based on the outcomes of AM experiments, a protected area manager selects the most preferred management action. CLAG does not get to choose which alternative future it wants to happen, although it can rank them in order of preference. CLAG evaluates a range of alternative futures precisely because it does not know which, if any of them, will materialize.

Second, it is not feasible for CLAG to conduct AM experiments on alternative futures. In other words, CLAG cannot implement alternative futures and measure their economic and ecological impacts on an ecosystem as required by AM experiments. Rather, CLAG must use the economic and ecological models discussed earlier to estimate the likely impacts of alternative futures.

Third, there is essentially only one source of uncertainty with management actions, namely, their economic and ecological impacts. In contrast, assessment of the alternative futures involves two sources of uncertainty, namely, the path development is likely to follow or which alternative future is most probable, and the economic and ecological impacts that are likely to occur with a particular alternative future. The first source is not relevant for management actions because protected area managers control management actions but not the alternative future that materializes. The second source of uncertainty is similar to that associated with management actions. In general, AM is not a suitable approach for handling the uncertainty associated with alternative futures.

How might CLAG handle the two sources of uncertainty associated with alternative futures? The first source of uncertainty (which alternative future is most likely to occur) can be handled three ways. First, CLAG admits it has no basis for determining the most likely alternative future and therefore evaluates a wide range of alternative futures. If the impacts vary significantly across alternative futures, then knowing the most likely alternative future makes a big difference and conducting further analysis to identify the most likely alternative future is warranted. Conversely, if the impacts do not vary significantly across alternative futures, then knowing the most likely alternative future is of little consequence provided, of course, CLAG is reasonably confident that the most likely alternative future has impacts that fall within the range of impacts for the selected alternative futures.

Second, CLAG can employ a Delphi approach in which consensus is reached on the probability of occurrence of each and every alternative future (Bakus et al. 1982). The most likely alternative future is the one having the highest probability of occurrence. Third, CLAG can determine the expected value of impacts by weighting each impact by the probability of its occurrence. For example, if the estimated impact of the mth alternative future on total output is TO_m, the consensus probability of the mth alternative future is p_m, and there are five alternative futures, then expected total output (ETO) is:

$$ETO = \sum_{m=1}^{5} p_m TO_m$$

where the sum of the probabilities equal one. Expected values for habitat suitability are calculated in a similar manner. While expected values of objectives are useful information, CLAG may be interested in identifying alternative futures that are likely to have highly adverse impacts on TO or HS even though their probability of occurrence is low. This is because alternative futures that decrease TO or HS are likely to be rejected.

The second source of uncertainty (what are the economic and ecological impacts of a particular alternative future) can be handled by using interval estimates rather than point estimates of impacts. An interval estimate provides a range of values for an impact that reflects the range of uncertainty about those impacts. A point estimate is a single value. For example, an interval estimate of HS is [50, 75], whereas a point estimate is 60. An interval estimate of how an alternative future affects HS can be derived by selecting a range of values for the variables that enter the landscape change model. These variables include the baseline land requirements per $1,000 of final expenditures in a sector, the method used to estimate transition probabilities for conversion of land from one cover class to another, the boundaries of management zones, the number of land-cover classes, the method used to allocate land requirements within management zones, and others.

Similarly, interval estimates of TO can be developed by using a range of input-output coefficients (c'_{ru}) in the IMPLAN model that reflects improvements in technology through time. While accounting for uncertainty regarding the ecological and economic impacts of alternative futures makes for a more realistic and robust evaluation, it complicates the assessment of alternative futures. Another way to deal with uncertainty in a MADM framework is to use fuzzy set theory (Munda et al. 1994).

There are other general approaches to handling uncertainty besides AM and the approaches suggested for use by CLAG. The *safe minimum standards* (SMS) approach is demonstrated for determining the minimum level of biological diversity to preserve in order to ensure sufficient functioning and resilience of ecosystems. Biodiversity can be assessed using various measures, such as minimum viable population sizes needed to ensure survival of various species. One SMS decision rule posits that when comparing alternative policies for conserving biodiversity under different states of nature, such as disallowing development that threatens biodiversity or allowing development that runs the risk of reducing biodiversity, the preferred policy is the one that minimizes the maximum loss that is likely to occur (minimax criterion) unless the social costs of that policy are unacceptably high (Bishop 1978).

Two possible states of nature are relevant to this decision problem: conserving biodiversity is of value in the future or conserving biodiversity is not of value in the future. Social costs of preserving current populations of species are the forgone net private benefit that would have occurred with development (difference between benefits of development minus expected benefits of conservation) plus the direct costs of preventing further declines in species (Crowards 1998). For instance, if the maximum loss with development is $20 million and the maximum loss with biodiversity conservation is $15 million, then the minimax criterion selects biodiversity conservation as the preferred policy unless its social costs are unacceptably high. What constitutes unacceptably high social costs is a political decision.

The above SMS-based decision rule has been criticized on the grounds that it (1) admits knife-edge decisions (the optimal policy may cause only slightly lower losses than the second-best policy), (2) assumes the costs of different states of nature are known with certainty and ignores the distribution of those costs between gainers and losers, (3) is static because the probabilities of different states of nature are unknown and do not influence the decision (no learning over time), and (4) allows the decision to conserve biodiversity to be reversed in a future time period (Van Kooten and Bulte 2002).

Another framework for making decisions that is especially useful when ecological thresholds are unknown and/or impacts of losing ecological services are irreversible is the *precautionary principle*. It says that a development option with potentially high and irreversible environmental consequences, such as oil development in the Arctic National Wildlife Refuge, should be rejected regardless of the expected loss in net private benefits.

Comparing a policy that conserves biodiversity and ecosystem functions with one that promotes development that threatens biodiversity is further complicated by the fact that the potential ecological benefits of preserving biodiversity are typically long term, whereas the economic costs are typically short term. This imbalance tends to favor policies that reduce biodiversity especially when benefits and costs are discounted. Discounting is inappropriate when losses in biodiversity result in the irreversible consequence of species extinction.

Optimization and Simulation

Optimization and simulation are useful in evaluating and comparing the economic and ecological impacts of alternative futures and management actions for protected areas. Optimization involves formulating a decision problem as a mathematical model and solving it to obtain a solution. Mathematical models range from very simple to very complex. For example, benefit-cost analysis is a simple optimization model for identifying the most economically efficient alternative future. The model requires calculating the net present values of all alternative futures and selecting the one that maximizes net present value. Ranking alternative futures based on MADM essentially involves a mathematical model that requires identifying efficient and sustainable alternative futures, specifying and estimating the economic and ecological attributes of alternative futures, and selecting a utility function for ranking alternative futures. However, MADM problems do not necessarily have optimal solutions. The Multiple Attribute Scoring Test for Capacity (MASTEC), described in chapter 7, is an optimization model that allows a protected area manager to identify the best management action for achieving compliance with established carrying capacities. The model is solved using a chance-constrained mathematical programming model.

Simulation generates values of a variable over space and/or time using a deterministic process model that incorporates mathematical relationships between input and output variables. The IMPLAN model simulates changes in total economic output, household income, and employment (output variables) associated with changes in final expenditures on goods and services by sector (input variables). It consists of a set of input-output coefficients that measure the total direct and indirect outputs required in all sectors to achieve a one-dollar increase in final expenditures in a primary sector. Input-output coefficients are estimated using historical data on money flows between sectors. Another example of simulation is to infer habitat suitability for grizzly bear (output variable) based on landscape structure metrics calculated from land-cover maps (input variables).

Other kinds of ecological models are used to simulate ecological impacts. Stream-habitat models have been used to simulate how stream-dwelling organisms respond to changes in siltation and chemical pollution (Fausch et al. 1988). Quantitative relations between fish/invertebrate communities and siltation of streams (Berkman and Rabeni 1987; Rabeni and Smale 1995) and between fish communities and dissolved oxygen and summer temperatures in streams (Smale and Rabeni 1991) have been used to assess the health of aquatic ecosystems. Ecological effects of proximate habitat conditions in a stream have been evaluated using the index of biological integrity (Karr et al. 1986). The CENTURY model has been used to simulate how net primary productivity, nutrient availability, and carbon/nitrogen fluxes respond to changes in climate and management (Parton et al. 1995). STELLA is a popular modeling environment for simulating interactions between ecological and economic processes (Matthias and Hannon 1997; Graham 2002). It is described in the next section.

Integrated Ecological Economic Modeling

CLAG needs to assess the economic and ecological impacts of alternative futures. This requires an understanding of the relationships among development, economic activity, landscape change, and grizzly bear habitat. Such understanding can be acquired from application of integrated ecological economic models (Barbier et al. 1995). Integrated ecological economic models are typically complex because ecological economic systems "are characterized by strong (usually nonlinear) interactions between the parts, complex feedback loops that make it difficult to distinguish cause from effect, and significant time and space lags, discontinuities, thresholds, and limits" (Costanza et al. 1993). Compared to the modeling of economic systems, modeling interactions between economic and ecological processes is nascent.

Ecological economic models are valuable in evaluating sustainability (Costanza 1991), building mutual understanding and soliciting input from stakeholder groups, and maintaining a substantive dialogue between stakeholders (Costanza and Folke 1997). A model helps stakeholders to understand complex systems and make collaborative decisions. Developing and applying integrated ecological economic models in a consensus-building framework involves 12 steps:

1. Defining the problem
2. Identifying stakeholders
3. Establishing a process for bringing stakeholders together, such as a roundtable
4. Seeking agreement on a facilitator
5. Defining stakeholder interests
6. Holding a roundtable
7. Undertaking a scoping exercise
8. Building and running a scoping model
9. Developing a detailed model
10. Presenting model results
11. Building consensus recommendations
12. Monitoring the development of the preferred action (Costanza and Tognetti 1996)

A generic ecological economic model for assessing sustainability is illustrated in figure 8.3. The model contains two modules, an ecosystem module having socioeconomic and ecological components and a sustainability assessment module. The socioeconomic component explains how the services of reproducible capital (machinery, buildings, and equipment), natural capital (land, water, air, and other natural resources), and human capital (not shown in the diagram) are combined to produce consumer goods and services. Consumption of the latter in-

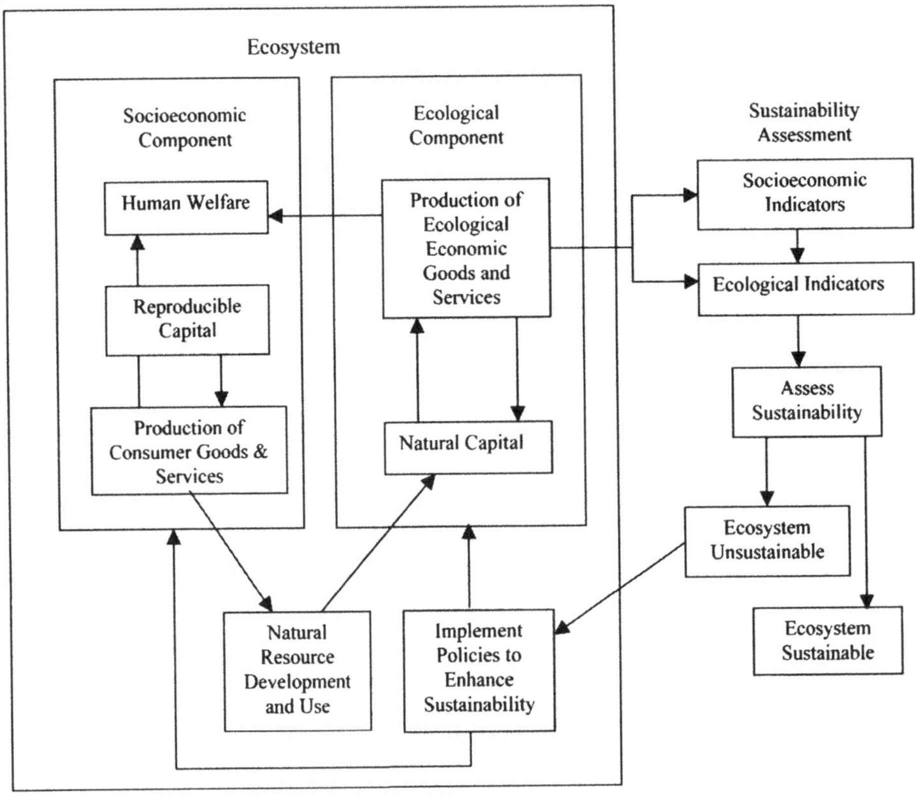

Fig. 8.3. Generic integrated ecological economic model for assessing sustainability.

creases human welfare. Production of consumer goods and services depreciates the stock of reproducible capital, which is replenished through investment (not shown). The ecological component shows that stocks of natural capital provide a range of ecological economic goods and services that sustain human welfare. Natural resource use and development deplete natural capital. Maintaining the stock of natural capital enhances the sustainability of the ecological system and typically requires the implementation of specific policies and programs.

The sustainability assessment module evaluates ecosystem sustainability in terms of socioeconomic indicators, such as total output, and ecological indicators, such as grizzly bear habitat. If the ecosystem is unsustainable, then policies are formulated to enhance sustainability. If the ecosystem is sustainable, then policy action is not needed. Policies influence ecosystem sustainability by altering one or more socioeconomic and/or ecological variables. For example, encouraging landowners to sell conservation easements on private land in prime grizzly

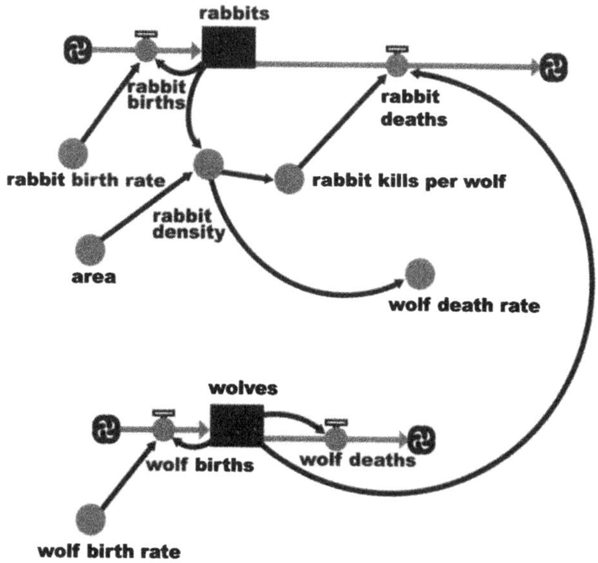

FIG. 8.4. STELLA model of predator (wolf)-prey (rabbit) relationships. (From Tilidetzke, undated)

bear habitat could increase protection of grizzly bears because conservation easements prohibit development.

Ecological economic models typically employ optimization and/or simulation techniques described in the previous section. STELLA (Matthias and Hannon 1997) is a popular dynamic modeling system for simulating interrelationships between ecological and economic processes. Application of STELLA requires a group of scientists and practitioners to combine its knowledge about how a particular ecosystem operates. Modeling involves a three-step procedure. The first step requires the group to develop a graphical model of the relationships between stocks and flows. Figure 8.4 is a STELLA model that describes the relationships between wolves (predator) and rabbits (prey) (Tilidetzke, undated). Populations of wolves and rabbits are stock variables designated by rectangles. Births and deaths of wolves and rabbits are flow variables designated by circles. Arrows between stocks and flows are connectors. Applying the birthrate to population determines the number of births, which adds to the population. The population of rabbits and wolves, and predation rates (rabbit kills per wolf), determine rabbit deaths, which indicates a feedback loop between population of wolves and death rate for rabbits. If births exceed deaths, then rabbit population increases. If deaths exceed births then rabbit population decreases. If births equal deaths, then rabbit population remains stable. Dynamic models are simulated using a particular time step, such as a month or year.

The second step involves using STELLA software to translate the graphical model into a set of equations and initial conditions that are used to stimulate the model. This phase requires detailed mathematical specification of the connectors. Connectors that would need to be specified in figure 8.4 are the direction and magnitude of the relationship between population and birthrates, food availability per animal, and food eaten per animal, between food eaten per animal and death rate, and so on. In the third step, the user views the simulation results in graphical, tabular, and animated form. Correlations between simulated and measured population values are used to assess the accuracy of a dynamic model. A high correlation indicates high accuracy, whereas a low correlation indicates low accuracy. When accuracy is low, the functional relationships among variables in the model need to be revised until an acceptable level of accuracy is achieved. This procedure is known as model calibration.

Dynamic models are useful in assessing impacts of policies. For example, suppose the model indicates that bear population decreases through time because deaths exceed births. A policy that attempts to reverse this trend is to increase food supply for bears by protecting habitat that contains important food sources. The success of this policy is determined by using the model to simulate how bear population responds to a lower rate of decline in lands containing food sources. Other policies are simulated in a similar manner.

Spatial Analysis

This section describes several information technologies relevant to protected area management in Cimarron National Park. Knowledge of landscape features is valuable for both nonanalytical and analytical purposes. On the nonanalytical side, for instance, most national parks generate maps that have useful information about roads, campgrounds, trails, visitor facilities, and other features. These maps help visitors and others to learn about the opportunities and facilities available in the park. Figure 8.5 is an example of such a map. It provides information about the trails in the vicinity of Logan Pass in Glacier National Park. While the map is helpful in locating trails, it is insufficient for analytical tasks like assessing the environmental impacts of reconstructing Going-to-the-Sun Road, which goes over Logan Pass. The latter would require detailed spatial information about soils, vegetation, and wildlife patterns and topography, as well as vehicular traffic. A more comprehensive map is needed for environmental impact assessment.

Landscape features are important in explaining biophysical processes. For example, consider the likely impacts on soil erosion and water quality of constructing a trail between a valley and pass in Cimarron National Park. If there is a steep slope between the valley and mountain pass, then constructing a series of switch-

FIG. 8.5. Logan Pass trails, Glacier National Park. (From Glacier National Park 2002.)

backs up the slope results in considerably less soil erosion and runoff than having the trail go straight up the slope. In addition, a switchback trail is less steep and easier for hikers to traverse. Having detailed spatial information about soil types, vegetative cover, and topography allows estimation of soil erosion rates and runoff with alternative trail locations.

Spatial analysis in the context of protected areas consists of applying information technologies to spatial data and information for the purpose of improving understanding of the relationships between landscape patterns and ecological economic processes. Landscape patterns encompass many elements including soil type, landform (characteristic geomorphic features of the landscape), climate (composite, long-term weather of a region), interactions among organisms, land use, and natural disturbance and succession (Turner et al. 2001). Ecological processes influenced by landscape patterns include soil erosion, sediment and water transport, energy and nutrient cycling, movement of fish and wildlife, spread of wildfire and disease, and others. The old way of doing spatial analysis, which is to interpret a map or picture using the eye and brain (Goodchild 2000), has been substantially enhanced by advancements in geographic information systems

(GIS), GIS-based environmental modeling, remote sensing, global positioning systems, interactive mapping, and computerized landscape visualization. This section describes these information technologies.

Geographic Information Systems

A GIS is a computer-based system for doing complex and comprehensive spatial analysis that involves entering, storing, analyzing, and displaying geographically referenced information (Haines-Young et al. 1993; U.S. Geological Survey 2002a). A GIS consists of (1) tools for data entry and manipulation, (2) a database manager, (3) smart maps that can be queried for information and printed, and (4) a graphical user interface that facilitates use of the GIS (U.S. Geological Survey 2002a).

Figure 8.6 illustrates eight map layers in a hypothetical GIS. Each layer describes a particular landscape feature, such as land use, soils, hydrography, and hypsography. Features are defined by location and a set of attributes. For example, the transportation layer contains information about road locations and attributes. Attributes indicate whether a road is paved, improved, or unimproved, and owned by federal, state, or private interests. The hydrography layer indicates where rivers, streams, and lakes are located and possibly water quality. A GIS permits many kinds of spatial analyses, such as determining the suitability of a landscape for grizzly bears based on the number, size, and location of roadless areas. Data used in a GIS can be purchased, donated by individuals, or downloaded from the Internet, and come from a variety of sources including digitized or scanned maps, tabular data, Global Positioning Systems, field surveys, remote sensing, and aerial photography.

Two data models are used in a GIS: raster or vector (U.S. Geological Survey 2002b). The raster model represents a landscape feature as a matrix of square or hexagonal equal-sized cells in continuous space. Each cell has a value and location. A raster-based model of land cover is shown in figure 8.7. This figure shows that each grid cell in a digital elevation model indicates the surface elevation for the area covered by that cell (say 30 m by 30 m) as well as the location of the cell. Using a smaller cell size (finer grid) increases resolution, but requires more storage space. Conversely, using a larger cell size (coarser grid) decreases resolution but requires less storage space. Cell size is normally based on the original map scale and minimum mapping unit. Satellite imagery and aerial photos are in raster format.

A vector-based GIS uses points, lines, and polygons to represent landscape features. Locations of features are determined using x,y coordinates or points on the Earth's surface. Lines use a series of coordinate pairs to represent features that have length like a road or river. Polygons are areas that have closed borders, such as the boundaries of a campground or lake. A vector-based data structure for land cover is illustrated in figure 8.8.

*Digital orthophoto quadrangle and satellite images.

FIG. 8.6. Map layers in a hypothetical GIS. (Created by Chris Fulcher, Center for Agricultural, Resource and Environmental Systems, University of Missouri-Columbia.)

1	1	1	3	3	3	3	3	3
1	1	1	3	3	3	3	3	3
1	1	2	2	2	2	3	3	3
1	1	2	2	2	2	2	3	3
1	2	2	2	2	2	3	3	3
3	3	3	2	2	2	3	3	3
3	3	3	3	2	3	3	3	3
3	3	3	3	3	3	3	3	3

1 Residential 2 Water 3 Farmland

FIG. 8.7. GIS raster-based model of land cover. (From U.S. Geological Survey 2002b)

Landscape features can be represented using either a raster or vector model. Discrete features like the location of a parking lot are easier to represent with a vector model. Vector or raster models are used to represent continuous features,

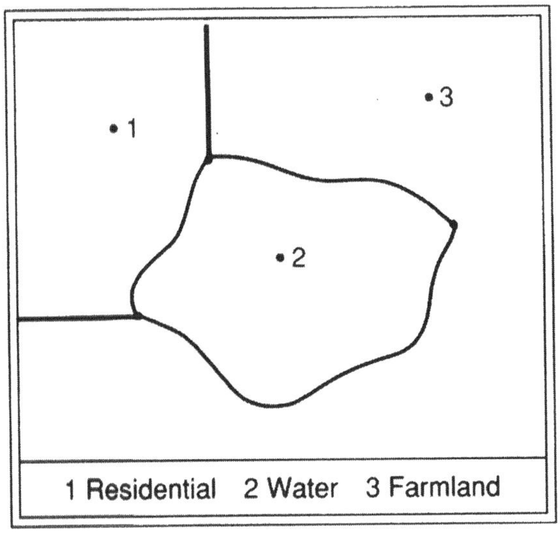

FIG. 8.8. GIS vector-based data model for land cover. (From U.S. Geological Survey 2002b)

such as vegetation type and elevation (GIS.com 2002). While computer manipulation of raster data is faster, vector data are generally more detailed and visually appealing because they more nearly approximate the appearance of objects.

Several kinds of data functions are employed in a GIS, including digitizing and editing of data, obtaining data from remote sites, changing map projections to increase data compatibility, enhancing visual display through advanced cartographic techniques, and exporting data (Goodchild 2000). A GIS is helpful in understanding the relationships between data layers and performing data queries. Data queries that are particularly relevant in land-use analysis are:

1. What patterns are evident in the distribution of the phenomenon? What patterns exist, for example, in the use of land in the study area? Do particular uses cluster in certain areas, and what pairs of uses tend to be neighbors?
2. How do different phenomena at a location influence each other? How is land use affected by or correlated with soil type, rainfall, or slope?
3. How is a phenomenon influenced by the same phenomenon at other locations? Do certain land uses tend to disperse themselves, or to cluster?
4. How is a phenomenon influenced by other phenomenon in its neighborhood? Is land use influenced by distance to the nearest urban area, for example?
5. What changes are occurring on the landscape, and how are they driven by processes of different types?

6. How well do the phenomena in an area fit the predictions of a specific model? What patterns are predicted to emerge in the future? For example, what trends are apparent in the land uses in an area, and what future patterns will they lead to? (Goodchild 2000)

A GIS was used to evaluate grizzly bear habitat in the Greater Yellowstone Ecosystem (GYE) (Noss et al. 2002). The assessment used static (habitat suitability) and dynamic (population viability) models to identify biologically significant and irreplaceable sites for focal species that are vulnerable to development. GIS data on vegetation, topography, climate, and land-use variables such as road density were used in habitat-suitability models or resource selection functions for focal species (Boyce and McDonald 1999). Population viability analysis, which tracks the population as individuals are born, disperse, and die, was conducted by incorporating GIS variables related to mortality risk and habitat productivity in the PATCH program (Schumaker 1998). The PATCH program determines survival and fecundity of individual animals. The modeling system developed for the GYE assesses likely impacts of landscape change scenarios (expressed in terms of factors such as road and population expansion) during the period 2000–2025 on population viability and identifies probable sources and sinks. One prediction made by the study is a relatively high probability of persistence for grizzly bears over the short to medium term, but a much smaller future population distribution and size. Discontinuing new road building in existing roadless areas on public lands in the GYE is expected to reduce this negative impact (Noss et al. 2001).

As development proceeds in greater ecosystems containing critical grizzly bear habitat, landscapes often become fragmented, grizzly bear habitat is degraded and encounters between humans and bears increase. The latter increases both the risk of injury or death of people and bear fatalities. Human-related bear fatalities jeopardize achievement of the recovery goals for grizzly bears. Ironically, while the presence of grizzly bears and other wildlife attracts visitors to national parks, increased visitation and associated economic growth places the grizzly bear at greater risk of extinction.

GIS-based Environmental Modeling

A powerful GIS application is in environmental modeling. Two examples of GIS-based environmental models are presented, one for calculating soil loss and the other for assessing habitat suitability for bobwhite quail. Soil erosion rates are estimated using the Revised Universal Soil Loss Equation (RUSLE) (USDA 2002). RUSLE estimates soil erosion rates based on five factors: soil erodibility, slope length, slope steepness, cover-management, and supporting practices. When RUSLE is applied to a large number of sites in a landscape, it is more efficient and accurate to extract the values of the first three factors from a vector-coded GIS. In particular, soil erodibility is an attribute in the soils layer, and slope length and slope steepness can be calculated from the hypsography layer. Erosion rates calculated with RUSLE can be displayed in a GIS.

The second environmental model is a web-based spatial decision support system (SDSS) for assessing winter habitat for bobwhite quail in the state of Missouri. The SDSS calculates a habitat suitability index (HSI) surface that indicates the probability of suitable winter habitat for bobwhite quail in selected areas of the state.

Habitat suitability of a point location is determined by the values of three landscape metrics in the bobwhite quail's mean home range. Winter home range for northern bobwhite quail is a 345-m radius area (or 38 ha). The three landscape metrics are the shape index of row crop fields (SRC), the edge density of woody patches (EDW), and the edge density of fields enrolled in the Conservation Reserve Program (EDC). Landscape metrics are calculated by applying FRAGSTATS to a land-cover map created using 30-m resolution satellite images for the period 1991–93.

Winter habitat (Y) is related to the three landscape metrics as follows (Burger et al. 1998):

$$Y = -13.9995 + 3.1823 * SRC + 0.0542 * EDW + 0.0629 * EDC$$

The HSI surface depicts probability values ranging from 0 to 1. For display purposes, the probability values are scaled to integer values between 0 and 100 as follows:

$$HSI = Exp(Y) / [1 + Exp(Y)]$$

The HSI surface is mapped using four classes for the rescaled probability values: 0–25 for not suitable, 26–50 for marginally suitable, 51–75 for moderately suitable, and 76–100 for highly suitable winter habitat.

The SDSS for assessing habitat suitability for bobwhite quail is illustrated in figure 8.9. The interface screen A appears when the model is started. At the top of the interface screen is a set of buttons for performing functions like zoom-in, zoom-out, pan, feature information, habitat suitability, and print map. To the left of the Missouri land-cover map is the legend for land-cover classes. To create the HSI surface, the habitat suitability analysis button is clicked, a polygon is delineated on the map using the pointer (see polygon in B), and a double click brings up a menu for selecting the cell size (resolution) for the HSI surface. The minimum cell size is 30 m because that is the resolution for the statewide land-cover map. After the desired cell size is chosen, the "OK" button is clicked and the resulting HSI surface is mapped as shown in B. Most of the selected polygon is highly suitable winter habitat for bobwhite quail (magenta color). A printed HSI surface is obtained by clicking the print button.

A

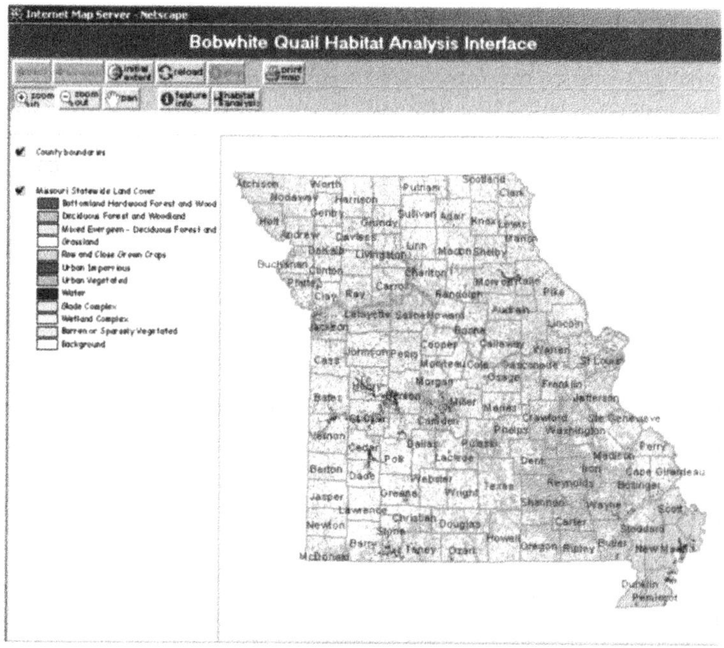

B

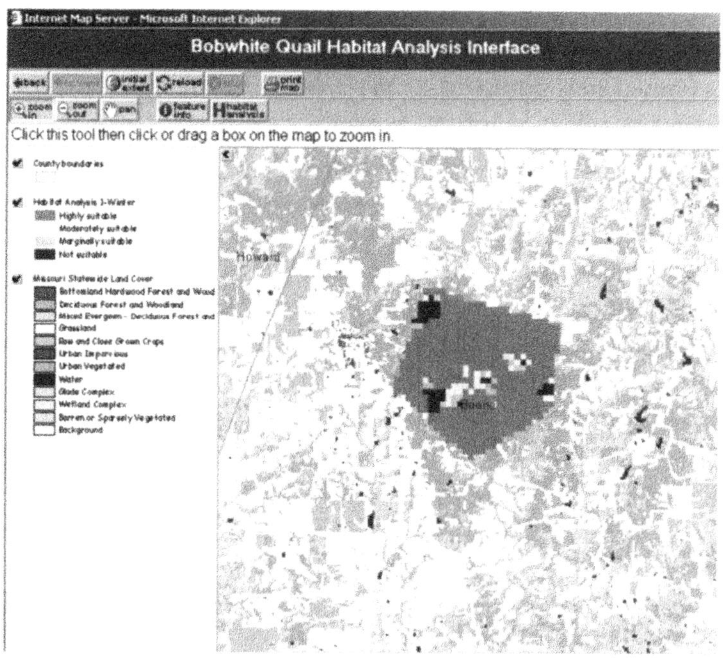

FIG. 8.9. Winter habitat suitability assessment for bobwhite quail in Missouri. (A): bobwhite quail habitat analysis interface; (B): habitat suitability surface. (From Center for Agricultural, Resource and Environmental Systems, http://www.cares.missouri.edu/projects/completed/irms/interface.html.)

Remote Sensing

Remote sensing is one of the most rapidly growing and useful information technologies for evaluating the status and condition of natural resources. Examples of phenomena that can be remotely monitored at relatively frequent time intervals and high spatial resolutions include the extent and likely impact of drought, retreat of glaciers, and quantification of human-induced landscape change. Remote sensing is a valuable tool for mapping, monitoring, and extracting information about habitat conditions (Ustin and Costick 2000), inventorying and assessing the spatial distribution of animals and ecosystem diversity (Forman 1995) (see the following sidebar for an application of remote sensing to monitoring elephant populations), understanding ecosystem form and pattern in mountain protected areas (Walsh et al. 1994), and developing and monitoring policies and management strategies (Jenik 1997). These uses of remote sensing are relevant to the habitat assessment that CLAG wants to perform.

Spatial Analysis of Elephants in Amboseli National Park

Kenya's Amboseli National Park (ANP) was established as a national park in 1974 for the exclusive use of wildlife and tourism. The 392 km^2 (151 mi^2) park lies immediately northwest of Mt. Kilimanjaro, a 5,951-m (19,340-ft)-high snowy volcanic peak, on the border with Tanzania. ANP is part of the 3,000 km^2 (1,158 mi^2) Amboseli ecosystem and is famous for its big game and scenic beauty. During the dry season, the park contains large concentrations of wildlife making it a popular destination for tourists. Five main wildlife habitats are found in ANP: open plains, acacia woodland, rocky thorn bush country, swamps, and marshland. The areas surrounding ANP are covered with dusty volcanic ash from Mt. Kilimanjaro causing ANP to be generally dusty and dry most of the year. However, the park contains springs, swamps, and marshes recharged by underground water originating as snowmelt from Mt. Kilimanjaro. Water and food attract a wide range of wildlife including elephants, zebra, wildebeests, giraffes, impala, leopards, lions, hippos, antelope, rhinos, wild dogs, hyenas, cheetah, buffalo, and more than 400 species of birds. ANP is considered one of Kenya's most popular national parks and one of the premier locations in the world for watching wild elephants. For more than four hundred years, Masai tribes have inhabited the Amboseli region. While the Masai initially protested the establishment of ANP by spearing elephants and rhinos, they have generally protected wildlife from poaching and have not tolerated elephant poaching. For this reason, Amboseli elephants are one of the few remaining undisturbed elephant populations in Africa.

Since 1972, ANP has been the site of the longest running study of elephants in the world—the Amboseli Elephant Research Project (AERP). The project team, led by Cynthia Moss, has photographically identified over 1,800 elephants according to date, time, and place. The AERP maintains data on 1,160 elephants currently living in and around the ANP. These data provide an invaluable source of baseline information about elephants. Data management in the early days of the project involved using paper maps with hand-recorded data. Satellite imagery is now used to analyze spatial and temporal changes in the movements of Amboseli elephants. For example, 61-cm (24-inch) resolution QuickBird satellite imagery from DigitalGlobe is used to identify current positions and even sizes of elephants throughout the ecosystem. This and other information have allowed AERP to determine travel patterns for Amboseli elephants. Availability of high-resolution satellite imagery allowed AERP to discontinue the traumatizing practice of tracking elephants with radio collars.

Thirty years of data have been integrated into a geographic information system (GIS) that allows AERP to answer questions about where elephants are distributed and why. ArcGIS Tracking Analyst was applied to the 30-year database to assess changes over time in family social interactions. ArcGIS Spatial Analyst is being used to determine how different groups of elephants utilize different areas of the park. Global positioning systems and mobile GIS software have proved very useful in field data collection and analysis. Other GIS applications allowed AERP to

1. Determine that a few male elephants stay in safe areas during the day and move outside the park at night to raid agricultural crops
2. Make and negotiate precise land-use planning decisions
3. Identify human-elephant conflicts by tracking changes over time in land cover, for example, from pastoral to agricultural uses
4. Accurately visualize spatial and temporal changes in the distribution of elephants, how elephants respond to changing human behaviors, and patterns of intergroup dominance among elephants

With the help of spatial information technologies, AERP has been able to produce one of the longest and most extensive data sets available for any population of large mammals.

Sources: ESRI (2002); Amboseli National Park (2003a,b).

Remote sensing is a particularly versatile and cost effective way to detect and analyze changes in land cover and landscape structure. It allows synoptic viewing, near real time mode of operation, repetitive coverage at short intervals, consistent image quality, low cost data acquisition, and easy integration with GIS layers (Sugumaran et al. 1994; Sugumaran 2002). Other advantages of remote sensing are the relatively low cost of acquiring coarse resolution data, such as TM images, greater availability of high resolution images, such as IKONOS, improved techniques for processing and interpreting satellite imagery, and the capacity to analyze landscape features (Quattrochi and Pelletier 1991). Remote sensing images for multiple time periods can be used to measure changes through time in landscape structure, such as the size, form, and pattern of patches, corridors, and background matrix. Furthermore, temporal changes in habitat suitability for target species can be evaluated by combining landscape structure changes measured using remotely sensed data with landscape-based habitat suitability criteria for those species (Griffiths and Lee 2000; Gulinck et al. 2000).

Remote sensing technologies involve use of devices ranging from handheld charge devices coupled to sensors mounted on field equipment. For instance, mounting a grain meter and global positioning system on an agricultural combine provides site-specific yield data for precision application of nitrogen to crop fields (Ustin and Costick 2000). Current and planned land observing systems fall into four categories based on the characteristics and resolution of the sensors, namely, multispectral systems, hyperspectral systems, hyperspatial systems, and radio/microwave systems (Stoney 1997).

Landsat Multispectral Scanner, Landsat Thematic Mapper (TM), SPOT, High Resolution Visible and Indian Remote Sensing, and other multispectral sensors used in satellites provide four to eight spectral bands and moderate (10–30 m) spatial resolution. The multiple spectral bands are well suited for mapping and analysis of watersheds, land use/cover, roads and infrastructure, vegetative cover and landscape fragmentation, and natural disturbance patterns (Ustin and Costick 2000).

Figure 8.10 illustrates how Landsat TM images were used to detect urban growth in St. Charles County, Missouri, in 1984, 1992, and 2000. The land-cover maps were developed using two TM scenes per year, one during leaves on and another during leaves off. To ensure that the pixels from different years overlap, the multiyear TM scenes were imported into ERDAS IMAGINE image processing software and coregistered with a root mean square error (RMSE) of 0.3 pixels. A supervised maximum likelihood algorithm was used to separately classify all six scenes. In order to improve accuracy, areas of clouds and urban areas were clipped from the images and classified separately. Training samples used to verify the accuracy of land-cover classifications came from various sources, including field survey, color infrared air photos, and USGS Digital Orthographic Quarter Quadrangles. Image accuracy ranged from 85 percent to 90 percent (Compas and Sugumaran 2004).

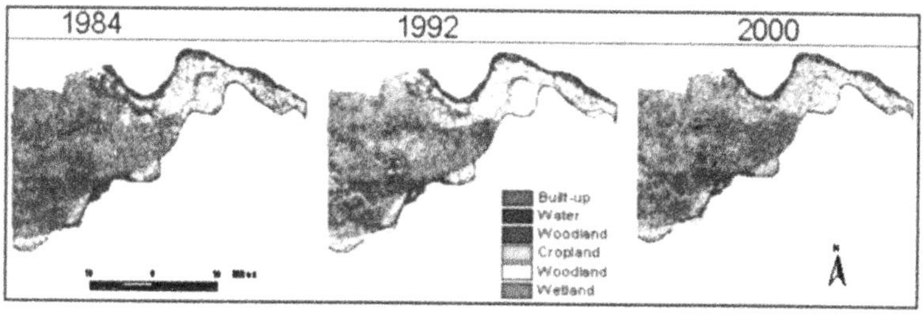

FIG. 8.10. Detecting land cover changes in St. Charles County, Missouri, with Landsat TM images. (Created by Ramanathan Sugumaran, University of Northern Iowa.)

Hyperspectral imaging systems are capable of achieving a high-resolution spectrum for each pixel (picture element) by employing large numbers of adjacent narrow wavelength spectral bands. The resulting data set, called an "image cube," has latitude, longitude, and altitude dimensions for all pixels. Pixel resolution is at least 30 m. Sensors in hyperspectral imaging systems are capable of identifying biogeochemical properties of plants, soils, and geologic materials such as the mineral, carbonate, organic matter, and iron content of soils (Ben-dor et al. 1999; Clark 1999) and plant properties like pigments, water, and cellulose (Ustin et al. 1999). This information is very useful in assessing ecosystem health and environmental pollution (Ustin and Costick 2000).

Hyperspatial imaging sensors are capable of providing imagery with very high spatial resolution similar to that of airborne photogrammetry. For example, IKONOS (Space Imaging Corporation) has a 1-m resolution panchromatic band sensor and a 4-m resolution multispectral band sensor (blue, green, red, and near infrared). Figure 8.11 illustrates an IKONOS multispectral image of the area surrounding Many Glacier Hotel in Glacier National Park.

Pointing capabilities of hyperspatial sensors make it possible to acquire stereo images that allow development of high-resolution digital elevation models and estimates of the heights of the land surface and tree canopies (Ustin and Costick 2000).

An example of radar/microwave systems is the NASA/JPL (National Aeronautics and Space Administration/Jet Propulsion Laboratory) airborne Synthetic Aperture Radar (SAR). SAR detects the water content and structural characteristics of grassland vegetation and, unlike satellite images, is not influenced by cloud cover. In addition, SAR discriminates between different grassland types provided each type has a distinguishable canopy structure (Hill 2000). LIDAR (light detection and ranging) or laser altimetry is similar to a radar-based remote sensing system. It fires a pulse of laser energy at the target surface and measures the dis-

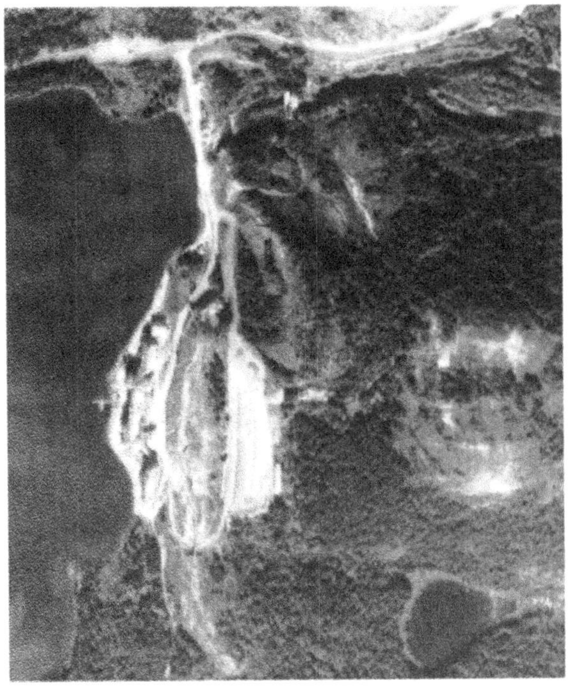

FIG. 8.11. IKONOS image of Many Glacier Hotel and vicinity in Glacier National Park, 2000. (From Center for Agricultural, Resource and Environmental Systems, University of Missouri-Columbia.)

tance between the sensor and the object based on the amount of time it takes the pulse to travel between the sensor and the object. LIDAR was used in three Apollo missions to measure the topography of the moon. Global topography has been mapped using small footprint, low-altitude, single-return LIDAR systems. Precise three-dimensional measurement of the Earth's surface can be made with medium-large footprint systems (laser beam diameters on the order of tens of meters). LIDAR provides estimates of tree canopy height and the vertical distribution of intercepting surfaces within the canopy. Forest stand age or successional stage can be estimated by combining canopy height (from LIDAR) with information on species composition and other characteristics (Dubayah et al. 2000).

As an application, consider how CLAG might use remote sensing technologies to evaluate the extent to which rapid economic growth in the GCE is fragmenting grizzly bear habitat. The approach considered here has three steps. In the first step, land-cover maps for multiple years are created by applying several classification algorithms (supervised maximum likelihood, artificial neural network, and a knowledge-based classifier) to cloud-free, multiseason (winter and summer), multiyear, and multispectral Landsat Thematic Mapper (TM) images having 30-m resolution. In the second step, landscape structure metrics are calculated using the land-cover maps for multiple years. Examples of landscape structure metrics include patch number, size, shape, and perimeter; patch size coefficient of variation, isolation, connectivity, relative richness, relative evenness, relative patchiness, matrix porosity, diversity, dominance, fractal dimension, nearest neighbor probability, contagion, edges, and vegetative cover (Forman and Godron 1986; Turner et al. 2001).

Landscape metrics are quite useful in assessing ecological impacts of landscape change. For example, fractal dimension has been used to measure the complexity of landscape patterns when comparing different landscapes, changes over time in a particular landscape, and different-sized patches (Turner et al. 2001). A patch is a nonlinear surface area that differs in appearance from its surroundings (Forman and Godron 1986). Fractal dimensions are often lower in human-dominated landscapes or cover types (Turner 1990; Mladenoff et al. 1993). Patch size appears to be positively correlated with species and/or habitat diversity (Burgass and Sharpe 1981). Patch area is a good predictor of the number of tree and bird species (Rudis and Ek 1981). The amount of edge and interior spaces affects the abundance and diversity of organisms and wetland functioning (Harris 1988; Quinn and Harrison 1988; Robinson and Quinn 1988).

Since one landscape metric is not sufficient to characterize habitat suitability, the third step involves using a set of relatively independent (uncorrelated) landscape metrics to identify landscape patterns that directly influence habitat suitability for grizzly bears. Landscape metrics can be used to track changes through time in "1) the proportion of the landscape containing suitable habitat, 2) the number of patches of suitable habitat above the minimum size required by the animal, and 3) the connectivity of suitable habitat as measured by interpatch distances" (Turner et al. 2001). Of particular interest in the third step is whether landscapes that favor survival of grizzly bears are being compromised by actual or future development. The grizzly bear habitat suitability map that results from this three-step evaluation method is similar to the one shown in figure 8.12 for probability of occurrence of suitable habitat for eastern timber wolves in the upper midwestern United States.

Global Positioning System

A global positioning system (GPS) is a technology for determining the position (latitude and longitude), height, and velocity of an object. A GPS consists of a constellation of satellites that emits specially coded radio signals and a receiver that processes those signals to determine the latitude, longitude, and height of the receiver. The satellite constellation consists of four satellites in each of six orbits around the Earth for a total of 24 satellites. A GPS receiver calculates the velocity of a moving object, such as a vehicle or airplane, based on repeated measurements

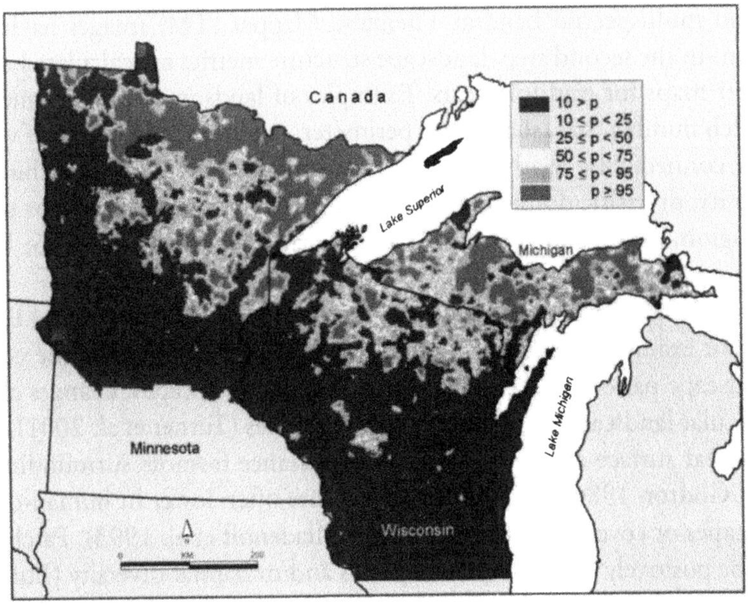

FIG. 8.12. Map of probability of occurrence of suitable wolf habitat in the upper Midwest. (From Turner et al. 2001, p. 215.)

of the object's position through time. Spatial resolution for GPS ranges from 100 m with lower-cost systems to sub-cm accuracy with higher-cost differential GPS. The most widespread application of GPS is in navigation of aircraft, ships, ground vehicles, and individuals on foot. Linking an automobile-mounted GPS receiver to an onboard computer-based road atlas allows a driver to identify alternative routes (fastest, shortest, etc.) to a selected destination and the estimated time of arrival. GPS was originally developed and extensively used by the military, although there has been tremendous growth in civilian applications (Dana 2002).

GPS technology is extremely useful in locating points, vectors, and polygons in a vector GIS. For example, a GPS could be used to determine the boundaries for different types of vegetation and burn areas and track the movements of grizzly bears and other species in the GCE. GPS technology is particularly useful when a user, such as CLAG, cannot afford to buy detailed satellite imagery (such as IKONOS) for an area. Another application of GPS is to verify GIS data obtained by other means. GPS technology has greatly improved the capacity to manage spatial variability in soil types and associated crop yields. Two applications are noteworthy. First, a map of spatial variability in yields for a crop field can be determined using the data collected by a combine-mounted GPS and grain yield meter. Second, attaching a GPS to a magnetic induction device provides grid-based measurements of soil depth throughout a crop field. These measurements are used to determine the optimal fertilization application rates for different areas of the field.

Interactive Mapping

Web-based interactive mapping is a very powerful way to create maps and identify and evaluate landscape features. Websites utilizing interactive mapping permit creation of custom maps for an area, application of interpretive functions to the map (zoom in/out, determine geographic coordinates of a point, clip selected layers, determine the area of a delineated polygon, etc.), creation of a list of attributes of map features (properties of a soil type, the size and perimeter of a lake, etc.), and map printing. Interactive mapping offers several benefits. First, the user does not need to purchase GIS software or know how to use it. Second, the user does not have to purchase or acquire spatial data. Third, the only computer software needed to do web-based interactive mapping is an Internet browser. Fourth, it is easy for the provider to update data used to produce maps. Fifth, the provider does not have to work with each user individually, which significantly reduces the cost of providing spatial information and analysis.

Interactive mapping allows CLAG to create maps of suitable habitat for grizzly bears under current development patterns and alternative futures. Several interactive mapping applications appear on the website for the Center for Agricultural, Resource, and Environmental Systems (www.cares.missouri.edu). The interactive mapping tool illustrated here is for Dardenne Creek Watershed, which is located west of St. Louis, Missouri. The tool employs three easy steps illustrated in figure 8.13A. In step 1, the first tab is clicked, which allows the user to specify the area of interest. In this example, Dardenne Creek Watershed is chosen. In step 2, the user selects the data layers to be mapped. Three data layers are selected for the illustration: streams and creeks (1:24,000), lakes and ponds (1:24,000), and land cover for 1999. In step 3, the user verifies the selected layers and makes a map by clicking the "make map" button. The resulting map is shown in figure 8.13B. The left panel contains the legends for the four attributes being mapped. The top of the window contains several buttons that allow various functions to be applied to the map. For example, the zoom-in button allows a user to zoom in on a subwatershed area.

Computerized Landscape Visualization

Information technologies described thus far allow CLAG to compare the biophysical and socioeconomic impacts of alternative futures. However, they do not allow users to fully visualize the appearance of a landscape under alternative futures. For example, GIS allows the creation of two-dimensional maps of land cover under alternative futures. Unfortunately, these maps do not portray the full visual impacts of a scenario. Remote sensing captures current and historical images but does not allow visualization of the likely impacts of alternative futures.

A

B

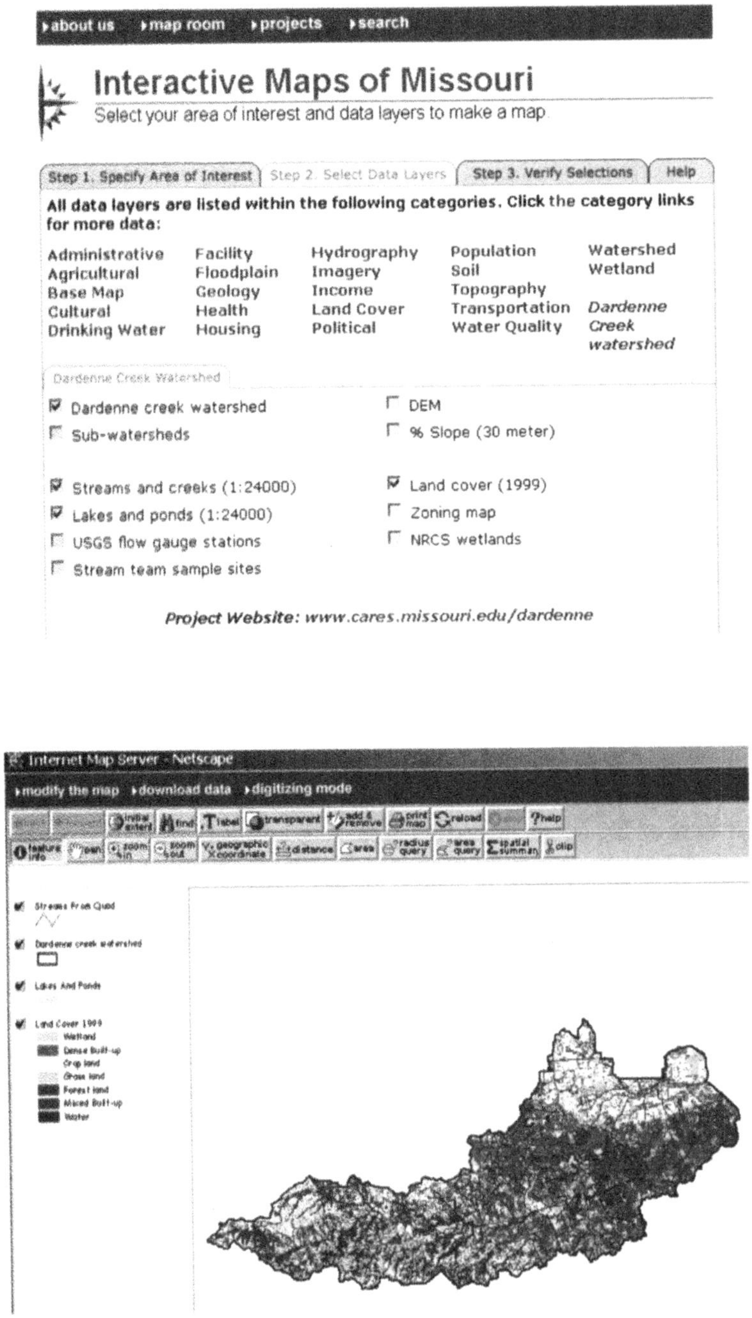

FIG. 8.13. Interactive mapping for Dardenne Creek watershed. (A): interactive map menu; (B): map of Dardenne Creek watershed. (From Center for Agricultural, Resource and Environmental Systems [www.cares.missouri.edu] at the University of Missouri-Columbia.)

Computerized landscape visualization helps overcome these limitations of GIS and remote sensing. Two visualization tools are described here. CommunityViz is a GIS-based computerized landscape visualization tool that incorporates three modules: SiteBuilder 3D, Scenario Constructor, and Policy Simulator (http://www.communityviz.com). The SiteBuilder 3D module allows the user to construct 3D digital images and do a virtual "fly-through" of a simulated landscape, such as one created under an alternative future. Scenario Constructor allows quantitative comparison of alternative futures based on a set of criteria chosen by the user. Policy Simulator uses agent-based modeling techniques to forecast the likely future economic and demographic impacts of alternative futures. CommunityViz is designed for a variety of applications including land-use planning, natural resource management, urban planning, community planning, and landscape architecture. It helps planning departments, citizens, and developers anticipate and rectify potential problems with proposed community development plans.

CommunityViz has been used in several western communities including Sante Fe, New Mexico, Lyons, Colorado, and Missoula, Montana (Larmer 2002). In Missoula, homeowners, planners, and wildfire professionals used CommunityViz to assess fire risk for homes and communities in the wildland-urban interface. The Firewise Scenario Constructor developed for Missoula allows users to interactively sketch wildfire risk scenarios and evaluate them based on community objectives and constraints. Other components include impact analysis, indicator tracking, comparison of alternatives, and benefit-cost analysis. CommunityViz is suitable for assessing alternative treatments such as tree and brush clearing near homes; prescribed burns; additional water sources; methods of construction for roof, siding, and decks; and strategic placement of fire protection resources. It also allows visualization of a proposed build-out or subdivision in the wildland-urban interface.

Computerized landscape visualization was used to visualize a central Iowa landscape in 2025 with three landscape policies that differ in terms of the relative importance of agricultural production, water quality, and biodiversity (Nassauer et al. 2002). Assumptions common to the three policies are (1) the landscape embodies profitable agricultural production by private landowners, public concerns for water quality, and biodiversity; (2) the landscape is affected by global markets, international and national agricultural policy, and national, state, and local environmental policy; and (3) agricultural and environmental policy reflect societal perceptions, values, and concerns.

Assumptions and landscape consequences unique to each policy are as follows:

Existing Landscape: Land-cover patterns have been shaped by 1994 agricultural policy that supports crop prices, limits production, controls erosion, and improves wildlife habitat.

Landscape Scenario I: The dominant policy objective is profitable agricultural production, and profit is perceived as short-term economic return. Land-cover patterns respond to high world demand for Corn Belt grain crops with high use of production inputs.

Landscape Scenario II: Land-cover patterns evolve in response to federal policy that has enforced clear, measurable water quality performance standards for nonpoint source pollution of surface and groundwater on a farm-by-farm basis. Land-cover patterns and agricultural enterprises change in order to achieve mandated reductions in surface and groundwater pollution in the most efficient way.

Landscape Scenario III: Land-cover patterns are dramatically changed by policy interventions designed to increase indigenous biodiversity. Public investments in biodiversity are used to maintain and restore native plants and animals within a landscape that continues to support profitable agricultural production. Land-cover patterns change to increase habitat with innovative, profitable, agricultural production methods.

Digital images for the three landscape policy scenarios are illustrated in figure 8.14.

Computerized landscape visualization could be used in the GCE case study to visualize how alternative futures specified by CLAG are likely to change landscape appearance in areas surrounding the park. While landscape structure is more meaningful in assessing grizzly bear habitat than landscape appearance, the latter could influence the preferences of CLAG members for alternative futures and habitat protection strategies.

Summary

Several principles and methods are useful in planning and managing protected areas and in the assessment and ranking of future management actions and alternative futures for a greater ecosystem. Benefit-cost analysis (BCA) allows determination of the economic efficiency of alternative futures based on net present value, annual net benefit, benefit-cost ratio, and internal rate of return. A positive net present value and annual net benefit, a benefit-cost ratio greater than one, and an internal rate of return that exceeds the market interest rate indicates an alternative future is economically efficient. BCA does not account for negative impacts of development on nonpriced ecological services, such as grizzly bear habitat in a greater ecosystem. Cost effectiveness analysis arrays specific habitat protection strategies in order of increasing cost, but does not allow determination of the most economically efficient level of habitat protection.

Economic impact analysis employs models such as IMPLAN to estimate the impacts of future increases in final expenditures on goods and services (as specified by an alternative future) on total output, household income, and employment in the affected area.

Landscape change and ecological economic impacts can be modeled as a two-stage procedure. In the first stage, alternative futures are selected for evaluation. Each scenario specifies future changes in final sales on goods and services in all economic sectors. An alternative future is translated into future land uses in the ecosystem. Changes in land use entail the conversion of undeveloped land uses to developed land uses. In the second stage, future land requirements for an alternative future are allocated to locations on the landscape using transition probabilities estimated from historical land-cover maps for the ecosystem. Changes in land use are allocated to land parcels using techniques such as cellular automata, local transition probabilities, or multiple attribute evaluation. Impacts of future land cover/use changes on habitat suitability are evaluated by calculating landscape structure metrics using future land-cover maps and interpreting the implications of those metrics for habitat suitability.

Multiple attribute decision making (MADM) is valuable for ranking alternative futures based on their multiple attributes. Ranking employs a utility function that is a weighted sum of economic and ecological attributes of alternative futures. Application of MADM to a set of alternative futures requires selecting and quantifying the attributes of alternative futures, determining alternative futures that provide efficient combinations of attributes, eliminating unsustainable alternative futures, estimating attribute weights, and ranking alternative futures based on a utility function. Conflicts between stakeholders can arise when they have different rankings of alternative futures. Such conflict can be resolved by applying consensus-based decision making to the attribute weights or alternative futures, or voting for attributes or alternative futures.

AM posits that imperfect human understanding of ecosystems requires treating management actions as experiments for generating data and information to test hypotheses regarding the behavior of ecosystems. An AM approach can be implemented using a two-stage procedure that uses Bayes' rule to calculate posterior probabilities that are used to determine the most likely state of an ecosystem.

AM is not suited for handling uncertainty regarding which alternative future is most likely to occur. This source of uncertainty could be handled by first determining whether a range of alternative futures have significantly different ecological and economic impacts. If impacts differ, then the most likely scenario is determined using a Delphi approach. Another source of uncertainty surrounds the economic and ecological impacts of a particular scenario. This source can be handled using interval estimates rather than point estimates of economic and ecological attributes of alternative futures.

a

b

FIG. 8.14. Computerized landscape visualization for three central Iowa landscape scenarios: (a) existing landscape, (b) landscape scenario I, (c) landscape scenario II, (d) landscape scenario III. (From Nassauer et al. 2002.) *(continues)*

Optimization involves the mathematical formulation and solution of a decision problem. Benefit-cost analysis is a simple optimization problem. MADM problems do not necessarily have optimal solutions. Simulation generates the values of a variable over space and/or time using a deterministic process model such as IMPLAN or STELLA. Integrated ecological economic models are useful in explaining the relationships between economic and ecological processes and com-c

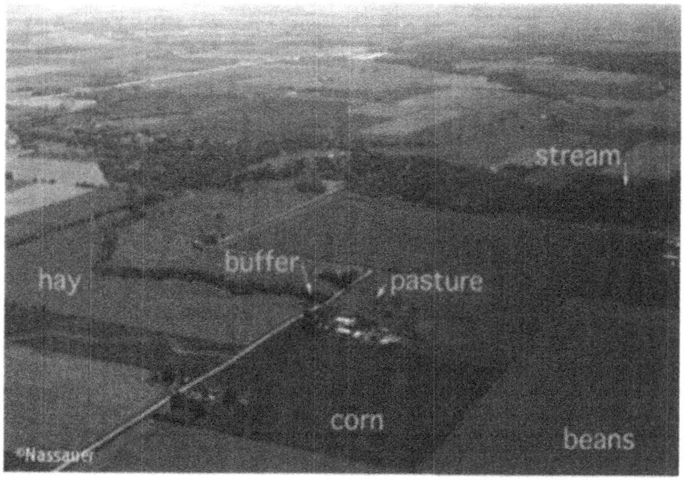

d

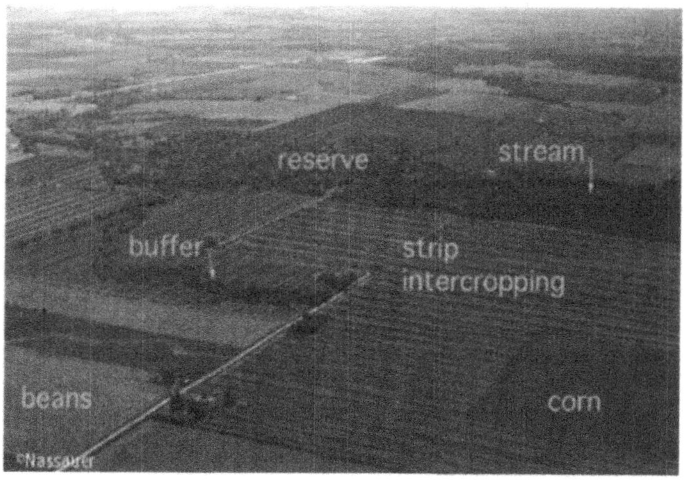

FIG. 8.14. *(continued)*

paring strategies for improving the sustainability of an ecological economic system.

Geographic information systems are excellent for complex and comprehensive spatial analysis of site-specific features and processes and environmental modeling. Rapid advancements in the availability, accuracy, and cost of remotely sensed data render this technology extremely useful in monitoring and evaluating the status and condition of natural resources, including habitat suitability. Many types of satellite- and airborne-based sensors are available with wide-ranging spectral capabilities, accuracies, and applications. A global positioning system (GPS) determines latitude, longitude, height, and velocity of objects such vehicles and ships. GPS is useful in natural resource planning and management for verifying the size and location of objects, such as the boundaries of a national park or burn area.

Interactive mapping allows users to create maps and evaluate map objects on the Internet. Operations available in interactive mapping include creation of custom maps, applying interpretive functions to the map, listing attributes of map features, modifying the map, and printing it. Computerized landscape visualization allows users to visualize the appearance of a landscape under different alternative futures and policy options using 3D digital images and virtual "fly-through."

References

Agarwal, C., G. L. Green, M. Grove, T. Evans, and C. Schweik. 2000. *A review and assessment of land-use change models: Dynamics of space, time, and human choice.* Bloomington, IN: Center for the Study of Institutions, Population, and Environmental Change.

Alig, R. J. 1986. Econometric analysis of the factors influencing forest acreage trends in the southeast. *Forest Science* 32: 119–134.

Amboseli National Park. 2003a. http://www.kws.org/amboseli.htm (accessed April 4, 2003).

———. 2003b. http://www.pbs.org/wnet/nature/echo/html/body_park.html (accessed April 4, 2003).

Baker, W. L. 1989. A review of models of landscape change. *Landscape Ecology* 2: 111–133.

Bakus, G. J., W. G. Stillwell, S. M. Latter, and M. C. Wallerstein. 1982. Decision making: With application for environmental management. *Environmental Management* 6: 493–504.

Barbier, E. B., J. C. Burgess, and C. Folke. 1995. Background and overview (chap. 1). *Paradise lost? The ecological economics of biodiversity.* London, England: Earthscan Publications, Ltd.

Ben-dor, E., J. R. Irons, and G. Epema. 1999. Soil reflectance. In *Manual of remote sensing for the earth sciences,* ed. A. Rencz, 3: 111–188. Bethesda, MD: American Society of Photogrammetry and Remote Sensing.

Bergerud, W. A., and W. J. Reed. 1998. Bayesian statistical methods. In *Statistical methods for adaptive management studies,* ed. V. Sit and B. Taylor, 89–104. Land Management Handbook No. 42. Research Branch, Ministry of Forestry, Victoria, British Columbia.

Berkman, H. E., and C. F. Rabeni. 1987. Effect of siltation on stream fish communities. *Environmental Biology of Fishes* 18: 285–294.

Bishop, R. C. 1978. Endangered species and uncertainty: The economics of a safe minimum standard. *American Journal of Agricultural Economics* 60: 10–18.

———. 1993. Economic efficiency, sustainability, and biodiversity. *Ambio* 22: 69–73.

Boyce, M. S., and L. L. McDonald. 1999. Relating populations to habitats using resource selection functions. *Trends in Ecology and Evolution* 14: 268–272.

Brown, D. G., B. C. Pijanowski, and J. D. Duh. 2000. Modeling relationships between land use and land cover on private lands in the Upper Midwest, USA. *Journal of Environmental Management* 59: 247–263.

Burgass, R. L., and D. M. Sharpe (Eds.). 1981. *Forest Island dynamics in man-dominated landscapes. Ecological studies: Analysis and synthesis.* New York: Springer.

Burger, L., E. Kurzejeski, and others. 1998. Habitat model to predict landscape use of northern bobwhite in Missouri. Unpublished report. Department of Wildlife and Fisheries, Mississippi State University, Mississippi State, MS.

Clark, R. N. 1999. Spectroscopy of rocks and minerals and principles of spectroscopy. In *Manual of remote sensing for the earth sciences,* ed. A. Rencz, 3: 3–38. Bethesda, MD: American Society of Photogrammetry and Remote Sensing.

Clarke, K. C., L. Gaydos, and S. Hoppen. 1997. A self-modifying cellular automation model of historical urbanization in the San Francisco Bay area. *Environment and Planning B* 24: 247–261.

Compas, E., and R. Sugumaran. 2004. Urban growth prediction model on the web: A decision support tool for community planners. *Proceedings of 27th Annual Applied Geography Conference,* Oct. 20–24. St. Louis, MO.

Conroy, M. J. 2000. Mapping biodiversity for conservation and land use decisions. In *Spatial information for land use management,* ed. M. J. Hill and J. Aspinall, 145–158. The Netherlands: Gordon and Breach Science Publishers.

Cooksey, R. W. 1996. *Judgment analysis: Theory, methods and applications.* Sydney, Australia: Academic Press.

Costanza, R., and C. Folke. 1997. Valuing ecosystem services with efficiency, fairness and sustainability as goals. In *Nature's services: Societal dependence on natural ecosystem,* ed. G. C. Daily, 49–70. Island Press, Washington, DC.

Costanza, R., and S. Tognetti (Eds.). 1996. *Integrated adaptive ecological and economic modeling and assessment—a basis for the design and evaluation of sustainable development programs.* Draft synthesis paper, Scientific Committee on Problems of the Environment (SCOPE), 5 Bld de Montmorency, 75016, Paris, France.

Costanza, R., L. Wainger, C. Folke, and K-G. Maler. 1993. Modeling complex ecological economic systems: Toward and evolutionary, dynamic understanding of people and nature. *BioScience* 43: 545–555.

Costanza, R. E. 1991. Assuring sustainability of ecological economic systems. In *Ecological economics: The science and management of sustainability,* ed. R. Costanza, 331–343. New York: Columbia University Press.

Crowards, T. M. 1998. Safe minimum standards: Costs and opportunities. *Ecological Economics* 25: 303–314.

Dana, P. H. 2002. Geographer's craft project. Department of Geography, The University of Colorado-Boulder. Global Positioning System Overview, http://www.colorado.edu/geography/gcraft/notes/gps/gps_f.html (accessed October 10, 2002).

Deadman, P., R. D. Brown, and H. R. Gimblett. 1993. Modeling rural residential settlement patterns with cellular automata. *Journal of Environmental Management* 37: 147–160.

Dubayah, R., R. Knox, M. Hofton, J. Bryan Blair, and J. Drake. 2000. Land surface characterization using Lidar remote sensing. In *Spatial information for land use management,* ed. J. Hill and R. J. Aspinall, 25–38. The Netherlands: Gordon and Breach Science Publications.

Ellison, A. M. 1996. An introduction to Bayesian inference for ecological research and environmental decision-making. *Ecological Applications* 6: 1036–1046.

ESRI. 2002. African elephant cause aided by the GIS community. *ARCNEWS* 24: 12.

Fausch, K. D., C. L. Hawkesand, and M. G. Parsons. 1988. Models that predict standing crop of stream fish from habitat variables: 1950–85. General Technical Report PNW-213, U.S. Forest Service, Washington, DC.

Fischer, G., M. Makowski, and J. Antoine. 1996. Multiple criteria land use analysis. Working Paper WP-96-006. Laxenburg, Austria: International Institute of Applied Systems Analysis.

Folke, C. 1999. Ecological principles and environmental economic analysis. In *Handbook of environmental and resource economics,* ed. J. Van den Bergh. UK: Edward Elgar Publishing.

Foltz, J. C., J. G. Lee, M. A. Martin, and P. V. Preckel. 1995. Multiattribute assessment of alternative cropping systems. *American Journal of Agricultural Economics* 77: 408–20.

Forman, R. T. T. 1995. *Land mosaics: The ecology of landscapes and regions.* New York/Cambridge: Cambridge University Press.

Forman, R. T. T., and M. Godron. 1986. *Landscape ecology.* New York: John Wiley & Sons.

Geoghegan, J., L. A. Wainger, and N. E. Bocksteal. 1997. Spatial landscape indices in a hedonic framework: An ecological economics analysis using GIS. *Ecological Economics* 23: 251–264.

GIS.com. 2002. *Your internet guide to GIS (Geographic Information Systems).* http://www.gis.com/ (acessed October 6, 2002).

Glacier National Park. 2002. *Logan Pass trails.* http://www.nps.gov/glac/maps/trailslp.htm (accessed October 1, 2002).

Goodchild, M. F. 2000. Spatial analysis: Methods and problems in land use management. In *Spatial information for land use management,* ed. J. Hill and R. J. Aspinall, 39–52. The Netherlands: Gordon and Breach Science Publications.

Goulder, L. H., and D. Kennedy. 1997. Valuing ecosystem services: philosophical bases and empirical methods. In *Nature's services: Societal dependence on natural ecosystems,* ed. G. C. Daily, 23–47., Washington, DC: Island Press.

Graham, R. *Modeling an ecosystem with STELLA.* 2002. http://archive.ncsa.uiuc.edu/Pubs/access/92.4/92.4Stella.html (Accessed August 26, 2002).

Griffiths, G., and H. J. Lee. 2000. Landscape pattern and species richness; Regional scale analysis from remote sensing. *International Journal of Remote Sensing* 21: 2685–2704.

Gulinck, H., H. Dufourmont, P. Coppin, and M. Hermy. 2000. Landscape research, landscape policy and Earth observation. *International Journal of Remote Sensing* 21: 2541–2554.

Haimes, Y. Y., and W. A. Hall. 1974. Multiobjectives in water resource systems: the surrogate trade-off method. *Water Resources Research* 10: 615–624.

———. 1977. Multiobjective analysis in the Maumee River Basin: A case study on level-b planning. Report SED-WRG-77-1. Cleveland, OH: Case Western University.

Haines-Young, R., D. R. Green, and S. Cousins. 1993. Landscape ecology and spatial information systems. In *Landscape ecology and GIS,* ed. R. Haines-Young, D. R. Green, and S. H. Cousins, 3–8. London: Taylor & Francis Ltd..

Halbert, C. 1993. How adaptive is adaptive management? Implementing adaptive management in Washington State and British Columbia. *Reviews in Fisheries Science* 1: 261–263.

Haley, K. B. 1990. Operational research and management in fishing. In *Operations research and management in fishing,* ed. A. G. Rodrigues. NATO ASI Series, Series E: Applied Sciences, Vol. 189. Dordrecht, Netherlands: The Kluwer Academic.

Hall, G. F., D. E. Strebel, and P. J. Sellers. 1988. Linking knowledge among spatial scales: Vegetation, atmosphere, climate and remote sensing. *Landscape Ecology* 2: 3–22.

Hansen, M. J., S. E. Franklin, C. G. Woudsma, and M. Peterson. 2001. Caribou habitat mapping fragmentation analysis using Landsat MSS, TM, and GIS data in the North Columbia mountains British Columbia, Canada. *Remote Sensing of Environment* 77: 50–65.

Harris, L. D. 1988. The nature of cumulative impacts on biotic diversity of wetland vertebrates. *Environmental Management* 12: 675–693.

Hilborn, R. 1992. Can fisheries agencies learn from experience? *Fisheries* 17: 6–14.

Hilborn, R., C. J. Walters, and D. Ludwig. 1995. Sustainable exploitation of renewable resources. *Annual Review of Ecology and Systematics* 26: 45–67.

Hill, M. J. 2000. Applications for spatial data in grassland monitoring and management. In *Spatial Information for Land Use Management,* ed. J. Hill and R. J. Aspinall, 113–128. The Netherlands: Gordon and Breach Science Publications.

Holling, C. S. 1978. *Adaptive environmental assessment and management.* Chichester, England: John Wiley & Sons.

Hulbert, S. H. 1984. Pseudoreplication and the design of ecological field experiments. *Ecological Monographs,* 54: 157–211.

Hunsaker, C. T., D. A. Levine, S. P. Timmins, B. L. Jackson, and R. V. O'Neill. 1992. Landscape characterization for assessing regional water quality. In *Ecological indicators,* ed. D. H. McKenzie, D. E. Hyatt, and V. J. McDonald, 997–1006. London, England: Elsevier.

Irwin, L. L., and T. B. Wigley. 1993. Toward and experimental basis for protected forest wildlife. *Ecological Applications* 3: 213–217.

Jenik, J. 1997. The diversity of mountain life. In *Mountains of the world: A global priority,* ed. B. Masserli and J. D. Ives. New York/London: The Parthenon Publishing Group.

Johnson, K. N. 1997. Science-based assessments of the forests of the Pacific Northwest. In *Creating a forestry for the 21st century: The science of ecosystem management,* ed. K. A. Kohm and J. F. Franklin, 397–409. Washington, DC: Island Press.

Johnson, V. K. 2001. Trends in rural residential development in the Greater Yellowstone ecosystem since the listing of the grizzly bear, 1975–1998. *Yellowstone Science* 9: 2–9.

Joubert, A. R., A. Leiman, H. M. de Klerk, S. Katua, and J. C. Aggenbach. 1997. Fynbos (fine bush) vegetation and the supply of water: A comparison of multi-criteria decision analysis. *Ecological Economics* 22: 123–140.

Kangas, J. 1994. An approach to public participation in strategic forest management planning. *Forest Ecology and Management* 70: 75–88.

Karr, J. R., K. D. Fausch, P. L. Andermeier, P. R. Yant, and I. J. Schlosser. 1986. Assessing *biological integrity in running waters: a method and its rational.* Illinois Natural History Survey, Special Publication 5, Champaign, IL.

Keen, P. G. W., and M. S. S. Morton. 1978. *Decision support systems: An organizational perspective.* Reading, MA: Addison-Wesley.

Keeney, R. L., and H. Raiffa. 1976. *Decisions with multiple objectives: Preferences and value tradeoffs.* New York: John Wiley & Sons.

Kohm, K. A., and J. F. Franklin. 1997. Introduction. In *Creating forestry for the 21st century: The science of ecosystem management,* ed. K. A. Kohm and J. F. Franklin, 1–5. Washington, DC: Island Press.

Landis, J. D. 1994. The California urban futures model: A new generation of metropolitan simulation models. *Environment and Planning B* 21: 399–420.

Larmer, P. 2002. A new planning tool takes flight. *High Country News* 34: 6.

Lee, K. N. 1993. *Compass and gyroscope: Integrating science and politics for the environment.* Washington, DC: Island Press.

Lindall, S., and D. Olson. 1993. *MICRO IMPLAN 1990/1985 database documentation.* St. Paul, MN: Minnesota IMPLAN Group.

Luque, S. S. 2000. The challenge to manage the biological integrity of nature reserves: A landscape ecology perspective. *International Journal of Remote Sensing* 21: 2613–2643.

Makowski, M. 1994. *Methodology and a modular tool for multiple criteria analysis of lp models.* Working Paper WP-94-102. Laxenburg, Austria: International Institute of Applied Systems Analysis.

Matthias, R., and Hannon, B. 1997. *Modeling dynamic economic systems.* New York: Springer-Verlag.

McClain, R. J., and R. G. Lee. 1996. Adaptive management: Promises and pitfalls. *Environmental Management* 20: 437–448.

McGarigal, K., and B. Marks. 1995. *FRAGSTATS: Spatial pattern analysis program for quantifying landscape structure.* General Technical Report PNW-GTR-351. USDA Forest Service, Pacific Northwest Research Station, Portland, OR.

Meffe, G. K., L. A. Nielsen, R. L. Knight, and D. A. Schenborn. 2002. *Ecosystem management: Adaptive, community-based conservation.* Washington, DC: Island Press.

Mendenhall, W., 1975. *Introduction to probability and statistics.* 4th ed. Belmont, CA: Wadsworth Publishing Co.

Miller, G. T., Jr., 1990. *Resource conservation and management.* Belmont, CA: Wadsworth Publishing Co.

Mladenoff, D. J., M. A. White, J. Pastor, and T. R. Crow. 1993. Comparing spatial patterns in unaltered old-growth and disturbed forest landscapes. *Ecological Applications* 3: 294–306.

Munda, G., P. Nijkamp, and P. Rietveld. 1994. Qualitative multicriteria evaluation for environmental management. *Ecological Economics* 10: 97–112.

Nassauer, J. I., R. C. Corry, and R. M. Cruse. 2002. Alternative future landscape scenarios: A means to consider agricultural policy. *Journal of Soil and Water Conservation* 57: 2044–2053. (http://www.snre.umich.edu/nassauer/lab_index.html (accessed October 22, 2002).

National Park Service. 1993. Science and the national parks II. Adapting to change. U.S. Department of the Interior, Technical Report NPS D-899, Washington, DC.

———. 2002. *Going-to-the-Sun Road Citizen's Advisory Committee.* http://www.nps.gov/glac/gtsr/advisory/cac.htm (accessed November 2, 2002).

Natural Resources Conservation Service (NRCS), U. S. Department of Agriculture. 1999. *Summary report 1997: Natural resources inventory*. Iowa State University Statistical Laboratory. http://www.nhq.nrcs.usda.gov/NRI.

Noss, R. F., C. Carroll, K. Vance-Borland, and G. Wuerthner. 2002. A multicriteria assessment of the irreplaceability and vulnerability of sites in the Greater Yellowstone ecosystem. *Conservation Biology* 4: 895–908.

Noss, R. F., G. Wuerthner, K. Vance-Borlund, and C. Carroll. 2001. *A biological conservation assessment for the Greater Yellowstone ecosystem*. Report to the Greater Yellowstone Coalition, Bozeman, MT. Conservation Science, Corvallis, OR. Available from www.conservationscience.com. (accessed December 2001).

O'Neill, R. V., J. R. Krummel, R. H. Gardner, G. Sugihara, B. Jackson, D. L. DeAngelis, B. T. Milne, M. G. Turner, B. Zygmunt, S. W. Christensen, V. H. Dale, and R. L. Grahm. 1988. Indices of landscape pattern. *Landscape Ecology* 3: 153–162.

Parma, A. M., and the NCEAS Working Group on Population Management. 1998. What can adaptive management do for our fish, forest, food, and biodiversity. *Integrative Biology* 1: 16–26.

Parton, W. J., D. S. Oijima, and D. S. Schimel. 1995. Models to evaluate soil matter storage and dynamics. In *Structure and organic matter storage in agricultural soils*, ed. M. R. Carter and B. A. Stewart. Boca Raton, FL: CRC Press.

Perrings, C. A. 1994. Biotic diversity, sustainable development, and natural capital. In *Investing in natural capital: The ecological economics approach to sustainability*, ed. A. M. Jannson, M. Hammer, C. Folke, and R. Costanza, 92–112. Island Press, Washington, DC.

Pfleiger, W. L., and T. B. Grace. 1997. *Changes in the fish fauna of the lower Missouri River, 1940–83*. In *Community and evolutionary ecology of North American stream fishes*, ed. W. J. Matthews and D. C. Heins, 166–177. Norman, OK: University of Oklahoma Press.

Pijanowski, B. C., S. H. Gage, D. T. Long, and W. E. Cooper. 2000. A land transformation model for the Saginaw Bay watershed. In *Landscape ecology: A top-down approach*, ed. J. Sanderson and L. D. Harris, 183–198. Boca Raton, FL: Lewis Publishers.

Prato, T. 1998. *Natural resource and environmental economics*. Ames: Iowa State University Press.

———. 1999. Multiple attribute decision analysis for ecosystem management. *Ecological Economics* 30: 207–222.

———. 2000a. Multiple attribute Bayesian analysis of adaptive ecosystem management. *Ecological Modelling* 133: 181–193.

———. 2000b. Multiple attribute evaluation of landscape management. *Journal of Environmental Management* 60: 325–337.

Prato, T., and M. Hamed. 1999. Restoring natural habitats: Economic impacts of the Big Muddy National Fish and Wildlife Refuge. *Impact Assessment and Project Appraisal* 17: 227–241.

Prato, T., and S. Hajkowicz. 1999. Selection and sustainability of land and water resource management systems. *Journal of the American Water Resources Association* 35: 739–752.

Quattrochi, D. A., and R. E. Pelletier. 1991. Remote sensing for analysis of landscapes: An introduction. In *Quantitative methods in landscape ecology*, ed. M. G. Turner and R. H. Gardner. New York: Springer Verlag.

Quinn, J. F., and S. P. Harrison. 1988. Effects of habitat fragmentation and isolation on species richness: Evidence from biogeographic patterns. *Oceologia* 75: 132–140.

Rabeni, C. F., and M. A. Smale. 1995. Effects of siltation on stream fishes and potential mitigating role of the buffering riparian zone. *Hydrobiologia*, 303: 211–219.

Ramankutty, N., and J. A. Foley. 1999. Estimating historical changes in global land cover: Croplands from 1700 to 1992. *Global Biogeochemical Cycles* 13: 997–1028.

Robinson, G. R., and J. F. Quinn. 1988. Extinction, turnover and species diversity in an experimentally fragmented California annual grassland. *Oecologia* 76: 71–82.

Rudis, V. A., and A. R. Ek. 1981. Optimization of forest island spatial patterns: methodology for analysis of landscape pattern. In *Forest Island dynamics in man-dominated landscapes,* ed. R. L. Burgess, D. M. Sharpe, 241–256. New York: Springer.

Saaty, R. W. 1987. The analytic hierarchy process—what it is and how it is used? *Mathematical Modelling* 9: 161–176.

Schumaker, N. H. 1998. *A user's guide to the PATCH model.* EPA/600/R-98/135/.U.S. Environmental Protection Agency, Environmental Research Laboratory, Corvallis, OR.

Smale, M. A., and C. F. Rabeni. 1991. The *effects of special area land treatment (SALT) agricultural management practices on the biological health of headwater streams.* A final report submitted to the Missouri Department of Natural Resources, Water Quality Division, Jefferson City, MO.

Smith, A. D. M., and C. J. Walters. 1981. Adaptive management of stock recruitment systems. *Canadian Journal of Fisheries and Aquatic Sciences* 38: 690–703.

Smith, G. R., 1997. Making decisions in a complex and dynamic world. In *Creating forestry for the 21st century: The science of ecosystem management,* ed. K. A. Kohm and J. F. Franklin, 419–435. Washington, DC: Island Press.

Smith, P. G., and J. B. Theberge. 1987. Evaluating natural areas using multiple criteria: Theory and practice. *Environmental Management* 11: 447–460.

Solecki, W. D. 2001. The role of global-to-local linkages in land use/land cover changes in South Florida. *Ecological Economics* 37: 339–356.

Stoltzenburg, W. 1996. Extinction for the record. *Nature Conservancy,* May/June, p. 6.

Stoney, W. E. 1997. Land sensing satellites in the year 2000. In *IGARSS'97. Proceedings of the International Geoscience and Remote Sensing Symposium, Singapore,* ed. T. L. Stein. IEEE Geoscience and Remote Sensing Society, TX.

Strassert, G., and T. Prato. 2002. Selecting farming systems using a new multiple criteria decision model: The balancing and ranking method. *Ecological Economics* 40: 269–277.

Sugumaran, R., G. Sandhya, K. S. Rao, R. N. Jadhav, and M. M. Kimothi. 1994. Potential of satellite data in delineation of wastelands (mainly salt affected land) and correlation with ground information. *Indian Journal of Remote Sensing* 22: 113–118.

Sugumaran, R. 2002. Development of a range management decision support system using remote sensing, GIS and knowledge based system. *Computers and Electronics in Agriculture* 37: 199–205.

Summers, G. F., and D. Field. 2000. Rural development: Meaning and practice in the United States. In *National parks and rural development,* ed. G. Machlis and D. Field. Washington, DC: Island Press.

Tecle, A., F. Szidarovszky, and L. Duckstein. 1995. Conflict analysis in multi-resource forest management with multiple decision-makers. *Nature and Resources* 31: 8–17.

Tecle, A., L. Duckstein, and P. Korhonen. 1994. Interactive, multiobjective programming for forest resources management. *Applications in Mathematics and Computation* 63: 75–93.

Terborgh, J., and M. E. Soule. 1999. Why we need megareserves: Large-scale reserve networks and how to design them. In *Continental conservation: Scientific foundations of regional reserve networks,* ed. M. E. Soule and J. Terborgh, 199–209. Washington, DC: Island Press.

Theobold, D. M., and N. T. Hobbs. 1998. Forecasting rural land-use change: A comparison of regression- and spatial transition-based models. *Geographical and Environmental Modelling* 2: 65–82.

Thierauf. R. J., and R. C. Klekamp. 1975. *Decision making through operations research.* New York: John Wiley and Sons.

Tilidetzke, B. Undated. *Modeling and simulation using Stella.* Charleston Southern University. http://archives.math.utk.edu/ICTCM/EP-13/C37/pdf/paper.pdf (accessed October 8, 2002).

Turner, B. L., and W. B. Meyer. 1994. Global land-use and land-cover change: An overview. In *Changes in land use and land cover: A global perspective,* ed. W. B. Meyer and B. L. Turner, 3–10. Cambridge, England: Cambridge University Press.

Turner, M. G. 1989. Landscape ecology: The effect of pattern on process. *Annual Review of Ecological Systems* 20: 171–197.

———. 1990. Spatial and temporal analysis of landscape patterns. *Landscape Ecology* 4: 21–30.

Turner, M. G., R. H. Gardner, and R. V. O'Neill. 2001. *Landscape ecology in theory and practice: Pattern and process.* New York: Springer-Verlag.

U.S. Department of Agriculture, Agricultural Research Service. 2002. *Revised universal soil loss equation, National Soil Sedimentation Lab.* http://www.sedlab.olemiss.edu/rusle/description.html (accessed October 8, 2002).

U.S. Fish and Wildlife Service. 2003a. *Big Muddy National Fish and Wildlife Refuge.* http://midwest.fws.gov/BigMuddy/ (accessed April 10, 2003).

———. 2003b. *Final environmental impact statement summary for proposed expansion of Big Muddy National Fish and Wildlife Refuge.* http://midwest.fws.gov/planning/bigmuddy/top.htm (accessed April 10, 2003).

U.S. Geological Survey. 2002a. *Geographic information systems.* http://www.usgs.gov/research/gis/title.html. (Accessed October 3, 2002).

———. 2002b. *Data structures.* http://www.usgs.gov/research/gis/work5.html. (Accessed October 8, 2002).

Ustin, S. L., and L. A. Costick. 2000. Multispectral remote sensing over semi-arid landscapes for resource management. In *Spatial information for land use management,* ed. J. Hill and R. J. Aspinall, 97–111. The Netherlands: Gordon and Breach Science Publications.

Ustin, S. L., M. O. Smith, S. Jacquemoud, M. M. Verstraete, and Y. Govaerts. 1999. Geobotany: Vegetation mapping in earth sciences. In *Manual of remote sensing for the earth sciences,* ed. A. Rencz, 189–248. Bethesda, MD: American Society of Photogrammetry.

Van Kooten, G. C., and E. H. Bulte. 2002. *The economics of nature: Managing biological assets.* Malden, MA: Blackwell Publishers, Inc.

Villa, F., L. Tunesi, and T. Agardy. 2002. Zoning marine protected areas through spatial multiple-criteria analysis: The case of Asinara Island National Marine Reserve of Italy. *Conservation Biology* 16: 515–526.

Villa, F., M. Ceroni, and A. Mazza. 1996. A GIS-based method for multi-objective evaluation of park vegetation. *Landscape and Urban Planning* 35: 203–212.

Walsh, S. J., et al. 1994. Form and pattern in the alpine environment: An integrated approach to spatial analysis and modeling in Glacier National Park, USA. In *Mountain environments and geographic information systems,* ed. M. F. Price and D. I. Heywood, 189–216. London, England: Taylor and Francis Group.

Walters, C., 1996. *Adaptive management of renewable resources.* New York: Macmillan and Co.

Walters, C. J., and C. S. Holling. 1990. Large-scale management experiments and learning by doing. *Ecology* 71: 2060–2068.

Walters, C. J., and R. Green. 1997. Valuation of experimental management options for ecological systems. *Journal of Wildlife Management* 61: 987–1006.

Walters, C. J., and R. Hilborn. 1978. Ecological optimization and adaptive management. *Annual Review of Ecology and Systematics* 9: 157–188.

Wilhere, G. F. 2002. Adaptive management in habitat conservation plans. *Conservation Biology* 16: 20–29.

Williams, F. 2001. Between towns and wilderness: protecting the buffer zones. In *Return of the wild: The future of our natural lands,* ed. T. Kerasote, 73–86. Washington, DC: Island Press.

Wolfson, L. J., J. B. Kadane, and M. J. Small. 1966. Bayesian environmental policy decisions: Two case studies. *Ecological Applications* 6: 1056–1066.

Woodley, S. 2002. Planning and managing for ecological integrity in Canada's National Parks. In *Parks and protected areas in Canada: Planning and management,* ed. D. Dearden and R. Rollins, 97–114. 2nd ed. Ontario, Canada: Oxford University Press.

Yaffee, S. L., and J. M. Wondolleck. 1997. Building bridges across agency boundaries. In *Creating a forestry for the 21st century: The science of ecosystem management,* ed. K. A. Kohm and J. F. Franklin, 381–396. Washington, DC: Island Press.

Yakowitz, D. S., L. J. Lane, and F. Szidarovszky. 1993. Multi-attribute decision-making: Dominance with respect to an importance order of the attributes. *Applied Mathematics and Computation* 54: 167–81.

Zuerlein, G., G. Mestl, and M. Brohman. 1995. *The Missouri River: A crown jewel in need.* Lincoln: Nebraska Game and Parks Commission.

CHAPTER

9

Case Studies in Protected Area Management—I

Given the complex nature of systems in which conservation operates, the urgent need for action, and the current lack of information as to how to best proceed, we take as given that effective conservation ultimately requires an adaptive management approach.

—Nick Salafsky, Richard Margoluis, Kent H. Redford, and John G. Robinson

The preceding eight chapters lay a foundation regarding the values, institutions, concepts, principles, and approaches relevant to natural resource planning and management for protected areas. Chapters 9 and 10 provide case studies that examine natural resource issues facing national parks and protected areas and management approaches being used to address those issues. Several of the management approaches involve application of the concepts, principles, and approaches presented in the first eight chapters. Each chapter contains two extended case studies and three abbreviated case studies. Extended case studies provide a synoptic treatment and abbreviated case studies a concise treatment of specific natural resource issues, threats, and impacts facing protected areas. The extended case studies in this chapter address reducing cumulative impacts to natural and cultural resources in Waterton-Glacier International Peace Park and enhancing grizzly bear recovery in the Greater Yellowstone Ecosystem (GYE). Abbreviated case studies address alleviating human and ecological impacts of snowmobile use in the GYE; reducing conflicts between wildlife, urban development, and transportation in Banff National Park; and improving air quality in national parks.

Cumulative Effects in Waterton-Glacier International Peace Park

Managing a national park to achieve the purposes of its organic act is difficult and challenging. In the United States, the National Park Service Organic Act (16

USC. §§ 431–33) requires national parks to be managed so as "to conserve the scenery and the natural and historic objects and the wildlife therein and to provide for the enjoyment of the same in such manner and by such means as will leave them unimpaired for the enjoyment of future generations" (Rockwood 1988). The ability of national park managers to achieve the dual mandate of conserving resources and providing for public use and enjoyment in perpetuity is stymied by the cumulative effects of external and internal threats. Cumulative effects are additive over time and/or space, and occur when seemingly minor management actions are made or minor events occur in an ecosystem that eventually have major social, economic, and ecological consequences (Meffe et al. 2002). Management actions for dealing with cumulative effects are best addressed in a long-term, regionally based, stakeholder inclusive context known as adaptive ecosystem management (AEM), which combines ecosystem management (EM) and adaptive management (AM) (see chapter 8 for more details; see the following sidebar for example of a cumulative effects model).

Southern Alberta Regional Strategy and ALCES Cumulative Effects Model

Maintaining a sound and growing economy and a diverse, healthy environment is a challenge and priority that must be shared by all who live in our province.

—Alberta Premier Ralph Klein

The Southern Alberta Regional Strategy (SARS) is an example of a planning program in which residents of Alberta, Canada, are developing a vision for the future environmental, social, and economic benefits provided by their province and a plan for achieving that vision. Leadership for SARS comes from Alberta Environment whose goal is to "develop and implement province-wide strategies that integrate industry, government and public uses of Alberta's land and resources" with an emphasis on sustainable development. SARS is part of Alberta Environment's Integrated Resource Management program. The Northern East Slopes Sustainable Resource and Environmental Management Strategy is another Alberta program that is similar to SARS. The significance of SARS for the purpose of protected area management is that it addresses sustainable development in a region that includes Banff National Park and other provincial parks.

The first phase of SARS involves assessing the current state of the region; identifying a vision, goals, and principles for sustainable development; making policy recommendations; and identifying and prioritizing key regional issues. Specific objectives for the first phase are:

1. To undertake a process of information gathering, modeling and analysis to provide a good strategic picture of human development and its effects on land, water and air in southern Alberta. This information will then be used to explore various potential development scenarios and to identify current and probable future issues and opportunities pertaining to sustainability and quality of life.
2. To undertake a program of extension to apprise stakeholders, decision-makers and the public of the findings and to seek to build awareness and appreciation for issues that must be addressed to ensure future sustainability and a continuing high quality of life.
3. To compile conclusions and recommendations in a final report that articulates a vision for a desired future, that identifies principles for addressing issues and makes recommendations to the provincial government and to key stakeholders for next steps (Phase II).

The modeling analysis mentioned in phase I involves application of ALCES (A Landscape Cumulative Effects Simulator). ALCES is a comprehensive natural resource management model for evaluating cumulative effects of land uses, including human settlement, protected areas, recreation, agriculture, forestry, and energy, on key environmental indicators under alternative management regimes and global trends in climate and commodity prices. Dr. Brad Stelfox of Forem Technologies developed ALCES. In essence, ALCES allows resource managers and other stakeholders to assess environmental responses to interactions between land uses and natural processes. The primary focus of ALCES is integrated resource management. Modeling results are being used to support development of regional resource and environmental strategies for the SARS region. ALCES is also being used in the Northern East Slopes Sustainable Resource and Environmental Management Strategy.

Data needed to execute ALCES include (1) current land uses in forestry, energy sector, agriculture and human activities; (2) current natural processes, such as fire management, vegetation growth and yield, and insect outbreak; (3) current land cover; and (4) sector specific development projections. Based on these data inputs, ALCES simulates future landscape changes, including cumulative effects of land uses and resource outputs, for a range of future regional development scenarios and policy choices, and the potential trade-offs between competing land uses.

Sources: Alberta Environment (2003a,b); CyberCervus International (2002)

The first extended case study discusses how to design and evaluate management actions to alleviate cumulative effects in Waterton-Glacier International Peace Park and the Crown of the Continent Ecosystem in which it is located. The AEM framework presented in the first case study provides a template for evaluating management actions to resolve problems and issues addressed in subsequent case studies.

Background

In 1932, the U.S. and Canadian governments passed legislation creating Waterton-Glacier International Peace Park, the world's first International Peace Park (hereafter referred to as the Peace Park). The Peace Park symbolizes "peace and goodwill between the United States and Canada [and] represents the need for cooperation and stewardship in a world of shared resources" (Waterton Lakes National Park Resource Guide 2002). The Peace Park consists of Waterton Lakes National Park in Alberta, Canada, and Glacier National Park across the border in Montana, United States (see fig. 9.1). Both parks are UNESCO Biosphere Re-

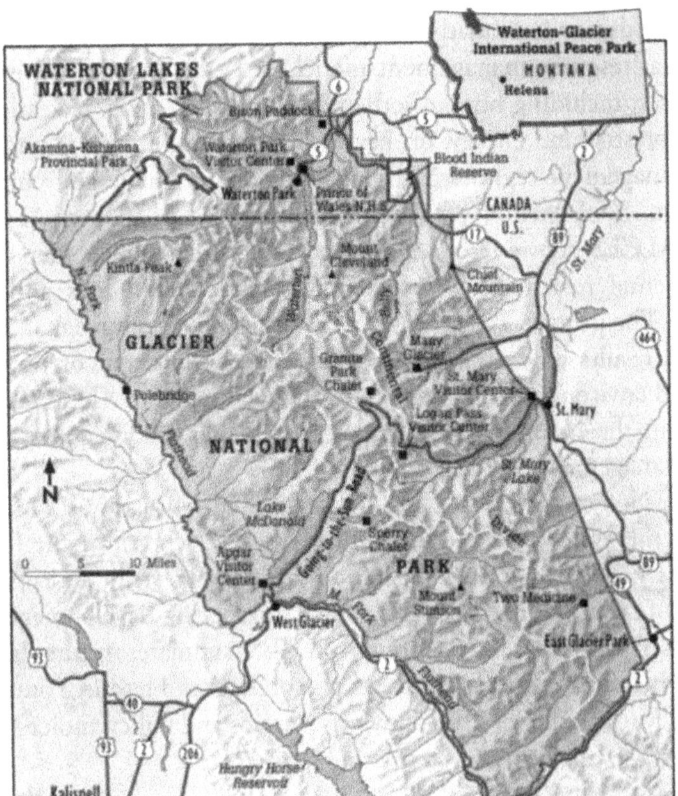

FIG. 9.1. Waterton-Glacier International Peace Park. (From National Parks Conservation Association 2002c)

Threats to Park Resources

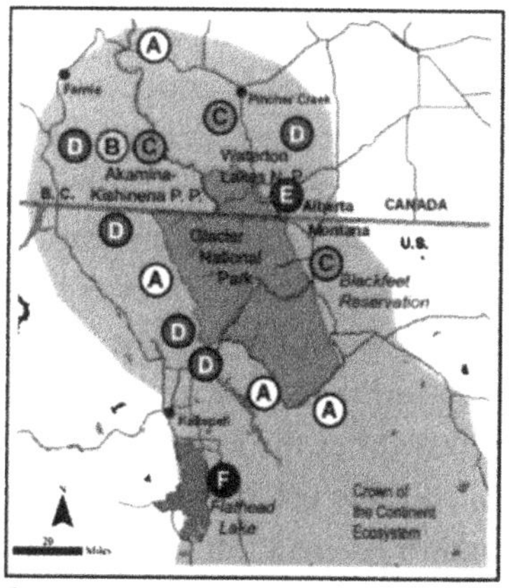

Examples of landscape alterations abount on land surrounding Waterton-Glacier International Peace Park. Most changes have negative impacts on the native species that find refuge in the park.

A - Proposed expansion of highways and paving of secondary roads may impede travel routes of wildlife.

B – Potential hard rock mining and open-pit coal mining would adversely affect water quality and wildlife populations such as bull trout.

C – High density road systems and other infrastructure associated with proposed logging, recreational development, and extraction of oil and gas along the Rocky Mountain Front would displace grizzlies and other wildlife.

D - Residential, commercial, and resort developments have encroached on seasonal range for many wildlife species.

E – Hunting of the gray wolf is legal in Alberta for nine months of the year.

F – Invasive non-native fish species and non-native weed species have migrated or been introduced into the park.

FIG. 9.2. External threats to Waterton-Glacier International Peace Park. (From National Parks Conservation Association 2002c.)

serves, and the Peace Park is a World Heritage Site (Flathead Basin Commission 2001; Montana's Flathead Valley 2001). Waterton Lakes National Park is 52,540 ha (129,728 acres) and Glacier National Park is 410,506 ha (1,013,594 acres). The Peace Park is the centerpiece of the Crown of the Continent Ecosystem (CCE), which includes the mountainous regions of northwestern Montana, southwestern Alberta, and southeastern British Columbia. The CCE includes the Peace Park, Bob Marshall-Great Bear-Scapegoat Wilderness complex, national forests and wildlife refuges, and lands owned by the Blackfeet and the Salish-Kootenai Indians (see in fig. 9.2). The headwaters of three major North American river systems are in the CCE, namely the Missouri/Mississippi, the Columbia, and the Saskatchewan/Nelson (Environment Canada 2002; Glacier National Park 2002).

In 1980, the National Park Service reported that national parks in the United States faced 4,300 threats (Dilsaver 1994). More than half of the threats to aesthetic qualities, cultural resources, air and water quality, plants, and wildlife came from sources outside the parks (fig. 9.2). The report documented only about 25 percent of the threats. The National Park Service recognizes the importance of having park managers use their authority to protect the resources and values of national parks. In this regard, NPS policy states that "strategies and actions beyond park boundaries have become increasingly necessary as the National Park Service strives to fulfill its mandate....Recognizing that parks are integral parts of larger regional environments, the Service will work cooperatively with others to...protect park resources and values and...address mutual interests...such as compatible economic development and resource and environmental protection" (U. S. Department of the Interior 2001). Along these same lines, Theberge and Theberge (2002) note that "if we consider parks and reserves as separate from the landscapes in which they are situated, then we fail to see the flow of species, energy, and materials that may confer long-term viability and persistence."

Glacier National Park faced almost four times the average number of threats as other national parks; 56 for Glacier vs. 13.5 for the 326 park units that existed at the time. The National Parks Conservation Association (NPCA) identified several internal and external threats to the Peace Park (National Parks Conservation Association 2002c). Threats include (1) proposed highway expansion; (2) conversion of working ranches and forests to residential, commercial, and resort developments; (3) clear-cut logging; (4) growth in sightseeing air tours; (5) invasions of nonnative species into parklands and waters; (6) legalized hunting for gray wolf in Alberta; (7) potential extraction of oil, gas, coal, and hard rock resources; (8) global warming and airborne chemical pollutants; and (9) shortage of personnel and operating funds for monitoring wildlife populations, completing archaeological research, maintaining historic structures and museum collections, and providing high-quality visitor services. Most of these threats are caused by events and circumstances external to the Peace Park. On a scale of 0 to 100, NPCA gave the Peace Park scores of 83 for natural resources, 52 for cultural resources, and 52 for stewardship capacity. Low scores assigned to cultural resources and stewardship capacity indicate that improvements are needed in these areas.

According to NPCA's report, impacts of threats to the Peace Park include (1) fragmented, degraded, and destroyed habitat for several wildlife species; (2) severe limitations on the movement of wide-ranging species like bears, wolves, deer, and elk; (3) reduced populations of native fish that are unable to compete with invasive nonnative species; (4) potential water quality degradation; (5) high likelihood of total absence of glaciers in 30 years and potentially severe impacts to alpine, floodplain, and wetland systems; and (6) degradation and loss of native vegetation. Table 9.1 lists the threats and impacts and figure 9.2 summarizes the nature, location, and impacts of the threats to the Peace Park. Other protected areas in North America face similar threats and consequences.

Principles for Alleviating Threats and Impacts

To make specific management prescriptions for threats to the Peace Park and other protected areas covered in the case studies requires detailed site-specific socioeconomic and ecological assessments. Such assessments are beyond the scope of

Table 9.1. Threats and Impacts to Waterton-Glacier International Peace Park and the Crown of the Continent Ecosystem

Threat
Proposed highway expansion (external)
Conversion of working ranch and forests to residential, commercial, and resort developments (external)
Clear-cut logging (external)
Growth in sightseeing air tours (external)
Invasions of nonnative species into parklands and water (internal and external)
Legalized hunting for gray wolf in Alberta (external)
Potential extraction of oil, gas, coal, and hard rock resources (external)
Global warming and airborne chemical pollutants (external)
Shortage of personnel and operating funds for monitoring wildlife populations, completing archaeological research, maintaining historic structures and museum collections, and providing high-quality visitor services (internal)
Impact
Fragmented, degraded, and destroyed habitat for several wildlife species
Severe limitations on the movement of wide-ranging species like bears, wolves, deer, and elk
Reduced populations of native fish that are unable to compete with invasive nonnative species
Potential water quality degradation
High likelihood of total absence of glaciers in 30 years and potentially severe impacts to alpine, floodplain and wetland systems
Degradation and loss of native vegetation

Source: National Parks Conservation Association (2002c).

this book, which discusses approaches and frameworks for systematic decision making. In addition, socioeconomic and ecological conditions in protected areas change often enough that management prescriptions have a short shelf life. For these reasons, the case studies use a more general approach based on application of the six principles listed in table 9.2 and discussed below.

1. *Wide-ranging threats to protected areas cannot be alleviated with a single management action.*

Increasing a park's budget and personnel devoted to monitoring wildlife populations and improving the maintenance of historic structures would do little to reduce landscape fragmentation in the CCE. Therefore, alleviating multiple threats or impacts requires a suite of well-designed and coordinated management actions. This does not mean, however, that management actions aimed at seemingly unrelated problems are totally independent. In the above example, part of the additional budget and personnel used to monitor wildlife populations could be used to monitor and evaluate how wildlife populations respond to habitat loss and degradation from landscape fragmentation.

2. *It is easier to alleviate internal threats than external threats to protected areas.*

The two national parks that make up the Peace Park could make a coordinated

Table 9.2. Principles for Designing Management Actions to Alleviate Threats to Protected Areas

Principle	Management implication
Wide-ranging threats/impacts cannot be alleviated with a single management action.	Alleviating multiple threats or problems requires a suite of well-designed and coordinated management actions.
It is easier to alleviate internal threats than external threats.	Alleviating external threats is likely to require more planning and engagement with stakeholders than alleviating internal threats. In dealing with external threats, planners and managers should focus on threats posing significant risks to ecological integrity.
Formulating and evaluating management actions in an EM framework accounts for interdependencies among social, economic, and ecological values.	An EM approach requires integrating scientific-based ecological knowledge and socioeconomic perspectives and values in a collaborative decision-making framework designed to enhance long-term sustainability of natural and cultural resources.
Adaptive management allows managers to handle uncertainty regarding the likely impacts of alternative management actions on social, economic, and ecological values.	Adaptive management requires more planning, research, and coordination than traditional management approaches.
Managers are likely to have greater success in alleviating the impacts of threats by working collaboratively with a broad range of stakeholders to establish and achieve common goals.	Collaboration requires managers to interact with stakeholders in a more comprehensive and interactive way than traditional top-down management
Success in alleviating threats is increased by integrating knowledge about: (1) cultural, social, economic, and ecological values (chapters 3 and 4); (2) institutional arrangements influencing management and operations (chapter 5); and (3) state-of-the-art decision-making approaches, concepts and methods (chapters 7 through 9).	AEM is a suitable integrating framework that allows consideration of these elements.

request to their respective agencies (Parks Canada for Waterton Lakes National Park and National Park Service for Glacier National Park) for additional personnel and equipment to monitor threatened and endangered species, complete archaeological research, maintain historic structures and museum collections, provide high-quality visitor services, and reduce the spread and adverse ecological impacts of nonnative species in the park. By comparison, it is more difficult for managers to alleviate external threats to the Peace Park from proposed highway expansion, continued fragmentation of working ranches and forests, and logging and mineral extraction on nearby lands. Managers have virtually no control over global warming and airborne chemical pollutants originating in far off places.

Differences in the ability of park managers to deal with internal vs. external threats arise because the environmental, socioeconomic, and demographic drivers of external threats are, for the most part, independent of the mandate to conserve park resources and provide for public enjoyment. These drivers are regional economic development, supply and demand for wood and minerals, global emissions of carbon and chemical pollutants, and other events beyond the control of park managers. This situation does not imply that Peace Park managers should focus only on reducing internal threats to the exclusion of external threats; quite the contrary. Most of the threats to the Peace Park are external (see table 9.1), and alleviating them is likely to require more planning and collaboration than internal threats. Although they would be using an indirect approach, managers could be effective in reducing external threats by aggressive education and outreach that makes clear to stakeholders the gravity of the threat.

3. *Formulating and evaluating management actions in an ecosystem management (EM) framework accounts for interdependencies among social, economic, and ecological values of protected areas.*

An EM approach requires integrating scientific-based ecological knowledge and socioeconomic perspectives and values in a collaborative decision-making framework designed to enhance long-term sustainability of protected areas. This requires threats/impacts to be addressed in a manner that balances social, economic, and ecological values at the CCE scale and accounts for the interdependencies among those values. For example, proposed highway expansion and commercial/resort development in gateway communities improve the quality of visitor services outside the Peace Park, but they threaten the park's ecological integrity. Similarly, using more of the Peace Park's limited budget to monitor wildlife populations and complete archaeological research and less to reduce habitat loss/fragmentation and controlling the spread of nonnative species involves trade-offs.

4. *Adaptive management (AM) allows managers to handle uncertainty regarding the likely impacts of alternative management actions on social, economic, and ecological values of protected areas.*

While outcomes of some management actions are fairly certain, outcomes of other actions are uncertain. Clear-cut logging and mineral extraction in areas adjacent to the Peace Park have a high likelihood of impairing the movement of wide-ranging species, increasing soil erosion and the spread of invasive species, and contaminating water. In contrast, impacts of global warming on alpine, floodplain, and wetland systems and impacts of legal hunting for gray wolf in Alberta on wolf recovery in the United States are more uncertain. The AEM approach is suitable in resource management settings where managers and stakeholders view uncertainty as an opportunity to learn more about protected area ecosystems. Using AM requires more planning, research, and coordination than traditional management approaches and might be resisted for political, social, and economic reasons.

5. *Protected area managers are likely to have greater success in alleviating the impacts of threats by working collaboratively with a broad range of stakeholders to establish and achieve common goals.*

A stakeholder is defined as "anyone who has an interest in the topic at hand and wishes to participate in decision making" (Meffe et al. 2002). Stakeholders for protected areas include private landowners and businesses, Native Americans, community planners, residential and commercial developers, resource extraction firms, public land management agencies, natural resource/environmental interest groups, and the general public. Effective cooperation among stakeholders requires establishing and maintaining collaborative relationships with stakeholders, and involving them in a more comprehensive and integrative way than is done in traditional management approaches.

Several organizations have been established to raise stakeholder awareness and understanding and facilitate collaboration on natural resource issues in the CCE. The Crown of the Continent Ecosystem Education Consortium (COCEEC) has as its goal to (1) encourage and support coordination and cooperation among individuals and organizations that educate about the human and natural resources of the CCE; (2) promote a sense of community among citizens of the region, a comprehensive view of the landscape, and an ethic centered on personal and community stewardship; (3) provide balanced educational leadership on emerging concepts of ecosystem stewardship, biological diversity conservation, and ecosystem sustainability; (4) encourage the development and dissemination of information and educational materials for presentation to diverse audiences in a variety of settings; and (5) cooperatively seek private and public support and partnerships for activities that further its mission (Crown of the Continent Ecosystem Education Consortium 2002).

The goal of the Glacier Field Office of the National Parks Conservation Association is to build grassroots community support for park conservation among diverse interests in the Waterton-Glacier region (National Parks Conservation Association 2002a). Other groups with different values work to keep forest roads open for logging, recreation, and other uses, which can negatively impact park conservation. A collaborative organization whose geographic domain includes the CCE is the Yellowstone-to-Yukon (Y2Y) Conservation Initiative. The goal of this initiative is to ensure that the region's rich and diverse wilderness, wildlife, native plants, and natural processes continue to function as an interconnected web of life, capable of supporting all the natural and human communities within it for present and future generations (Yellowstone to Yukon Conservation Initiative 2002). Chapter 7 and figure 7.10 provide more details about the Y2Y ecosystem and the Y2Y Conservation Initiative.

Transboundary cooperation between parks is especially important when the political jurisdictions have conflicting natural resource management policies. For example, hunting for gray wolf is not permitted in the U.S. portion of the CCE because the wolf is a protected species. Hunting for wolf is permitted across the border in Alberta, Canada. Such divergent policies can slow the recovery of gray wolves in the United States.

6. *Success in alleviating threats to protected areas is increased by integrating knowledge about the (1) cultural, social, economic, and ecological values (chapters 3 and 4); (2) institutional arrangements influencing management and operations (chapter 5); and (3) state-of-the-art decision-making approaches, concepts, and methods (chapters 6 through 8).*

Complex dynamic processes operating in natural-resource-based protected areas such as the Peace Park make those areas difficult and challenging to manage. In order to alleviate threats to the Peace Park and other protected areas, managers need to continually develop and evaluate management actions that balance multiple social, economic, and ecological values. This requires integration of knowledge about the multiple values of the park, the institutional setting in which resource management and other decisions are made, and various approaches for making decisions. Such integration can be accomplished using an AEM approach that incorporates the six principles proposed above.

Designing and Evaluating Management Actions

Well-designed management actions for the Peace Park can alleviate one or more threats to the Peace Park as summarized in table 9.1. Management actions are discussed first for external threats/impacts and second for internal threats/impacts. It is important to recognize that park management actions aimed at reducing one external threat/impact can influence another external threat/impact as well as social, economic, and ecological values. For example, a management action that bans clear-cut logging in national forests in the CCE might not only increase the movement of wide-ranging species across their home range, but also improve habitat for endangered fish species outside the park by reducing sediment delivery to streams. Banning clear-cuts is likely to have a negative impact on regional income and employment in gateway communities if it is more profitable than other forms of logging and the ban causes some timber operations to shift to areas outside the CCE that allow clear-cutting. Any reduction in regional income and employment attributable to the ban is a social cost and needs to be weighed against the fish and wildlife benefits of the ban.

Similarly, management actions to reduce internal threats are likely to generate benefits and costs in the regional economy. For example, a management action that reduces congestion in park campgrounds and roads by limiting the number of vehicles admitted to the park would improve the quality of visitor experiences (internal benefit), but reduce gate receipts (internal cost) and decrease regional income and employment in gateway communities (external cost). Decreases in regional income and employment have adverse economic consequences on gateway communities. Other examples of management actions aimed at reducing internal threats to the Peace Park that have external impacts are reconstructing Going-to-the-Sun Road and revising the commercial services plan for Glacier National Park. Both actions ultimately improve the quality of visitor services, but have potentially significant economic impacts on gateway communities. For this reason, protected area managers should approach management problems in a holistic manner.

The preceding discussion illustrates the advantages of evaluating management actions for the Peace Park using an EM approach. An EM approach requires protected area managers to consider the long-term impacts of management actions on the greater ecosystem in which the park resides. EM requires managers to consider how a ban on clear-cut logging affects not only wildlife species that have a significant portion of their home range in the park, but also water quality in streams outside the park and income and employment in gateway communities. Ecological economic impacts to the greater ecosystem are just as important to consider as ecological impacts within the arbitrarily defined boundaries of the park. Taking a regional perspective is just as important to EM as evaluating management actions in a long timeframe. Hence, the long-term social, economic, and ecological trade-offs involved in, say, replacing clear-cut logging (even-aged management) with selective cutting (uneven-aged management) or reconstructing Going-to-the-Sun Road using the quickest, lowest cost alternative rather than longest, highest cost alternative need to be considered.

Taking an adaptive approach to EM recognizes the uncertainty regarding the social, economic, and ecological consequences of management actions. For example, it is reasonable to expect that disallowing clear-cut logging in national forests in the CCE would benefit large carnivores because they are area- and edge-sensitive species. However, this benefit could be partially offset by other factors that influence carnivore populations such as food supply, natural disasters, weather, and climate change. Also, ecological effects of logging methods and climate change are confounded and hence difficult to separate, and logging releases carbon to the atmosphere, which contributes to climate change.

Implementing AEM

The Crown Managers Partnership (CMP) is a group of public and land resource management agencies in the CCE whose goal is to develop "management tools, data management and science (research/inventory/monitoring) at the ecosystem scale in cooperation with academic institutions" (Crown of the Continent Managers Partnership 2002). Consider applying AEM to the following scenario. Suppose CMP is given the authority and responsibility to implement AEM in the CCE and creates an Adaptive Management Council (AMC) for the purpose of developing and implementing management actions that alleviate threats/impacts based on the six principles listed in table 9.2.

Membership in the AMC is best determined based on the criteria of inclusivity, self-selection, and diversity of representation. Inclusivity means that all interested stakeholders or their representatives are invited to participate in AMC. Being inclusive helps "people with different views recognize and understand their common interest in working together" (Meffe et al. 2002). Self-selection means AMC members are free to choose how involved they become in particular issues. Degree of stakeholder involvement in an issue depends on how that issue affects a stakeholder and the manner in which AMC addresses the issue. In other words, AMC members will not be equally interested in all threats/impacts. Diversity of representation requires that AMC members reflect a broad spectrum of interests and concerns. Meffe et al. (2002) provide an excellent treatment of successful community-based conservation based on these and other principles. Based on these criteria, members of the AMC are likely to include scientists, private landowners, Native Americans, businesspersons, developers, community planners, environmental groups, and other stakeholders.

The primary objectives of the AMC are to (1) identify desirable outcomes for the ecosystem and management actions for achieving those outcomes; (2) compare the social, economic, and ecological consequences of management actions using adaptive management; and (3) selecting preferred management actions. AMC should identify management actions that are likely to provide sustainable and efficient (desirable) social, ecological, and economic outcomes using the approach discussed in chapter 7 and illustrated in figures 7.7 and 7.8. For instance, a desirable ecological outcome for carnivores is to provide sufficient habitat to sustain viable populations, especially for threatened and endangered carnivores that have a portion of their home range in the Peace Park.

One of the principles stated in table 9.2 is that wide-ranging threats/impacts cannot be alleviated with a single management action. In other words, desirable outcomes (alleviating multiple threats/impacts listed in table 9.1) cannot be achieved with one management action. This does not mean that a single management action influences only one threat/impact. For instance, buying conservation easements on ranchland might not only alleviate adverse ecological impacts of landscape fragmentation, but also improve movement of carnivores and water quality and reduce losses in native vegetation. AMC needs to identify several management actions to alleviate each threat/impact.

AMC can identify desired outcomes using "willing conformity." Willing conformity means that members of AMC internalize common values and shared meanings of the ecosystem to the point where they act in "accordance with those values at the expense of their own self-interest without the presence of coercion" (McLain and Lee 1996). Achieving willing conformity requires "mutual recognition of the legitimacy of the interests of others" (Shannon 1987; Moore 1994; McLain and Lee 1996).

Management actions that provide desirable outcomes can be compared using a multiple attribute decision-making (MADM) approach in which the social, economic, and ecological values of management actions are attributes (see chapter 8). For example, appropriate social, economic, and ecological attributes for management actions designed to alleviate adverse impacts of landscape fragmentation on carnivore populations are (1) local and regional income and employment, (2) landscape patterns that enhance habitat suitability for carnivores, (3) movement of carnivores, and (4) water quality. Impacts on income and employment can be estimated using the IMPLAN model discussed in chapter 8. Other management actions that AMC might consider are land donations, land purchases, land trusts, land exchanges, land zoning, and others (Brown 1999).

Selecting preferred management action using multiple attribute evaluation requires knowledge of the relative importance of social, economic, and ecological attributes of management actions to stakeholders. Not all stakeholders will attach the same relative importance to attributes. Based on the principles discussed in chapter 8, the ranking of management actions according to members' preferences requires (1) calculating stakeholders' utility scores for management actions using a simple additive utility function that combines the values of attributes for management actions and the relative importance (weights) of attributes, (2) ranking management actions based on those scores, and (3) resolving stakeholder disagreements about preferences for management actions using group decision-making techniques. Ananda and Herath (2002) identify several group decision-making techniques including focus groups, in-depth groups, consensus conferences (De Marchi and Ravetz 2001), citizens' juries (Crosby 1996; Aldred and Jacobs 2000), value integration surveys (Gregory 2000), policy workshops (McDaniels and Roessler 1998), multiparty negotiation (Ehtano et al. 2001), and negotiation forums (Eastman et al. 1998).

Suppose AMC is able to distinguish between management actions or elements of management actions that have relatively certain outcomes and those that have relatively uncertain outcomes in terms of alleviating the threats/impacts to the Peace Park and CCE. Management actions with relatively certain outcomes should be addressed using passive AM. If feasible, management actions with uncertain outcomes should be assessed using active AM.

Consider how AMC could use AM to evaluate management actions to reduce adverse effects of landscape fragmentation on carnivore species in the CCE. Table 9.3 explains the difference between passive and active AM for this evaluation. Events and conditions given in table 9.3 do not necessarily reflect actual events and conditions in the Peace Park and CCE. Passive AM is appropriate when AMC is relatively certain about how landscape fragmentation influences carnivore populations. With passive AM, a preferred management action is selected by comparing expected outcomes of alternative management actions to actual outcomes of current management actions with consideration given to public comments, applicable statutory requirements, and management objectives for public lands in the CCE. In the United States, management actions that are likely to have a significant impact on the human and natural environment must be evaluated prior to making a decision. This requires preparation of an environmental impact statement (EIS) or environmental impact assessment (EIA) that compares the impacts of alternative management actions and evaluates public comments on those alternatives. After completion of the final EIS/EIA, the authorizing official for the agency taking the action (usually the secretary of the department that manages the affected area) issues a record of decision that announces the preferred alternative. The EIS/EIA process is compatible with passive AM.

Active AM explicitly recognizes the uncertainty regarding the relationships between management actions, landscape fragmentation, and carnivore populations. For example, to evaluate conservation easements with active AM requires the AMC to design experiments to test hypotheses regarding how management actions involving different configurations (size and location of tracts) of conservation easements influence landscape fragmentation, carnivore populations, and social and economic attributes. Since it can take several years for carnivore populations to respond to management actions, AM experiments need to be long term. To the maximum extent possible, analysis of experimental results should account for other factors influencing carnivore populations such as food availability, climate, human contact, and land use/management within fragments. Confirmed hypotheses are used to identify the most likely effects of management actions on landscape fragmentation, carnivore populations, and social and economic attributes.

Other matters that AMC needs to address in implementing active AM include the number and scale of experimental units; inability to spatially replicate some management actions when larger scale environmental processes, such as climate change, affect localized experimental responses; accounting for complex transient responses (delays, sharp increases followed by slow decline, cycles, etc.); and unwillingness of stakeholders to accept experiments that could have unacceptable social and economic consequences (Walters and Holling 1990).

Two specific management actions are used to illustrate implementation of passive and active AM. The first management action is a ban on sightseeing air tours in the Peace Park. This action would eliminate the visual and auditory disturbances caused by sightseeing air tours, the option of seeing the park from the air, and the income and employment generated by air tours. While there is some uncertainty regarding the extent to which banning sightseeing air tours would enhance the experiences of on-the-ground visitors, most elements of this management action have relatively certain outcomes. Hence, AMC should evaluate the elimination of sightseeing air tours and similar management actions using passive AM.

The second management action is to have a land trust buy conservation easements on working ranches in the vicinity of the Peace Park in an effort to reduce

Table 9.3. Contrast Between Passive vs. Active Adaptive Management for Alleviating Impacts of Landscape Fragmentation on Carnivore Species

Data	Evaluation	Management action
Passive Adaptive Management		
Historical data indicate that carnivore populations in the CCE have been declining at the same time economic development and landscape fragmentation have been increasing.	Based on historical data and expert knowledge, a preferred management action is selected for dealing with fragmentation. The preferred management action assumes there is a particular relationship between landscape fragmentation and carnivore populations, e.g., reducing landscape fragmentation by 20 percent would increase carnivore populations by 30 percent.	The preferred management action is implemented. Monitoring is used to determine the effectiveness of the management action. This approach ignores uncertainty regarding the responses of carnivore populations to landscape fragmentation.
Active Adaptive Management		
Historical data indicate that carnivore populations in the CCE have been declining and economic development and landscape fragmentation have been increasing. Experiments are conducted to test hypotheses about how landscape fragmentation influences carnivore populations.	Based on historical data and simulation, alternative ecological models of carnivore population responses to landscape fragmentation and other factors are selected. Each model is based on a unique hypothesis concerning this relationship: H_1: Current rate of fragmentation does not adversely affect carnivore populations (no action required). H_2: Reducing landscape fragmentation by 30 percent increases carnivore populations by 20 percent. H_3,..., H_n (other hypotheses). Alternative management actions are implemented in a statistically valid experimental design and results used to test hypotheses.	Alternative management actions specify particular strategies for reducing landscape fragmentation, e.g., 30 percent for H_2, etc. Alternative management actions are compared to a control (H_1). Confirmed hypotheses are used to determine the most likely influence of landscape fragmentation on carnivore populations.

further landscape fragmentation. Uncertainty regarding this management action stems from two sources: whether the location and spatial extent of easements will be sufficient to reduce landscape fragmentation and whether the likely reductions in landscape fragmentation will significantly improve wildlife habitat. Due to these uncertainties, AMC should evaluate conservation easements and similar management actions using active AM.

AMC recognizes it is difficult or impossible to alleviate threats/impacts to the Peace Park and CCE for which it has little control. For instance, there are no management actions AMC can implement to reduce adverse impacts of climate change on glacier recession and alpine, floodplain, and wetland systems. Climate change is beyond the control of AMC and park managers. On the other hand, AMC can be supportive of research on the potential biophysical impacts of climate change and airborne chemical pollutants on the Peace Park. Finally, AMC needs to decide whether it is socially and politically feasible to remove certain threats, for example, hunting of gray wolf in Alberta.

In summary, AMC is created based on the principles of inclusivity, self-selection, and diversity of representation of stakeholders. It establishes desired outcomes or future conditions for the CCE and evaluates management actions to achieve those outcomes in an AEM framework. Elements of management actions having relatively certain outcomes are evaluated using passive AM and, if feasible, elements of management actions having uncertain outcomes are evaluated using active AM.

Grizzly Bear Recovery in the Greater Yellowstone Ecosystem

The second extended case study applies AEM to grizzly bear recovery in the Greater Yellowstone Ecosystem (GYE) (chapter 7 describes the GYE). The GYE's rich history of grizzly bear management is punctuated by two major turning points. The first one came in 1968 when Yellowstone National Park (YNP) adopted a natural regulation policy. The second turning point occurred in 1982 when the U.S. Fish and Wildlife Service announced its first Grizzly Bear Recovery Plan (U.S. Fish and Wildlife Service 1993). The plan came seven years after the grizzly bear was listed as a threatened species in the lower 48 states. This section provides background on grizzly bear management in YNP and GYE and discusses how AEM might have been used to implement natural regulation and recovery plans for grizzly bears in the GYE.

Background

Prior to European settlement, approximately 50,000 grizzly bears lived in prairies, forests, shorelines, and foothills from the Great Plains to the California coast in the United States and south into Mexico. There are thriving grizzly bear populations of 10,000 to 11,000 animals in British Columbia and 30,000 to 35,000 animals in Alaska. Grizzly bears are not a threatened or endangered species in Canada and Alaska. Human settlement, habitat destruction, and extensive shooting and trapping caused the grizzly bear population to fall to 1 percent of its historic level and 2 percent of its historic range in the lower 48 states (U.S. Fish and Wildlife Service 2001).

There are approximately 1,100 grizzly bears in the lower 48 states in five separate populations located in Montana, Idaho, Wyoming, and Washington (see fig. 9.3). The GYE has a population of 400 to 600 bears that is increasing at an annual rate of 2 percent to 4 percent. The Northern Continental Divide ecosystem (same as the Crown of the Continent Ecosystem) has a population of between 400 and 500 bears that is stable or increasing. The remaining three populations are 40 to 50 bears in the Selkirk Ecosystem, 30 to 40 bears in the Cabinet-Yaak Ecosystem, and approximately five bears in the Northern Cascades ecosystem (U.S. Fish and Wildlife Service 2001). In June 2000, the U.S. Department of the Interior announced a record of decision to reintroduce a minimum of 25 grizzly bears over a five-year period into Idaho's Selway-Bitterroot Wilderness. The decision required bears to be designated as "nonessential, experimental population" under Section 10(j) of the Endangered Species Act and managed by a Citizen Management Committee (Thompson 2000). In June 2001, the service proposed to withdraw its plan to reintroduce grizzly bears into the Selway-Bitterroot Wilderness (U.S. Fish and Wildlife Service 2001). As of this writing, a final decision has not been made on this proposal.

Construction of roads, campgrounds, and commercial facilities (hotels, service stations, grocery stores, gift shops, etc.) in YNP allows visitors to access and en-

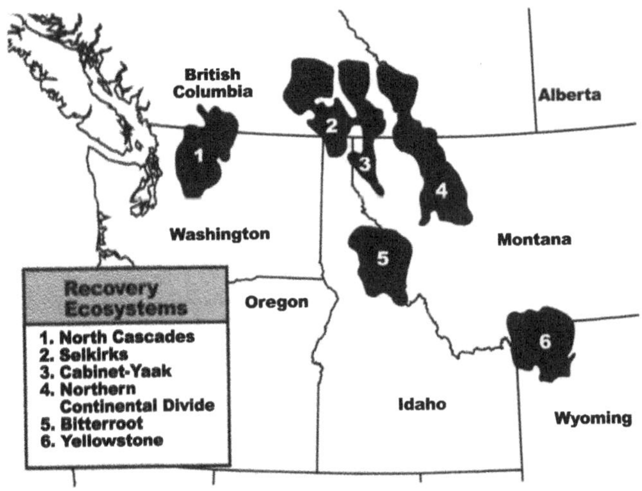

FIG. 9.3. Grizzly bear populations and recovery areas in the lower 48 states. Grizzly recovery in the Bitterroot area has been placed on hold. (From Interagency Grizzly Bear Committee 2002a.)

joy the park's scenic and wildlife resources. Unfortunately, they destroy wildlife habitat, deprive wildlife from accessing much of the park's limited riparian habitat, and increase wildlife habituation to humans. Roads, in particular, reduce grizzly bear access to important streamside-lakeside habitat and increase encounters with humans and bear mortality (Craighead 1991).

During the summers of 1959 through 1968, grizzly bears were permitted to feed on human garbage dumps at Rabbit Creek and Trout Creek in YNP. Dump feeding compensated for some of the loss of bear habitat caused by construction of roads, campgrounds, and commercial facilities. Watching the bears eat garbage was a major visitor attraction. In 1968, park officials implemented a natural regulation policy that among other things led park managers to close the garbage dumps within a one-year period.

The Craighead brothers disagreed with the way the natural regulation policy was being implemented, particularly the abrupt closure of garbage dumps. They believed natural regulation was not a viable management policy because the park's boundaries were too restrictive and the numbers of visitors too large relative to the megafauna. Additionally, they said that removing the dumps would not reduce grizzly bear habituation to people as maintained by park managers, but rather eliminate grizzly bear habituation to the dumps. The Craigheads recommended Yellowstone "be managed as an entity in which human presence and impact are properly recognized" and favored a phased closure of the garbage dumps over a 10-year period during which time the bear population would be closely monitored (Craighead 1991). The Yellowstone population of grizzly bears declined for several years after the dumps were closed (Schullery 1997).

In summarizing their opposition to natural regulation and rapid closure of the garbage dumps, the Craigheads stated that "if Yellowstone is to remain the 'jewel in the crown' of our national park system, its natural resources must be actively managed. For this to be done properly, the GYE will have to be biologically defined and delineated, and management policies formulated for the entire area" (Craighead 1991). This position is compatible with AEM.

In 1975, the grizzly bear was listed under the federal Endangered Species Act (ESA) of 1973 as a threatened species in the lower 48 states. According to the ESA, designation of a species as threatened, endangered, or recovered is based solely on the best available science; socioeconomic and political factors are not relevant considerations until recovery plans are developed. The Fish and Wildlife Service is responsible for recovery of grizzly bears and other declining species under the ESA. It completed the first Grizzly Bear Recovery Plan in 1982. Recovery activities include improving management of grizzly bears on public lands, genetic research, population monitoring, public education, and implementing the recovery plans for each population (Interagency Grizzly Bear Committee 2002a; U.S. Fish and Wildlife Service 2001). The major recovery programs for grizzly bears are in the Northern Continental Divide Ecosystem in northwestern Montana and the GYE in Wyoming, Idaho, and Montana.

The Interagency Grizzly Bear Committee (IGBC) was created in 1983. The purpose of the IGBC is to achieve recovery of grizzly bear populations in the lower 48 states through coordination of management and research actions of federal and state agencies having responsibility for the species (Interagency Grizzly Bear Committee 2002a). Members of the IGBC are the National Park Service; U.S. Fish and Wildlife Service; USDA Forest Service; the states of Idaho, Montana, Washington, and Wyoming; and British Columbia. Recovery plans developed for the five populations of grizzly bears focus on reducing the potential for human-bear encounters and related bear mortality and providing secure habitat in which females can raise their young. The recovery goal is to ensure adequate numbers of bears and suitable habitat for sustaining recovered grizzly bear populations. A recovered population is one for which there is a reasonable probability of existence into the foreseeable future, and the factors underlying its original listing under the ESA have been successfully addressed (U.S. Fish and Wildlife Service 2002). These factors are (1) the present or threatened destruction, modification, or curtailment of habitat or range; (2) overuse of habitat for commercial, recreational, scientific, or educational purposes; (3) disease or predation; (4) the lack or inadequacy of regulations; and (5) other natural or human factors that threaten continued existence of the population. Recovery activities include public education, reduction in bear access to human food and garbage, evaluation of road densities, research on availability of grizzly foods, and studies of bears and their habitat (U.S. Fish and Wildlife Service 2000b). Recovery is achieved when grizzly bears and their critical habitat are viable, self-sustaining, and no longer need protection (Interagency Grizzly Bear Committee 2002b).

The GYE's Grizzly Bear Recovery Zone consists of a Primary Conservation Area (PCA) of 23,833 km^2 (9,209 mi^2) and a 15-km (10-mile) wide buffer zone of 13,641 km^2 (5,271 mi^2) (see fig. 9.4). The PCA extends beyond the boundaries of YNP for two reasons. First, grizzly bears have an extensive home range of between 1,553 and 2,588 km^2 (600–1,000 mi^2) for males and up to 647 km^2 (250 mi^2) for females. It is not unusual for grizzly bears to travel 52 to 104 km (20–40 miles) in one day (U.S. Fish and Wildlife Service 1993). Second, as the population has expanded, bears have established habitat areas outside YNP; between one-third and one-half of the Yellowstone grizzly bear population resides outside the park (Wilkinson 2002). The grizzly bear is an umbrella species; maintaining sufficient habitat for grizzly bears benefits many other species. The PCA contains seasonal habitat that supports a recovered population and is managed to maintain habitat for a recovered grizzly bear population and minimize conflicts between grizzly bears and humans. A higher priority is given to protecting female bears than male bears. Should grizzly bear recovery conflict with other land uses,

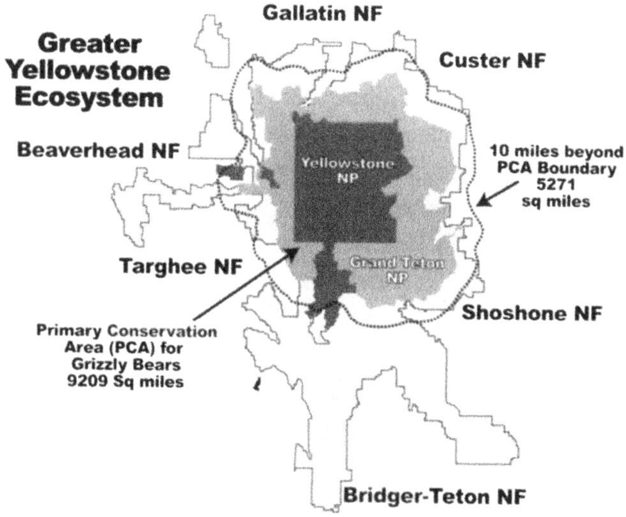

FIG. 9.4. Primary recovery area for grizzly bears in the Greater Yellowstone ecosystem. (From U.S. Fish and Wildlife Service 1993.)

recovery is favored. Grizzly bears are often removed from areas in which there are conflicts between bears and people (U.S. Fish and Wildlife Service 1993).

Bear recovery in the GYE has been very successful. During the period 1998–1999, the number and distribution of grizzly bears exceeded recovery targets. Adult females increased from a low of less than 30 in 1983 to about 100 in 2000. Success has been achieved in reducing grizzly bear mortalities and achieving habitat standards consistent with a recovered population. A conservation strategy is being developed to guide management of grizzly bears by Idaho, Montana, and Wyoming after the species is removed from the federal threatened species list and management is turned over to the states (U.S. Fish and Wildlife Service 2000a). There is considerable disagreement regarding whether the management plans developed by the states will provide adequate protection for grizzly bears.

AEM and Natural Regulation

Suppose it were possible to go back to the late 1960s. How might the natural regulation policy and closure of garbage dumps be implemented in YNP using AEM? The Leopold Report recommended that natural processes be allowed to operate and dominate in national parks and that parks be managed as vignettes of primitive America (Leopold et al. 1963; Keiter 1996). There are two opposing views regarding whether natural regulation is an appropriate management paradigm for YNP.

The first view maintains that natural regulation is a passive "hands off, let nature take its course" form of management that is appropriate and superior to active management based on precise goals (Knight 1991). In this view, active management is not appropriate in large national parks because it preempts the opportunity to learn about how ecosystems function. This concept of active management is not the same as active AM; the latter uses experiments to learn about ecosystems. The second view expressed by the National Research Council (2002) is that "true natural regulation (i.e., letting nature take its course with no human intervention) has not been possible for more than a century, nor is it likely to become possible in Yellowstone's foreseeable future."

The second view is that the Leopold Report did not advocate passive management and the reestablishment of primitive conditions (vignette of primitive America) in the park. Since the boundaries of YNP do not encompass an intact natural ecosystem, "biotic communities [cannot] regulate themselves within artificial boundaries and in areas where man is a massive intruder" (Craighead 1991). Nonintervention in park management poses a greater risk to park values than intervention (Berry et al. 2001). This view is appropriate for wide-ranging species like grizzly bears whose primary conservation area extends beyond the boundaries of the park (see fig. 9.4). In the second view, natural regulation requires hands-on active management using advanced management techniques (Craighead 1991) and specific management actions that restore natural conditions degraded by human activities (National Park Service 2001). The authors of this book favor the second interpretation of natural regulation.

Is passive and/or active AM compatible with the second interpretation of natural regulation as it pertains to grizzly bears? If restoring more natural conditions for grizzly bears is the desired outcome and abrupt closure of the dumps achieves that outcome, then closing the dumps would qualify as passive AM. Why? Because park managers assumed the preferred management action (rapid closure of the dumps) would result in more natural feeding conditions for bears. What was assumed about the impact of dump closure on grizzly bear population? It is not clear whether or how managers addressed this impact. Perhaps managers were willing to accept the trade-off that more natural feeding conditions might decrease the population of grizzly bears. Additionally, managers might have argued that any decline in grizzly bear population from rapid closure of the dumps would be temporary and that eventually the population would rebound to its pre-dump-closure level. It is fair to say that at the time the decision was made to close the garbage dumps, managers did not know with certainty how closing the dumps would influence the grizzly bear population. In other words, there was uncertainty regarding the relationship between bear population and the rate at which the garbage dumps were closed. History shows the grizzly bear population in Yellowstone declined after the garbage dumps were closed.

There are several ways active AM could have been used to implement natural regulation for bears. What follows is a plausible scenario for implementing active AM based on the principles enunciated in the first case study. Suppose an AMC is created for grizzly bears in the same manner as the AMC proposed for the Peace Park and the Citizen Management Committee proposed in connection with reintroduction of grizzly bears into Idaho's Selway-Bitterroot Wilderness. AMC would evaluate the Leopold Report and determine what natural regulation means for bear management and whether it is a suitable policy for the park. If natural regulation is deemed suitable, then AMC would select desired outcomes for grizzly bear recovery and preferred management actions for achieving those outcomes. Suppose AMC decides that it is desirable to achieve more natural conditions in the park (as did YNP managers) but not at the expense of having the bear population fall below some threshold level. The latter could be lower than the bear population that existed at the time the decision was made to close the garbage dumps.

AMC recognizes there is uncertainty about how the grizzly bear population responds to the rate at which the garbage dumps are closed. For this reason, AMC decides to phase out the garbage dumps using an active AM approach. Ideally, AMC should conduct independent experiments to learn how different dump-closure schedules affect grizzly bear population. Since there is only one Yellowstone grizzly bear population, it is not possible to test different closure schedules (treatments) on different Yellowstone bear populations and replicate treatments. Furthermore, it is not possible to use one of the two dumps as a control site because sites are not statistically independent; the rate at which one dump is closed is likely to affect the number of bears feeding at the other dump. For these and other reasons, it is not feasible to make strict application of active AM.

AMC decides to implement active AM using a second-best approach that involves experimenting with a flexible dump-closing schedule. The latter is designed to achieve the desired outcome (more natural feeding conditions for bears) without causing unacceptable declines in the grizzly bear population. An unacceptable decline occurs when bear population falls below a threshold population level set by AMC. Closing garbage dumps based on a flexible schedule reduces the risk that grizzly bear population falls below the threshold population. AMC starts the experiment with a slow (conservative) schedule and speeds it up as long as the bear population remains above the threshold. This approach is expected to allow time for bears to transition from dependence on human garbage to total dependence on natural foods.

AMC can establish threshold values for social and economic impacts of closing the dumps, although such impacts are not expected to be large. Social, economic, and ecological values are then combined in a MADM framework that allows integrated assessment of the impacts of dump closure. The flexible scheduling approach has some limitations. For example, it is not possible to control for other factors that influence grizzly bear population, such as the availability of other (nonhuman garbage) food sources, weather, disease, and natural hazards. AMC needs to consider these factors when assessing how dump closing affects the population of grizzly bears.

How does a flexible scheduling approach compare to the one recommended by the Craigheads? They recommended closing the dumps over a 10-year period, which is a fixed schedule, and monitoring bear population. Both approaches require careful monitoring of grizzly bear population. The second-best approach phases out the dumps according to a flexible schedule rather than fixed schedule. With the flexible schedule, the time it takes to phase out the dumps is not known until after the experiment is completed.

In summary, the decision to close the garbage dumps in YNP over a one-year period constitutes passive AM with the desired outcome of achieving more natural bear feeding conditions in the park. It was not feasible to use active AM to determine an optimal schedule for closing the garbage dumps. The second-best approach proposed here determines a schedule for closing the dumps that balances achievement of more natural feeding conditions and the risk of driving the population of grizzly bears below some threshold level.

AEM and Grizzly Bear Recovery

Suppose it were possible to go back to the early 1980s when the grizzly bear recovery program for GYE was being developed. How different might the program be if it were based on AEM? Seven years after the grizzly bear was listed as a threatened species, the Fish and Wildlife Service issued its first Grizzly Bear Recovery Plan. A year later, the IGBC was established. The IGBC is composed of federal and state agencies and the province of British Columbia. These entities currently manage grizzly bears and/or areas that provide habitat for grizzly bears. State and provincial agencies are included in the IGBC because the grizzly bear's range extends into state, provincial, and private lands and state agencies will have responsibility for managing grizzly bears when they are delisted or removed from the federal threatened species list.

Developing and implementing effective recovery plans for grizzly bears is complex and challenging and requires federal, state, and private land managers to make compromises and trade-offs between social, economic, cultural, biological, and political values. For this reason, it is not surprising that recovery plans have been challenged and criticized. In 1995, a federal judge declared that the Grizzly Bear Recovery Plan developed by the service was deficient because it relied too much on defining recovery in terms of bear population and did not establish habitat improvement goals (Willcox and Ellenberger 2000). In 1999, the service provided to the U.S. District Court for the District of Columbia updated and additional information on five specific grizzly bear issues. At that time, the service and agency partners agreed to investigate new methods to estimate total population size for grizzly bears in Cabinet-Yaak, Selkirk, and Northern Continental Divide Ecosystems. After scientific peer review, the service plans to use the new methods to estimate total population size and set the sustainable mortality limits for grizzlies in the GYE (U.S. Fish and Wildlife Service 1999).

Other criticisms levied at the recovery plan for Yellowstone grizzlies are that it:

1. Considers only the number of litters born and ignores litter size and total cub numbers,
2. Only accounts for human-caused mortality as a percentage of estimated population size instead of all deaths, including those attributable to natural causes,
3. Uses crude estimates of total population size (cub litter abundance times the ratio of total population size to the number of cub litters (10.5) measured more than a decade ago,
4. Ignores whether births exceed death over a substantial period of time and whether this condition is likely to persist in the future, and
5. Relies on crude indicators of recovery like the number of "Bear Management Units" in the recovery zone in which mother-cub families are detected rather than long-term trends in habitat size, quality, utilization, accessibility, carrying capacity, and impacts (Stringham 2002).

When the recovery plan was developed, there was insufficient knowledge about the minimum range size for grizzly bears and the maximum density of people beyond which the frequency and lethality of bear mortalities increases. Consequently, the recovery plan does not account for the impacts of human density on bear recovery (Mattson and Merrill 2002). Long-term survival of grizzly bears in the United States is enhanced when the available range exceeds 20,000 km^2 (7,722 square miles), the ratio of perimeter to area is less than 2, and human density is less than 7 persons per km^2 (18 humans per mi^2). By comparison, the GYE is 72,844 km^2, and in 2000 had a population density of 4.8 human per km^2. There is also a better understanding of how other factors affect grizzly bear survival including human behavior and attitudes that favor the presence of bears, greater topographic relief particularly where important food sources disperse bears away from people, presence/abundance of bison carrion, extensive oak communities, nuts from whitebark pine trees, army cutworm moths, spawning cutthroat trout, and others (Mattson and Merrill 2002). Unfortunately, many traditional food sources for grizzly bears are being threatened. Whitebark pine in many areas of the West (although not Yellowstone) is being devastated by blister rust. The population of cutthroat trout is being adversely affected by predation from nonnative lake trout. Global climate change may make the future presence of army cutworm moths questionable. Livestock interests want to curb the park's bison herd to control the transfer of brucellosis from bison to domestic cattle (Wilkinson 2002; also see the following sidebar for discussion of bison management in Greater Yellowstone Ecosystem).

Bison Management in Greater Yellowstone Ecosystem

Bison contribute to the aesthetic, biological, cultural, and ecological values of Yellowstone National Park and the Greater Yellowstone Ecosystem and are considered by many to be a national icon and symbol of the American West. Unfortunately, Yellowstone National Park does not provide sufficient habitat for bison, especially during severe winters. As a result, bison migrate out of the north side of the park into adjacent public and private lands in Montana.

Migrating bison carrying brucellosis can transmit it to domestic cattle. Brucellosis is a contagious bacterial disease that has no cure and for which vaccines are less than 100 percent effective. A pregnant cow infected with brucellosis typically aborts its first calf after the fifth month of gestation. Aborted fetuses would reduce the income of ranchers, and cause the state of Montana to lose its brucellosis-free status. Should this occur, Montana would not be allowed to ship its cattle to other states, resulting in even further income losses to the livestock industry and economic losses to the state's economy. Other impacts of brucellosis in cattle include (1) decreased weight gain by calves, (2) delays in calf production, (3) increased rates of culling and replacement, and (4) increased testing and vaccination costs.

After evaluating several alternatives for dealing with this potential problem, federal and state agencies (Montana Department of Livestock, Gallatin National Forest, Yellowstone National Park, and USDA-Animal and Plant Health Inspection Service) adopted the Interagency Bison Management Plan (IBMP) in December 2000. Development of the IBMP rests on two primary objectives. The first objective is to "maintain a wild, free-ranging population of bison plus its genetic integrity and its ecological function in the Yellowstone area." The second objective is to "address the risk of brucellosis transmission to protect the economic interest and viability of Montana's livestock industry." Major features of the plan are as follows:

- Establish special management areas outside the park where bison will be allowed during the winter when cattle are not present.
- Vaccinate bison when a safe and effective vaccine is available.
- Vaccinate and monitor cattle in specific areas near Yellowstone National Park.

- Haze bison on to appropriate public lands during winter or back into the park in the spring to keep bison separated from cattle.
- Adapt the plan as more knowledge and experience are acquired about the management of brucellosis, bison, and cattle.

A cornerstone of the IBMP is an adaptive management approach that treats bison management actions as opportunities to learn more about brucellosis, bison, cattle, and their management (see chapter 7 for a discussion of adaptive management). New knowledge gained from an adaptive management approach would be used to improve bison management. A major uncertainty that adaptive management would shed light on is whether bison and domestic cattle can be successfully separated outside the park.

A controversial provision of the IBMP is to limit the size of the Yellowstone bison herd population during the late winter/early spring to 3,000 animals in an effort to reduce the risk of brucellosis transmission from bison to cattle; the current population is about 3,800. According to the IBMP and IBMP Operating Procedures, when the bison population exceeds 3,000 animals in late winter/early spring, animals that leave the North Entrance of the park and enter lands where cattle are being grazed must be hazed to keep them inside the park. If this proves unsuccessful, then they are captured in the Stephens Creek capture facility and sent to slaughter. In the first winter season (2001–2002) after IBMP was adopted, 202 bison near or just outside the northern border of Yellowstone National Park were captured and slaughtered. In the second winter season (2002–2003), more than 231 bison were rounded up and sent to slaughter. The annual cost to taxpayers of the winter bison slaughter is over $2 million. Bison were also slaughtered even before the IBMP was implemented. In the unusually severe winter of 1996–1997, 1,083 bison searching for food on public lands outside the park's northern boundary were slaughtered.

Environmental groups view the slaughter of Yellowstone bison as unnecessary and undesirable because there is no documented case of brucellosis being transferred from wild bison to domestic cattle and the IBMP does not require bison to be tested for brucellosis before being slaughtered. While the Montana Department of Livestock supports bison killing as a means of safeguarding Montana's brucellosis-free status, environmental groups argue that it is a ploy to protect cattle grazing rights and reduce grazing competition between bison and cattle on public lands. A Citizens' Plan for management of Yellowstone's bison herd was developed by a coalition of environmentally oriented groups. In the Citizens' Plan, bison are treated as wildlife and given safe haven on America's public lands; bison hazing and slaughter are discouraged.

Managing Yellowstone bison raises a complex set of issues involving the relationship between sustaining the wildlife and ecological values of a premier natural park and protecting Montana's livestock industry. Only time will tell whether the IBMP is successful in balancing these competing values.

Sources: U.S. Department of the Interior, National Parks Conservation Association (2003d); U.S. Department of Agriculture (2000); Yellowstone Association (2003); Yellowstone Science (2003).

The Sierra Club criticized the recovery plan for not having a large enough recovery area and not linking the recovery area to other core habitat areas for grizzly bears. An alternative strategy for long-term recovery proposed by the Sierra Club involves (1) expanding the recovery area, (2) connecting the Yellowstone grizzly bear population to other populations, (3) protecting remaining wildlands, (4) restoring important but degraded habitat, (5) increasing public involvement in the recovery process, and (6) expanding efforts to reduce bear-human conflicts through improved sanitation and public education (Willcox and Ellenberger 2000; Stringham 2002).

Demographic context and geographic coverage have an important bearing on AEM for grizzly bear recovery. First, during the period the recovery plan for Yellowstone grizzly bears was developed and implemented, human population in the GYE was growing rapidly and significant amounts of the landscape were being converted to rural residential development. From 1980 to 2000, the population of the GYE grew at twice the national rate (Glick and Haggerty 2000). While some of this increase was dispersed around population centers, much of it was concentrated in more remote rural areas of the GYE (Glick and Haggerty 2000; Hansen et al. 2002). From 1970 to 1997, rural residential development in the Montana and Wyoming portions of the GYE jumped by more than 400 percent. One study showed that 320 of the 400 new homes randomly sampled in Gallatin County were constructed in prime wildlife habitat (Glick and Haggerty 2000). In addition, rapid growth in park visitation increased the likelihood of encounters between people and grizzly bears.

Based on these trends, AEM of grizzly bears requires coordinated management of public and private lands. This necessitates cooperation among federal land management agencies, state, county, and city governments, private developers and landowners, planning and zoning authorities, special interest groups, and others. Hansen et al. (2002) point out that a coordinated management approach organizes assessments and policies around the different and possibly conflicting management objectives of different land managers and stakeholders. A properly constituted and executed AMC can achieve the required coordination. While giving AMC the authority to coordinate management of public and private lands might seem politically, legally, and institutionally unrealistic, it is not without precedent. For example, in the Montana portion of the GYE, high-elevation summer and low-elevation winter range for elk are being cooperatively managed by federal, state, and private landowners (Schullery 1997). Similarly, a Citizens Management Committee has been proposed to manage the proposed reintroduction of grizzly bears into Idaho's Selway-Bitterroot wilderness.

An AM approach was not used to develop and implement the Grizzly Bear Recovery Plan. The plan was developed by the Fish and Wildlife Service with agency coordination handled by the IBGC. The latter includes only representatives from federal and provincial/state agencies. There was no direct, cooperative involvement of other stakeholders.

Second, while the size of the GYE is large enough and the human population density low enough to sustain a viable local population, some argue that the ultimate fate of grizzly bears depends on cooperative management of a "sprawling mosaic of federal, state, and private lands" that extends beyond the boundaries of YNP and GYE. A large biogeographic area could support 2,000 grizzly bears, the number some say is required to ensure sufficient genetic diversity and minimize the threats of random disasters (Schullery 1997). AEM of grizzly bears would require linking bear management in the GYE to changes in human population size, density, and distribution; land cover; and landscape fragmentation in a much larger biogeographic area. The Yellowstone-to-Yukon Conservation Initiative envisions management of large carnivores over continental scale (biogeographic) areas. Cooperative management of bears over a large biogeographic area is much more complex and challenging administratively than managing bears in the GYE. Even if separate recovery plans were developed and implemented for distinct populations of grizzly bears, as is currently done by the service, an AEM approach could alleviate some of the criticisms levied on the current grizzly bear recovery program.

Managing grizzly bear recovery in an active AM framework has the advantage of allowing managers to learn about the impacts of experimentally based management actions on grizzly bear habitat and populations. That knowledge could be used to improve management actions. Recall that one of the criticisms of the bear recovery plan is that it is not responsive to new scientific knowledge about the consequences of falling below a minimum range size for bears and/or exceeding a maximum human density. Scientific knowledge can be incorporated in bear recovery plans implemented in an active AM framework because management actions are reevaluated in the light of new knowledge (learning occurs) and revised as needed. A recent report by the National Research Council on ecological dynamics in YNP recommended that all management in the park be done as AM because of "insufficient scientific knowledge...to predict the consequences of different management actions." The report goes on to say that "long-term scientific investigations and experiments are needed to provide solid scientific evidence for evaluating management options" (National Research Council 2002).

While managing wildlife resources in the GYE using an AM approach is complex and challenging, it has the potential to achieve long-term desired outcomes and alleviate some of the criticisms levied at traditional top-down management approaches.

Snow Machine Use in the Greater Yellowstone Ecosystem

National parks and protected areas provide wonderful opportunities for outdoor recreation. While most park visitation occurs in the summer months, there has been significant growth in winter visitation. For instance, winter relative to nonwinter visitation in national park units within the United States is now higher than in the past. From 1985 to 2001, summer visitation (June, July, and August) as a percentage of total annual visitation decreased 8 percent, whereas winter visitation (December, January, and February) as a percentage of total annual visitation increased 19 percent (see fig. 9.5). Perhaps the most controversial recreational winter use activity in U.S. national parks is snowmobiling in Yellowstone and Grand Teton National Parks in the GYE. YNP in particular has become very popular for snowmobiling; West Yellowstone claims to be the "Snowmobile Capital of the World" (*Daily Inter Lake* 2002). In 1997, the animal rights group Fund for Animals sued the National Park Service for not evaluating the impacts of snowmobiling on the park's wildlife. Air pollution from snowmobiles has become so bad that fresh air was pumped into gatekeepers booths at the West Yellowstone entrance to reduce headaches, dizziness, and nausea suffered by rangers (*Daily Inter Lake* 2002). In November 2000, the secretary of the interior issued a record of decision to phase out recreational snowmobile and snowplane use by the winter of 2003/2004 in Grand Teton National Park, YNP, and the eight-mile John D. Rockefeller, Jr., Memorial Highway connecting the two parks. This decision favored protection of air and wildlife resources and more environmentally benign recreational activities, such as snowshoeing and cross-country skiing. Public comments were overwhelmingly in support of banning snowmobile use in these areas. Local communities, particularly West Yellowstone, opposed the ban because snowmobiling generates significant income and employment during the winter months. The snowmobile industry objected to the proposed ban on the grounds

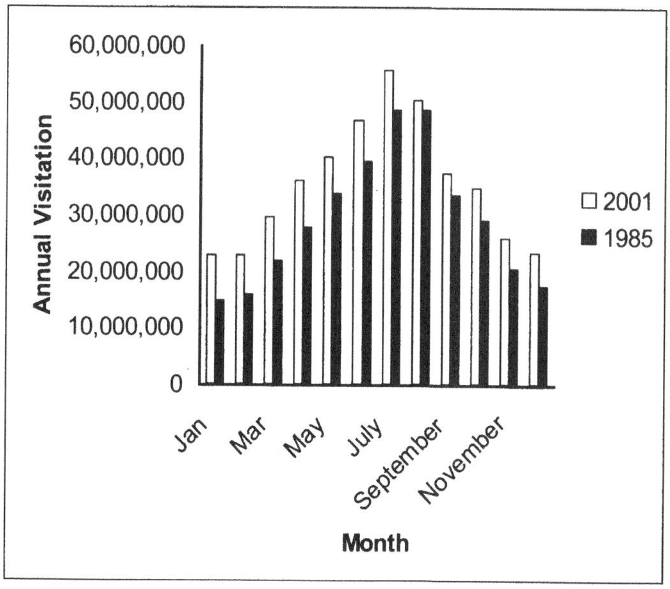

FIG. 9.5. Monthly visits to national parks in the United States, 1985 and 2001. (From National Park Service 2002c.)

that it infringed on its right of access to national parks and had questionable wildlife benefits.

Following the record of decision, the International Snowmobile Manufacturers Association, state of Wyoming, and others brought a lawsuit against the Department of the Interior asking the decision to be set aside. The lawsuit was settled when the Interior Department agreed to do a supplemental environmental impact statement (SEIS) for winter use in the three areas. The draft SEIS proposed three alternatives, solicited public comments on those alternatives, and considered new information not available at the time the original environmental impact statement was prepared. Of the 350,000 comments received on the draft SEIS, 80 percent favored a ban on recreational snowmobiles and snowplanes (alternative 1). Alternatives 2 and 3 employ adaptive management to "mitigate impacts on visitor experience and access, wildlife, air quality and natural sound while allowing snowmobile access on all existing oversnow routes" (National Park Service 2002d).

Alternative 2 in the draft SEIS is based on the traditional definition of carrying capacity (maximum number and type of visitors an area can sustain without causing irreversible deterioration of the physical environment and appreciable loss of visitor satisfaction; see chapter 7). It establishes interim visitor use limits for each of the six road segments in the GYE until a long-term visitor capacity study is completed. Interim use limits are predicted to keep impacts of snowmobile use within acceptable limits. The latter are defined in terms of standards for visitor experiences and park resources. Monitoring is done to determine whether use impacts violate the standards. Should the latter occur, managers would decrease use limits and adjust related management activities to comply with the standards (National Park Service 2002d).

Alternative 3 in the draft SEIS uses AM and the VERP carrying capacity method described in chapter 7 to assess whether snowmobile use is in compliance with carrying capacities. Carrying capacities are defined in terms of the quantity and quality of visitor experiences and park resources in each of seven road segments in the GYE. Indicators of visitor experiences and park resources are identified, monitored, and compared to corresponding standards for each road segment. Management actions are initiated when indicators violate standards (National Park Service 2002d). Alternative 3 moves directly to a visitor-carrying-capacity-based approach rather than establishing interim use limits until a long-term visitor capacity study is completed.

The framework used to develop alternatives 2 and 3 incorporates one or more of the following elements:

1. Imposing daily limits on numbers of snowmobiles in road segments with an average total daily use in YNP of 1,100 snowmobiles (during holidays and busy weekend use is as much as 1,650 per day)
2. Requiring snowmobiles to have Best Available Technology (BAT) to reduce carbon dioxide and hydrocarbon emissions and noise, for example, require snowmobiles to have cleaner and quieter four-stroke rather than two-stroke engines
3. Using an AM approach that utilizes an "open forum" strategy to make monitoring results, technical expertise, monitoring methods, and results of peer review available to the public
4. Requiring commercial and noncommercial snowmobile guides to receive training in an attempt to reduce negative wildlife impacts
5. Phasing in daily limits on snowmobile use, BAT, training and guiding requirements, and monitoring programs over a three-year period (2003–2004, 2004–2005, and 2005–2006)
6. Developing a new generation of snowcoaches that eventually meet BAT standards
7. Acquiring sufficient funds for the Park Service to implement all aspects of the program (Dougherty 2002; National Park Service 2002a,b)

The draft SEIS states: "The first step in adaptive management is to develop and implement a management scenario based on the best available information" (National Park Service 2002d). This implies passive AM. Specifically, alternative 2 defines interim use limits and standards for carrying capacity based on expert knowledge, which implies managers know with reasonable certainty how use limits (management actions) influence visitor experiences and resource conditions. A particular use limit is retained unless measured impacts violate established standards for visitor experiences and resource conditions. Presumably, use limits are decreased when this occurs, although it is not clear how the amount of decrease would be determined. Alternative 3 is passive AM because it implicitly assumes that managers know the best management action to comply with carrying capacities for visitor experience and resource conditions.

If active AM were used, then experiments would be devised to test hypotheses regarding the impacts of alternative management actions on visitor experiences and resource conditions, and a best management action would be selected based on results of the tests. How feasible is it to experiment with different snowmobile management actions? Numerous elements can be varied in active AM experiments for snowmobile use, including maximum daily use limits, percentage of tours that are guided, percentage of snowmobiles that are BAT, and others. While many levels and combinations of these elements are possible, not all are worth testing. In particular, it appears that provisos like requiring BAT in snowmobiles and snowcoaches and guided tours with trained tour guides would have many benefits in terms of reducing air and water pollution, abating noise, and alleviating adverse impacts on wildlife and other natural resources. Suppose alternatives 2 and 3 phase in these provisos over a three-year period as envisioned in draft SEIS. The remaining issue is how to set interim use limits (alternative 2) or judge the extent to which carrying capacities for visitor experience and resource conditions are being violated in each road segment (alternative 3).

Since the most significant differences in social, economic, and ecological impacts are likely to be between disallowing snowmobiles (alternative 1) and allowing snowmobiles with the above provisos (alternatives 2 or 3), it is reasonable to implement active AM with alternative 1 as the control and alternatives 2 and 3 as treatments with different snowmobile use levels. Provided snowmobile use is relatively independent across the road segments in the GYE (use in one segment does not influence use in another segment), then the control condition (no snowmobiles) can be randomly assigned to one or two road segments and the two treatments randomly assigned to the remaining five road segments. While this approach would be less expensive and time-consuming, it is not as statistically robust as randomly assigning the control (ban snowmobiles) and treatments (alternatives 2 and 3) to each road segment for a prescribed number of years (e.g., alternative 3 for the first five years, control for second five years, and alternative 2 for the third five years). Given the limited budget of the National Park Service and the social and political challenges of implementing active AM, it might be possible and feasible to implement only passive AM.

For the most part, the above discussion focuses on how different levels of snowmobile use are likely to influence visitor experiences, which is fundamentally a social phenomenon, and the quality of air, water, and wildlife resources, which are fundamentally biophysical or ecological phenomena. While social and ecological impacts of alternatives are important, so are economic impacts. The three alternatives evaluated in the draft SEIS for winter use in the GYE are likely to have substantially different impacts on recreational use in the winter months. The difference is especially pronounced between alternative 1 (no snowmobiles) and alternatives 2 or 3 (snowmobiles allowed). Different levels of snowmobile use in the GYE will result in different levels of income and employment in the region and to a lesser extent the state. The differences in recreation use, income, and employment should be considered in comparing alternative winter-use plans. As mentioned in the first case study and chapter 8, income and employment impacts of changes in demand (in this case for the recreational sector) can be estimated at the county and state levels using the IMPLAN model. Furthermore, social, economic, and ecological impacts of alternative management actions can be integrated using a MADM framework.

The final SEIS developed a fourth or preferred alternative that has the following components:

1. Imposing daily limits on numbers of snowmobiles in road segments with an average total daily use in YNP of 950 snowmobiles
2. Requiring snowmobiles to have Best Available Technology (BAT) with minimum reductions in hydrocarbon emissions of 90 percent and carbon monoxide emissions of 70 percent compared to a standard two-stroke snowmobile (four-stroke snowmobiles satisfy these limits)
3. Using an adaptive management approach that monitors and evaluates the short- and long-term impacts of management actions on park resources and values, and adjusts management actions based on evaluation results (National Park Service 2003)
4. Requiring all snowmobilers in the park to be accompanied by an NPS-approved guide in an effort to reduce negative wildlife impacts (commercially guided tours will make up 80 percent of the entries and noncommercially guided tours will comprise 20 percent of the entries)
5. Developing a new generation of snowcoaches that eventually meet BAT standards and become the key to winter transportation (Yellowstone Association 2003)

Reducing Conflicts between Wildlife, Urban Development, and Transportation in Banff National Park

As mentioned in the first case study, present and future urban development and transportation routes in and around national parks and protected areas are major threats to their ecological integrity. These threats are especially acute when urban development and major transportation routes occur within a national park or protected area as is the case in Banff National Park (BNP) in Alberta, Canada, and elsewhere. BNP was established in 1885 as Canada's first national park. The park is 6,641 km^2 (2,564 mi^2) in size and is part of the Central Rockies Ecosystem and a UNESCO World Heritage Site (Canadian Rockies.net 2002; Parks Canada 2001).

Banff-Bow Valley is located in the park and contains the Town of Banff, Hamlet of Lake Louise, Banff-Radium Highway, Bow Valley Parkway, Icefields Parkway (which connects Banff and Jasper National Parks), Trans-Canada Highway, Canadian Pacific Railway, and other park roads. There is a ski area at Lake Louise. The valley has over 8,000 permanent residents. BNP has about 4.7 million visitors per year, most of whom arrive by private vehicle. An additional 4 million people use the Trans-Canada Highway on their way to other locations. Traffic is about 20,000 vehicles per day on the Trans-Canada Highway, and 30 to 35 trains per day on the Canadian Pacific Railway. In the decade of the 1990s, traffic on the Trans-Canada Highway increased at between 2 percent and 3 percent per year. Urban development and transportation corridors through the park fragment wildlife habitat, restrict wildlife movement, eliminate or alter natural communities, contribute to wildlife/vehicle collisions and associated wildlife and human injuries and mortalities, reduce secure areas for wildlife, change predator-prey relationships, and others (Parks Canada 2001; Banff National Park 2002).

Wildlife-human conflicts are especially acute in the Banff-Bow Valley because it is not only a major location for townsites, campgrounds, visitor accommodations, and services and transportation routes, but also an important source of food and shelter for wildlife. In addition, the valley contains corridors that allow wildlife to access different parts of their range; an especially important element for long-term survival of elk, grizzly bears, lynx, wolverines, and wolves. In an effort to reduce damages caused by collisions between wildlife and vehicles, fences were constructed along a portion of the Trans-Canada Highway in the park. In addition, two wildlife overpasses were constructed along the portion of the highway in the park. Fences and overpasses have reduced wildlife injuries and mortalities. However, motor vehicles and trains continue to have adverse impacts on wildlife populations. Between 1981 and 1999, collisions between wildlife and vehicles on the Trans-Canada Highway resulted in the death of 16 wolves, 14 black bears, and 4 cougars. Most of the wildlife mortalities caused by trains involve elk. Other strategies Parks Canada has identified for reducing vehicle-related wildlife impacts include reducing speed limits, educating motorists through signage, limiting private-vehicle access to certain frontcountry areas, providing alternative transportation modes to private vehicles, and improving critical wildlife corridors (Parks Canada 2001).

Increased traffic in and between Banff and Jasper National Parks is having other adverse impacts including traffic congestion and associated deterioration in visitor experiences. The Banff-Bow Valley Task Force concluded that continuation of current trends would "lead to the destruction of the conditions in the Banff-Bow Valley that are required for a National Park" [and that] changes must come quickly if the park is to survive (Banff-Bow Valley Study 1996). A regional transportation study is being done for Bow Valley that aims to develop "a sustainable transportation strategy that encourages the efficient and effective movement of people and goods to and through the region, recognizing as paramount the need to ensure protection of the ecological values of the region while maintaining attractive visitor experiences" (Parks Canada 2001).

The 1988 amendments to the Canada National Parks Act state that the primary management goal for national parks is "maintenance or restoration of ecological integrity, through the protection of natural resources and natural processes." The Panel on the Ecological Integrity of Canada's National Parks (2000) determined that "an ecosystem has integrity when it is deemed characteristic for its natural region, including the composition and abundance of native species and biological communities, rates of change and supporting processes." An ancillary goal is "to maintain national parks unimpaired for the enjoyment of future generations" (Parks Canada 2001).

The 1997 management plan for BNP acknowledges that urban development and transportation corridors are permanent features of the park. Nevertheless, Parks Canada, the agency that manages Canada's national parks, is committed to implementing a human use management strategy for BNP. Human use management has become necessary due to "major concerns about the capacity of existing infrastructure to support the growing number of visitors," and increased noise and air pollution; traffic congestion; crowding; demands for parking and other park services; human-wildlife conflicts; and wildlife mortality, habituation, and displacement. In other words, increased visitation is threatening both the quality of visitor experiences and wildlife populations in the park (Banff National Park 2003). This threat has led to the concept of responsible human experience or use without abuse (Wilkinson 2000).

The goal of the BNP management plan is to provide high-quality visitor experiences while reducing the stress that human activity places on ecological integrity. The plan addresses issues and recommends actions "at a site-specific, landscape management unit, and park-wide scale, primarily during the nonwinter period." The plan is being developed using a passive adaptive management approach that is expected to generate new knowledge and understanding (learning) about "the complex relationship between ecological integrity and human use" (Parks Canada 2001). Knowledge gained from AM will be used by park managers to improve management actions for reducing adverse human and ecological impacts of visitation. Elements of the AM approach include (1) setting clear human use management objectives for the park; (2) establishing social (visitor experience), economic, cultural, and ecological performance indicators and standards (targets) for those objectives at various spatial and temporal scales; (3) monitoring for compliance with standards; and (4) adjusting management actions when indicators violate standards. This approach is very similar to the VERP (carrying capacity) method described in chapter 7. Development and implementation of the research and monitoring program is being guided by an advisory committee composed of experts. The public, stakeholders, and park communities will have opportunity to comment on draft objectives established for ecological management areas within the park and major human use management proposals (Parks Canada 2001).

It appears that the AM approach being used in BNP does not involve the kind of public and stakeholder involvement envisioned for the AMC discussed in the previous case studies. In particular, the public, stakeholders, and others are allowed to comment on objectives and management proposals, but are not directly involved in proposing and evaluating management actions. There is an important difference between AM efforts for BNP and potential AM efforts for the Peace Park; threats are mostly internal for BNP and external for the Peace Park. In this respect, it should be easier to alleviate transportation threats to BNP than to the Peace Park, other things equal. However, the transportation threats are potentially more harmful to BNP than to the Peace Park. Proposed expansion of Highways 2 (US) and 3 (Canada) and paving of the North Fork Road are likely to have less detrimental ecological impacts on wildlife in the Peace Park than the impacts of the Trans-Canada Highway on wildlife in BNP because the former threats originate outside the Peace Park whereas the latter originate within BNP. Similarly, the Burlington Northern-Sante Fe railroad line lies entirely outside Glacier National Park, whereas the Canadian Pacific Railway goes through BNP.

Despite these and other differences AM for BNP has many of the same elements discussed in earlier case studies, namely, (1) establishing objectives (desired outcomes) for dealing with particular management concerns; (2) taking an ecosystem scale perspective that extends beyond the boundaries of the park; (3) recognizing there is uncertainty about how management actions influence social, economic, and ecological values; (4) establishing performance standards or indicators for social, economic, and ecological attributes; (5) conducting research and monitoring to determine whether measured attributes of management actions are within limits imposed by established standards; (6) adjusting management actions when monitoring indicates standards are being violated; and (7) involving experts, the public, stakeholders, and communities in planning and evaluation.

Improving Air Quality in National Parks

Many national parks in the United States exceed air quality standards established by the Environmental Protection Agency (Mansfield 2002). Ozone, sulfur dioxide, and carbon dioxide are the three primary kinds of air pollutants that harm plants, animals, lakes, streams, soils, and visibility in national parks. Ground-level ozone (O_3) is the principal toxic ingredient in smog. It forms when nitrogen oxides and hydrocarbons from power plants and motor vehicle exhaust react in sunlight (National Park Service 2003c). High concentrations of ground-level ozone obscure scenic views, impair visitor health, and harm plants in national parks.

In terms of ozone's impairment of visibility, the smog that routinely engulfs the Great Smoky Mountains National Park reduces visibility of distant mountains from 115 km (77 miles) to an average of 21 km (14 miles) in the summer (Mansfield 2002). Annual visibility in Sequoia and Kings Canyon National Parks is just 39 miles compared to natural visibility of 120 to 180 miles in the western United States (National Parks Conservation Association 2003a). Poor visibility degrades visitor enjoyment and satisfaction (Ross 1988). Average visibility in Shenandoah National Park in Virginia has decreased from about 136 km (90 miles) to less than 30 km (20 miles) (National Parks Conservation Association 2003b,e).

In terms of ozone's health hazards, figure 9.6 indicates the number of unhealthy ground-level ozone days in the five national parks in the United States having the highest ozone levels. An unhealthy ozone day occurs when the daily maximum eight-hour average concentration of ground-level ozone exceeds 85 parts per billion. People with heart and respiratory ailments and other health problems, children, and the elderly are especially vulnerable to unhealthy con-

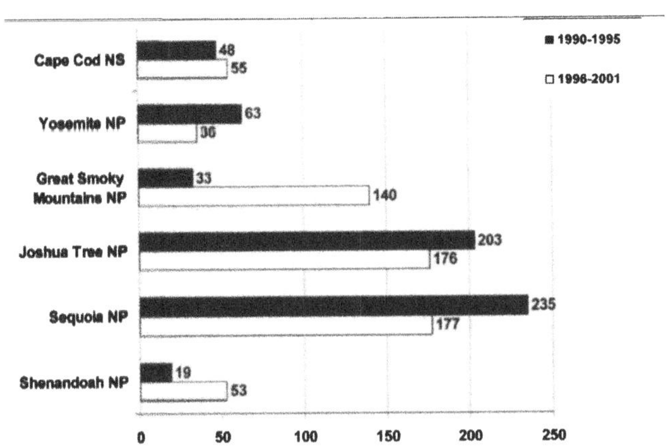

FIG. 9.6. Number of days with unhealthy ozone levels for five national parks having the highest levels, United States, 1990–2001. (From National Park Service 2003a.)

centrations of ozone (National Park Service 2003d). High ozone concentrations can induce asthma in healthy children (Jeffords 2002). Sequoia and Kings Canyon National Parks had 61 days in 2001 in which the air was unhealthy to breath due to high concentrations of ground-level ozone, which were more health hazard days than any other national park (National Parks Conservation Association 2003a). Ozone problems in the three California park units (Sequoia, Joshua Tree, and Yosemite) are caused by rapid growth in California's population, urban development, number of motor vehicles and power plants, and nonenforcement of clean air standards. The average annual ozone exposure level in Great Smoky Mountains National Park (133,200 ppb-hours) exceeds levels in Atlanta, Georgia, Knoxville, Tennessee, and Charlotte, North Carolina (Hymon 2001). Unhealthy ozone concentrations occur in other national parks (see National Park Service 2003a).

In terms of ozone's ecological impacts, research plots indicate that 39 percent of the Jeffrey and ponderosa pine trees in Sequoia and Kings Canyon National Parks display visible damage from ozone. When ozone levels exceed ambient levels by 50 percent, photosynthesis and growth in giant sequoia seedlings is suppressed (National Park Service 2003d).

Sulfur dioxide emissions from power plants that burn fossil fuels contribute to acid deposition. Wet (acidic rain, fog, and snow) and dry (acidic gases and particles) acid deposition occurs when a mild solution of sulfuric acid and nitric acid is formed in the atmosphere by the reaction of sulfur dioxide (SO_2), nitrogen oxides (NO_x), water, oxygen, and other chemicals. Acid deposition in soils and water can slow plant growth, injure or kill forests, and kill fish in lakes (Green Nature 2002a). Environmental degradation from acid deposition is especially acute in the northeastern region of North America and the southwestern United States. High-elevation forests and lakes in the Appalachian Mountains, including Adirondack State Park in New York, Shenandoah National Park in Virginia, and Great Smoky Mountains National Park in Tennessee, have been degraded by acid deposition (Green Nature 2002b).

Use of fossil fuels by electrical power-generating plants, industrial facilities, motor vehicles, and the burning of forests produce emissions of carbon dioxide (CO_2), which contribute to the greenhouse effect and global climate change. Slash-and-burn agriculture, timber harvesting, and the tilling of farm fields contribute to global climate change by releasing to the atmosphere carbon stored in vegetation and soils. Global climate change is adversely impacting national parks and protected areas, especially those at higher latitudes. For example, less than one-third of the glaciers that existed in Glacier National Park in 1850 are present today, and most remaining glaciers are mere remnants of their previous size (U.S. Geological Survey 2003). If current temperature trends continue, all glaciers in Glacier National Park will disappear in 30 years (Hall and Fagre 2003). Future changes in temperature and precipitation caused by global climate change are expected to alter the amount and seasonal distribution of water, shift vegetation patterns, and change the spatial distribution of wildlife in national parks and protected areas. Impacts will be greater in some protected areas than others. Mercury is another potent air pollutant that is threatening the ecosystems at Everglades, Big Bend, and Arcadia National Parks (National Parks Conservation Association 2002b).

Park managers are able to control the amount of air pollution caused by snowmobiles in Yellowstone National Park by restricting the numbers and types of snowmobiles allowed in the park. Park managers cannot directly control ozone, sulfur dioxide, and carbon dioxide emissions from power plants and motor vehicles. Most of the pollution caused by these sources originates outside the parks. For example, much of the air pollution in Great Smoky Mountains National Park comes from nearby coal-fired power plants (National Parks Conservation Association 2003b). Air quality in Joshua Tree National Monument would be impaired by a proposed landfill and the six power plants under construction or proposed for the area (Hymon 2001). Furthermore, while AM experiments can be conducted to improve understanding of the air pollution impacts of snowmobile use in Yellowstone National Park (third case study), it is not feasible to conduct AM experiments for pollution sources external to the park.

In essence, the operation of power plants and motor vehicles imposes an external cost on national parks in the form of damages from air pollution. What attempts have been made to reduce these damages? The U.S. Congress has been legislating air quality since 1970 when the Clean Air Act was passed (see chapter 5 for details). In an effort to reduce adverse impacts of air pollution on Class I areas, Congress amended the Clean Air Act in 1977 and 1990. The 1977 amendments added a section on "prevention of significant deterioration [in order] to preserve, protect, and enhance the air quality in national parks, national wilderness areas, national monuments, national seashores, and other areas of special national or regional natural, recreational, scenic, or historic value" (42 USC. Sec. 7470). National parks and wilderness areas larger than 2,429 ha (6,000 acres) and national wilderness areas and national memorial parks larger than 2,024 ha (5,000 acres) in existence on August 7, 1977, were designated as Class I areas (National Park Service 2003b). Class I is the highest air quality designation permitted by law. National parks designed as Class I areas are depicted in figure 9.7. The Clean Air Act amendments of 1977 also exempted old sources or facilities (power plants and refineries constructed prior to 1977) from the New Source Review (NSR) program. The NSR program requires new sources of air pollution and old sources making major upgrades that result in pollution increases to install the best available pollution control technology.

The 1990 amendments to the Clean Air Act added a new section that authorizes the Environmental Protection Agency, National Park Service, and other federal agencies to conduct research on visibility, evaluate clean air corridors, and

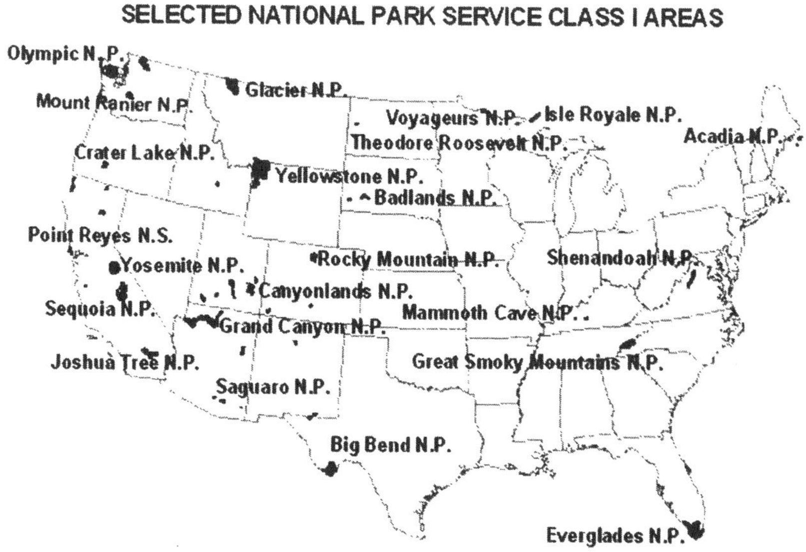

FIG. 9.7. Selected national parks in the United States designated for Class I air quality. (From National Park Service 2003b.)

identify emission sources and source regions impairing visibility in Class I areas (National Park Service 2003b). Unfortunately, the amendments do not ensure achievement of Class I air quality standards. Many parks are in violation of the standards and there is little they can do to achieve compliance.

In 2002, the Environmental Protection Agency proposed to drop the requirement that expansion of old facilities comply with NSR program. This proposal would allow 17,000 old facilities to operate indefinitely with limited or no pollution controls. If implemented, this proposal is likely to worsen air quality in national parks located near old facilities. EPA justified its proposed change on grounds that President Bush's Clear Skies Initiative (CSI), which sets pollution limits for nitrogen oxides, sulfur dioxides, and mercury, but not carbon dioxide, eliminates the need to require old facilities to comply with the NSR program. Meanwhile, the Clean Power Act of 2002 was introduced. It proposes greater emission reduction for nitrogen oxides, sulfur dioxides, carbon dioxide, and mercury in a shorter period of time than CSI while retaining the requirement that upgraded old facilities satisfy the NSR program. One study claims the Clean Power Act of 2002 would result in cleaner air than the CSI (National Parks Conservation Association 2003c).

Policy relating to air quality is dynamic making it difficult to say whether future air quality will improve or worsen in protected areas. Progress toward improving air quality in the United States and other countries is greatly influenced by the political process. Since improvement in air quality is likely to be viewed by many as a lower priority in protected areas than in metropolitan and rural areas where people live, it appears that air quality trends in protected areas will follow national trends with few exceptions. Since air quality influences public enjoyment and ecological integrity of protected areas, management agencies need to continue their air quality research and monitoring programs and stay involved in guarding air quality. Even though federal land management agencies cannot lobby Congress on legislation like the Clean Power Act, they are routinely involved in proposing ways to protect air quality in the vicinity of protected areas. In addition, management agencies give expert testimony to Congress relating to proposed air quality legislation. Environmental interest groups like the National Parks Conservation Association, National Wildlife Federation, and Sierra Club are effective in lobbying Congress and others to pass legislation that improves the quality of air and other natural resources and as litigants in court cases involving natural resource and environmental issues.

Summary

Two extended and three abbreviated case studies are presented here that demonstrate the complexity of management issues facing national parks and protected areas and apply in an AEM framework the values, institutions, concepts, principles, and approaches described in earlier chapters. The first case study describes how AM can be used to design and evaluate management actions to alleviate internal and external threats to natural and cultural resources in the Peace Park, which is part of the CCE. AM is best implemented by a working group of protected area managers, scientists, community planners, environmental groups, developers, and other stakeholders, and constituted based on inclusivity, self-selection, and diversity of representation of stakeholders.

AM accounts for uncertainty regarding the social, economic, and ecological consequences of management actions. Passive AM is appropriate when AMC is relatively confident about how ecosystems respond to management actions. A preferred management action is selected based on a comparison of alternative and current management actions, which can be done using multiple attribute evaluation. Active AM is appropriate when AMC does not know how ecosystems respond to management actions. Experiments are used to test hypotheses about these responses and results are used to improve management. Passive and active AM are well suited to identifying management actions for alleviating internal and external threats to the Peace Park.

It would have been possible to use AEM to balance achieving more natural feeding conditions for grizzly bears (natural regulation) with the risk that too rapid a closure of human garbage dumps would decrease grizzly bear population in the Greater Yellowstone Ecosystem (GYE). Active AM for this purpose would have to be implemented using a second-best approach because it is not possible to test different dump closure schedules (treatments) on different Yellowstone bear populations and replicate treatments. The second-best approach experiments with a flexible dump-closing schedule that varies the rate of dump closure depending on how bear population levels are responding. An advantage of this approach is that it allows bears to transition from dependence on human garbage to total dependence on natural food sources.

The SEIS for winter use in parks in the GYE employs passive AM to determine whether snowmobile use complies with carrying capacities. Alternative 2, which is based on the traditional definition of carrying capacity, defines interim-use limits or standards for snowmobile use based on expert knowledge. It requires managers to know with reasonable certainty how use limits (management actions) influence visitor experiences and resource conditions. Monitoring is done to determine whether imposition of use limits keeps impacts from violating standards. If standards are violated, then use limits are decreased and related management activities adjusted accordingly. Alternative 3, which is based on the VERP carrying-capacity method, uses AM to determine whether the quantity and quality of visitor experiences and park resources are compatible with established standards. If the latter are violated, then management actions are initiated. Alternatives 2 and 3 are passive AM. It would be possible to implement active AM with alternative 1 as the control and alternatives 2 and 3 as treatments with different snowmobile-use levels.

The BNP management plan is using passive AM to improve understanding of how human use affects visitor experiences and ecological integrity in the park and to alleviate wildlife-human conflicts and degradation of visitor experiences. This application of passive AM involves setting clear human use management objectives for the park, establishing various performance indicators and standards for those objectives, monitoring for compliance with standards, and adjusting management actions when standards are violated.

Reducing adverse impacts of air pollution is fundamentally different from alleviating many of the internal and external threats facing national parks and protected areas. It is very different from alleviating many of the ecological threats to the Peace Park, recovering grizzly bears in the GYE, alleviating adverse impacts of snowmobile use in the GYE, and reducing conflicts between wildlife, urban development, and transportation in BNP. Air pollution and global climate change are external threats sustained by social, economic, and political forces far removed from the milieu in which national parks and protected areas operate. Fortunately, protected area managers have undertaken research and monitoring programs to better understand the levels and consequences of air pollution in protected areas; and executive, legislative, and judicial branches of government have taken action to protect air quality in the vicinity of national parks. Nevertheless, air pollution and global climate change are not suitable for adaptive management because of their national and global origins.

References

Alberta Environment. 2003a. *A new regional strategy in Southern Alberta.* http://www3.gov.ab.ca/env/regions/southern/strategy.html (accessed April 22, 2003).

———. 2003b. *Using ALCES (A Landscape Cumulative Effects Simulator) in regional strategies.* Integrated Resources Management Branch. http://www3.gov.ab.ca/env/irm/alces.html (accessed April 22, 2003).

Aldred, J., and M. Jacobs. 2000. Citizens and wetlands: Evaluating the Ely's citizen' jury. *Ecological Economics* 34: 217–232.

Ananda, J., and G. Herath. 2002. Assessment of wilderness quality using the Analytic Hierarchy Process. *Tourism Economics* 8: 189–206.

Banff-Bow Valley Study. 1996. *Banff-Bow Valley at the crossroads, summary report of the Banff-Bow Valley Task Force.* Auditor General for Canada, Ottawa, Canada.

Banff National Park. 2002. *The Banff-Bow Valley Study.* http://raysweb.net/banff/pages/bowstudy.html (accessed December 31, 2002).

———. 2003. *Banff National Park management plan summary.* http://www.worldweb.com/parkscanada-banff/mp_trane.html (accessed January 10, 2003).

Berry, J., D. Decker, J. Gordon, R. Heitschmidt, D. Huff, D. Knight, W. Romme, and D. Swift. 2001. Rocky Mountain National Park science-based assessment of vegetation management goals for elk winter range. Unpublished report to the National Park Service.

Brown, P. 1999. Tools for ecological stewardship. In *Ecological stewardship: A common reference for ecosystem management,* vol. III, ed. W. T. Sexton, A. J. Malk, R. C. Szaro, and N. C. Johnson, 463–496. Oxford, UK: Elsevier Science Ltd.

Canadian Rockies.net. 2002. *Banff National Park, major highlights.* http://www.canadianrockies.net/banff/index.html (accessed December 31, 2002).

Craighead, J. J. 1991. Yellowstone in transition. In *The Greater Yellowstone ecosystem: Redefining America's wilderness heritage,* ed. R. B. Keiter and M. S. Boyce, 27–39. New Haven, CT: Yale University Press.

Crosby, N. 1996. *Correcting an authentic voice of the people: Deliberation on democratic theory and practice.* Chicago, IL: Midwest Policy Science Association.

Crown of the Continent Ecosystem Education Consortium. 2002. http://nrmsc.usgs.gov/coceec/default.htm (accessed December 3, 2002).

Crown of the Continent Managers Partnership. 2002. Unpublished concept paper.

CyberCervus International. 2002. *An evaluation of ALCES, a landscape cumulative effects simulator for use in integrated resource management in Alberta.* New Sarepta, Alberta. http://www3.gov.ab.ca/env/irm/docs/alcesregional.pdf (accessed April 22, 2003).

Daily Inter Lake. 2002. US may back off ban on snowmobiles in Yellowstone. February 16.

De Marchi, B., and J. R. Ravetz. 2001. Participatory approaches to environmental evaluation. Policy Research Brief No. 10. In *Concerted action: Environmental evaluation in Europe,* ed. C. L. Spash and C. Carter. United Kingdom: Cambridge Research for the Environment. http://www.landecon.cam.ac.uk/eve/publ.html#PRB (accessed January 8, 2003).

Dilsaver, L. M. (Ed.). 1994. *America's National Park System: The critical documents.* Lanham, MD: Rowman & Littlefield Publishers.

Dougherty, R. 2002. Snowmobiles to stay in Yellowstone. *National Parks* 76: 8.

Eastman, R. J., H. Jiang, and J. Toledana. 1998. Multi-criteria and multi-objective decision making for land allocation using GIS. In *Multiplecriteria analysis for land-use management,* ed. E. Beinat and P. Nijkamp, 227–251. Dordrecht, The Netherlands: Kluwer Academic Publishers.
Ehtano, H., E. Kettunen, and R. P. Hamalainen. 2001. Search for joint gains in multi-party negotiations. *European Journal of Operations Research* 130: 54–69.
Environment Canada. 2002. *Crown of the Continent Ecosystem data atlas.* http://www.mb.ec.gc.ca/COMMUNITY/ecoACTION/funding/ba01s28.en.html (accessed November 19, 2002).
Flathead Basin Commission. 2001. *Land use trends.* http://www.montanaweb.com/fbc/trends.htm (accessed November 19, 2002).
Glacier National Park. 2002. The Crown of the Continent Ecosystem. http://www.nps.gov/glac/resources/bio1.htm#Crown (accessed December 3, 2002).
Glick, D., and M. Haggerty. 2000. Land use planning in Greater Yellowstone: Is it making the grade? *Greater Yellowstone Report,* Winter.
Green Nature. 2002a. *Acid rain and forests.* http://greennature.com/article26.html (accessed January 7, 2003).
Green Nature. 2002b. *An introduction to acid rain.* http://greennature.com/article24.html (accessed January 7, 2003).
Gregory, R. S. 2000. Valuing environmental options: A case study comparison of multi-attribute and contingent evaluation survey methods. *Land Economics* 76: 151–173.
Hall, M. P., and D. B. Fagre. 2003. Modeled climate-induced glacier change in Glacier National Park, 1850–2100. *Bioscience* 53(2):131–140.
Hansen, A. J., R. Rasker, B. Maxwell, J. J. Rotella, J. D. Johnson, A. W. Parmenter, U. Langner, W. B. Cohen, R. L. Lawrence, and M. P. V. Kraska. 2002. Ecological causes and consequences of demographic change in the New West. *BioScience* 52: 151–162.
Hymon, S. 2001. Smog blankets many western parks. *Daily Inter Lake* (Kalispell, MT), July 18.
Interagency Grizzly Bear Committee. 2002a. *Grizzly bear recovery.* http://www.fs.fed.us/r1/wildlife/igbc/ (accessed December 19, 2002).
———. 2002b. *United States Department of Agriculture and United States Department of the Interior memorandum of agreement to establish the Interagency Grizzly Bear Committee.* http://www.fs.fed.us/r1/wildlife/igbc/Information/Charter.pdf (accessed December 19, 2002).
Jeffords, J. J., III. 2002. The clean air challenge. *National Parks* 76(November/December): 16–17.
Keiter, R. B. 1996. Ecosystem management: Exploring the legal-political framework (chap. 5). In *National parks and protected areas: Their role in environmental protection,* ed. R. G. Wright, 63–88., Cambridge, MA: Blackwell Science.
Knight, D. H. 1991. The Yellowstone fire controversy. In *The Greater Yellowstone ecosystem: Redefining America's wilderness heritage,* ed. R. B. Keiter and M. S. Boyce, 87–103. New Haven, CT: Yale University Press.
Leopold, A. S., et al. (Advisory Board on Wildlife Management). 1963. Wildlife management in the national parks. *American Forests* 69: 32–35, 61–63.
Mansfield, D. 2002. Smokies the nation's most polluted park, says study. Associated Press release, September 24, Environmental News Network, http://enn.com/news/wire-stories/2002/09/09242002/ap_48504.asp (accessed January 6, 2003).
Mattson, D. J., and T. Merrill. 2002. Extirpations of grizzly bears in the contiguous United States, 1850–2000. *Conservation Biology* 16: 1123–1136.
McDaniels, T. L., and C. Roessler. 1998. Multiattribute elicitation of wilderness preservation benefits: A constructive approach. *Ecological Economics* 27: 299–312.
McLain, R. J., and R. G. Lee. 1996. Adaptive management: Promises and pitfalls. *Environmental Management* 20: 437–448.
Meffe, G. K., L. A. Nielsen, R. L. Knight, and D. A. Schenborn. 2002. *Ecosystem management: Adaptive, community-based conservation.* Washington, DC: Island Press.
Montana's Flathead Valley. 2001. http://www.fcvb.org/ (accessed November 19, 2002).

Moore, S. A. 1994. Interaction processes and the resolution of environmental disputes: Case studies from public land planning in the United States and Australia. Ph.D. dissertation. University of Washington, Seattle.

National Parks Conservation Association. 2002a. *Across the nation.* http://www.npca.org/across_the_nation/npca_in_the_field/northern_rockies/northern_rockies.asp (accessed December 3, 2002).

———. 2002b. Bush's clear skies plan bad for parks. *National Parks* 76 (September/October): 12.

———. 2002c. *State of the parks, a resource assessment: Waterton-Glacier International Peace Park.* http://www.npca.org/across_the_nation/park_pulse/glacier/glacier.pdf (accessed November 19, 2002).

———. 2003a. *Fact sheet: Sequoia and Kings Canyon National Parks.* http://www.npca.org/across_the_nation/visitor_experience/code_red/fact_sheets/sequoia.asp (accessed January 7, 2003).

———. 2003b. *Great Smoky Mountains National Park description.* http://www.npca.org/across_the_nation/ten_most_endangered/great_smoky.asp (accessed January 7, 2003).

———. 2003c. *Media center: Park advocates call Bush clear skies initiative a smokescreen for polluter-friendly regulatory policy.* http://www.npca.org/media_center/PressReleaseDetail.asp?id=100 (accessed January 7, 2003).

———. 2003d. *Parklines: Help stop the slaughter of Yellowstone bison.* http://www.npca.org/take_action/action_alerts/ActionAlert.asp?strAction=fetch&G=1 (accessed April 24, 2003).

———. 2003e. *Regional haze in national parks.* http://www.npca.org/media_center/factsheets/haze.asp (accessed January 8, 2003).

———. 2003f. *Wildlife protection: Bison belong.* http://www.npca.org/wildlife_protection/threats/bison_belong/default.asp (accessed April 24, 2003).

National Park Service. 2001. *Management policies 2001.* Washington, DC: National Park Service.

———. 2002a. FSEIS released to cooperators. *Yellowstone Science* 10 (Winter): 29.

———. 2002b. New framework for winter use of parks announced. *Yellowstone Science* 10 (Summer): 25.

———. 2002c. *NPS visitation database reports, Public Use Statistics Office.* http://www2.nature.nps.gov/stats/ (accessed December 30, 2002).

———. 2002d. *Winter use plans: Yellowstone and Grand Teton National Parks and John D. Rockefeller, Jr. Memorial Highway.* http://www.nps.gov /grte/winteruse/intro.htm (accessed December 27, 2002).

———. 2003a. *Air quality and deposition monitoring, Air Resources Division.* http://www2.nature.nps.gov/ard/gas/index.htm (accessed January 7, 2003).

———. 2003b. *Clean Air Act.* http://www2.nature.nps.gov/ARD/caamalk.htm (accessed January 7, 2003).

———. 2003c. *Ozone-the invisible poison.* http://www.nps.gov/seki/ozone.htm (accessed January 7, 2003).

National Research Council. 2002. *Ecological dynamics on Yellowstone's Northern Range Committee on Ungulate Management in Yellowstone National Park.* Washington, DC: National Academy of Science.

Panel on the Ecological Integrity of Canada's National Parks. 2000. *Unimpaired for future generations? Maintaining ecological integrity with Canada's national parks: Volume I—A call to action.* Ottawa: Department of Canadian Heritage.

Parks Canada. 2001. *Record of 2001 forum. Appendix B: Discussion paper: Working towards a strategy for human use management in Banff National Park.* http://www.worldweb.com/ParksCanada-Banff/management_planning/record_of_2001_Forum/AppendixB_e.htm (accessed December 31, 2002).

Rockwood, T. H. 1988. Organizational responses to a conflicting mandate: Legitimacy and structural adaptations in the National Park Service. MA Thesis, Washington State University, Pullman.

Ross, D. 1988. Effects of visual air quality on visitor experience (chap. 3). In *Air quality in the national parks: A summary of findings.* Natural Resources Report 88-1, Air Quality Research and Monitoring Program, National Park Service. http://www2.nature.nps.gov/ard/vis/visitexp.html (accessed January 8, 2003).

Schullery, P. 1997. *Searching for Yellowstone: Ecology and wonder in the last wilderness.* New York: Houghton Mifflin Co.

Shannon, M. A. 1987. Forest planning: Learning with people. In *Social science in natural resource management systems,* ed. M. L. Miller, R. P. Gale, and P. J. Brown. Boulder, CO: Westview Press.

Stringham, S. F. 2002. Smokey and mirrors: The war between science and pseudoscience in grizzly bear conservation. *Wild Earth* 12: 36–41.

Theberge, J. C., and J. B. Theberge. 2002. Application of ecological concepts to the management of protected areas. In *Parks and protected areas in Canada: Planning and management,* ed. D. Dearden and R. Rollins, 70–96. 2nd ed. Ontario, Canada: Oxford University Press.

Thompson, S. 2000. *Impacts of current and projected national forest management on grizzly bear recovery in the Bitterroot ecosystem. Whitefish, MT.* http://www.nwf.org/nwfwebadmin/binaryVault/grizzlyreport.pdf (accessed December 17, 2002).

U.S. Department of the Interior. 2001. *National Park Service, management policies 2001, Section 1.5 external threats and opportunities.* http://www.nps.gov/refdesk/mp/chapter1.htm (accessed July 22, 2003).

U.S. Department of the Interior and U.S. Department of Agriculture. 2000. *Record of decision for final environmental impact statement and bison management plan for the State of Montana and Yellowstone National Park.* www.planning.nps.gov/document/yellbisonrod.pdf (accessed April 22, 2003).

U.S. Fish and Wildlife Service. 1993. Grizzly bear recovery plan. Missoula, MT.

———. 1999. News release. U.S. Fish and Wildlife Service to improve scientific methodology to assess grizzly bear population in Cabinet Yaak, Selkirk and Northern Continental Divide Ecosystems. May 25, Lakewood, CO. http://www.r6.fws.gov/pressrel/99-20.html (accessed December 26, 2002).

———. 2000a. *Draft conservation strategy for the grizzly bears in the Yellowstone ecosystem.* Interagency Conservation Strategy Team. http://www.r6.fws.gov/endspp/grizzly/Yellowstone_constrat2000/yelcs2000.pdf (accessed December 17, 2002).

———. 2000b. *Grizzly bear recovery: Overview and update,* Mountain-Prairie Region, Lakewood, Colorado.

———. 2001. News Release. *Public comment analysis now available on Bitterroot grizzly bear proposal.* October 22, Lakewood, CO.

———. 2002. *Grizzly bear recovery, Questions and answers about grizzly bears.* http://www.r6.fws.gov/endspp/grizzly/Yellowstone_constrat2000/q&agb.htm (accessed December 19, 2002).

U.S. Geological Survey. 2003. *Global change research—A focus on mountain ecosystems, Northern Rocky Mountain Science Center.* http://www.nrmsc.usgs.gov/research/global.htm (accessed January 7, 2003).

Walters, C. J., and C. S. Holling. 1990. Large-scale management experiments and learning by doing. *Ecology* 71: 2060–2068.

Waterton Lakes National Park Resource Guide. 2002. http://www.watertonpark.com/reference/special_status.htm (accessed November 19, 2002).

Wilkinson, P. F. 2000. Protecting for ecological integrity in Canada's national parks: Allowable and appropriate visitor use. Fourth Science and Management of Protected Areas Association Conference (SAMPAA), University of Waterloo, Waterloo, Ontario, May 14–19, 2000.

Wilkinson, T. 2002. A grizzly future. *National Parks* 76: 22–25.

Willcox, L., and D. Ellenberger. 2000. *The bear essentials for recovery: An alternative strategy for long-term restoration of Yellowstone's great bear.* Sierra Club Grizzly Bear Ecosystems Project, Bozeman, MT. www.sierraclub.org/wilderness/grizzly (accessed December 19, 2002)

Yellowstone Association. 2003. *Issues in Yellowstone: Bison management in the Yellowstone area.* Spring.

Yellowstone Science. 2003. Bison capture operations outside north entrance. *News and Notes* 11: 39–40.

Yellowstone to Yukon Conservation Initiative. 2002. http://www.rockies.ca/y2y/ (accessed December 3, 2002).

CHAPTER

10

Case Studies in Protected Area Management—II

The axiom for protecting the national park system is to consider that it is dedicated country, hallowed ground to leave as beautiful as we have found it, and not country in which man should be so impressed with himself that he tries to improve God's handiwork.

—David Brower

The six case studies presented in this chapter examine natural resource issues facing national parks and protected areas and management approaches to resolve those issues. Some of the case studies apply the concepts, principles, and approaches presented in the first eight chapters. Managing ungulates (elk and bison) in Yellowstone National Park and water management in Everglades and Kruger National Parks are presented as expanded case studies. Three abbreviated case studies are also included that involve preserving marine protected areas (Wilson's Promontory National Park in Australia, Basin Head Lagoon in Prince Edward Island, Canada, and Prince William Sound–Copper River Watershed and Glacier Bay National Park and Preserve in Alaska), reducing landscape fragmentation in Indiana Dunes National Lakeshore, and the importance of research in managing Ozark National Scenic Riverways. This chapter continues the case studies begun in chapter 9.

Managing Ungulates in Yellowstone National Park

Management of ungulates (elk and bison) in Yellowstone National Park (YNP) is fraught with challenges and controversy. To fully appreciate the debates over elk management in YNP, it helps to understand the evolution of elk management practices in the park. From the inception of YNP in 1872 through the early 1900s, elk and bison were treated as desirable species and predators like grizzly bear, wolf, coyote, and lion as undesirable species (see the following sidebar regarding wolf recovery in the Greater Yellowstone Ecosystem). Wildlife management during this period consisted of supplemental feeding of desirable species and eradication of predators. During the period 1911–13, government biologists concluded that the killing of predators, reduction in the seasonal movement and dispersal, and lack of aboriginal hunting allowed the elk population to reach unprecedented levels that warranted culling (Wagner 1994).

Wolf Recovery in the Greater Yellowstone Ecosystem

When Yellowstone National Park was established in 1872, northern Rocky Mountain wolves (*Canis lupus*) inhabited the park. A program of predator control in the late 1800s and early 1900s reduced the wolf population to the point that no wolf packs were sighted in the 1940s and there was no verifiable evidence of wolves from the 1970s to the mid-1990s. The gray wolf is listed as endangered throughout its historic range in the lower 48 states; in Minnesota it is listed as threatened. In 1978, John Weaver recommended that wolves be introduced to Yellowstone from Canada or Minnesota in an effort to restore a viable population. Nine years later in 1987, the U.S. Fish and Wildlife Service proposed that an experimental nonessential population of wolves be established in the park; the recovery plan considered the entire range including the park and portions of Wyoming, Idaho, and Montana. Designation as experimental nonessential allowed killing and removal of wolves when necessary to protect local residents from undue hardship. The original wolf recovery plan was approved in May 1980 and the revised version in August 1987. Lead agencies for the recovery effort are the U.S. Fish and Wildlife Service and the National Park Service.

After Congress provided funding in October 1991, the U.S Fish and Wildlife Service (USFWS) in consultation with the National Park Service and U.S. Forest Service began preparation of an environmental impact statement for restoring wolves to Yellowstone National Park and central Idaho. Opinions regarding wolf reintroduction were varied. A survey of the general public living in counties surrounding Yellowstone National Park showed 52 percent opposed reintroduction. A survey of members of the Wyoming Stock Growers Association indicated that 91 percent disapproved. In contrast, a survey of the national general public, Wyoming members of Defenders of Wildlife, and the Wyoming Wildlife Federation showed approval rates of 49 percent, 89 percent, and 67 percent, respectively.

Several objections were raised to wolf recovery. They included (1) wolves would kill livestock; (2) livestock operations might be restricted by government regulations aimed at protecting wolves; (3) wolves are dangerous to humans; (4) wolf management around Yellowstone would be prohibited by the Endangered Species Act; (5) wolves that escape from the park could not be caught; (6) mobility of park visitors would be restricted in order to protect the wolves, such as excluding visitors from denning areas; (7) wolves would ruin hunting outside the park and reduce the number of big game that migrate out of the park; (8) wolves would eliminate or seriously reduce prey herds in Yellowstone; I(9) wolves would kill or compete with grizzly bears; and (10) wo!ves introduced to Yellowstone would spread throughout the rest of the West causing the federal government to impose unreasonable restrictions on the public, such as closing roads, to protect them.

Several of these objections cannot be substantiated and others can be handled by appropriate government action. This is not to say that wolf reintroduction would have no ecological and human impacts. For example, a report prepared for Congress by scientists from the University of Wyoming predicted a 15 percent to 25 percent reduction in elk, and 5 percent to 15 percent reductions in bison, moose, and mule deer. Another study predicted that the moose population would decline by 10 percent to 15 percent and mule deer population by 20 percent to 30 percent. Coyotes would likely decrease and red foxes increase with only minor impacts on grizzly bears and mountain lions.

Several years later, the secretary of the interior signed the Record of Decision on the Final Environmental Impact Statement for reintroduction of gray wolves to both Yellowstone and Idaho. Regulations specifying how wolves would be managed as a nonessential experimental population took effect in November 1994. In November 1995, after 20 years of planning, public comment, and study, 14 wolves were translocated from Canada to Yellowstone National Park. Seventeen more Canadian wolves were released to the park in 1996. Wolves were first placed in acclimation pens for 10 weeks and then "soft" released into the wild. All but one released wolf pack settled in the vicinity of their acclimation pens.

Tens of thousands of park visitors have observed wolves since their return in 1995, providing opportunities to learn about predator-prey relationships, endangered species restoration, and the importance of maintaining intact ecosystems. Wolf-recovery goals have been attained (10 breeding pairs in each recovery area for three consecutive years). In particular, Yellowstone National Park has 14 wolf packs in the park with 12 breeding pairs in the park and 1 pair north of the park. Average pack size is 9.5. Gray wolves are expected to be taken off the endangered list, after which management will be transferred to the states of Montana, Idaho, and Wyoming.

Sources: Bath (1997); Mech (1997); National Park Service (1999b, 2003e); Phillips and Smith (1997).

In an effort to reduce the elk population, the National Park Service (NPS) embarked on a 38-year (1930–1967) program in which thousands of elk were killed, trapped, or removed from the park for private hunting. Between the early and late 1960s, the elk population decreased from about 10,000 to below 5,000. The latter was the perceived carrying capacity of the park. Over 4,000 elk were shot and trapped in the winter of 1961–62. Most of the elk reduction occurred on the northern range, which includes the Lamar Valley.

The culling program was generally accepted by academics at Montana State University and The Wildlife Society, and supported, in 1963, by an Advisory Board on Wildlife Management created by the U.S. Department of the Interior. The board recommended "direct reduction" (killing) of ungulates in national parks unless park managers could control their numbers by other means.

At the time, the direct reduction policy seemed right to natural resource managers. As John Good observed, "Every range manager, every wildlife manager, every Forest Service executive, every person connected with state agencies agreed that there were too damn many elk in Yellowstone and the park service had to knock 'em down" (Good 2000). One of the reasons park officials wanted to reduce the elk population was to reverse the extensive damages to aspen caused by elk grazing.

In 1967, political events led NPS to abandon the elk reduction program. The killing of thousands of elk by government hunters was strongly opposed by outfitters, guides, and the state of Wyoming. Outfitters and guides in particular wanted to bring private hunters into the park to kill elk, an idea that the public opposed. Two governors from Wyoming sued in federal court to stop the killing; they lost the case on the grounds that the state of Wyoming had no standing in terms of how the park was managed. Enter Senator Gale McGee (D-Wyoming). He was a member of the Senate appropriations subcommittee that set funding levels for the NPS and up for reelection in 1968. Senator McGee opposed the elk reduction program and expressed to NPS director George Hartzog that future funds for the NPS were in jeopardy unless the killing ceased. Prior to a meeting to discuss the elk problem, Director Hartzog announced that he had been convinced by Senator McGee that the killing of elk needed to stop and that he would order Park Superintendent John McLaughlin "to stop the elk slaughter immediately" (Wagner 1994; Good 2000). Subsequent studies of national regulation in YNP indicate that the direct reduction program for elk was unjustified (Houston 1982; Despain et al. 1986).

Shortly after the elk reduction program was terminated, the NPS announced a natural regulation policy (see also chapter 7). Natural regulation allows natural processes to operate and dominate, and is based on the notion that density-dependent changes (declines in birthrates and survivorship as population increases) cause wildlife populations to achieve an equilibrium density (Boyce 1991). Critics of this position argue that plant-herbivore interactions result in complex ecosystem dynamics that undermine the idea that wildlife populations converge to some equilibrium density (Caughley and Lawton 1981; Schaffer 1988). A dynamic equilibrium population for elk in YNP was estimated to be 19,000 in the winter of 1987–88 following five relatively mild winters, and 15,000 in an average winter (Merrill and Boyce 1991). In the period 1995–96, elk population in the park increased to 20,000 (*U.S. News & World Report* 1996).

Craighead (1991) argued that YNP was too small and had too many visitors to support the ecological processes of an intact natural system and that natural regulation resulted in overpopulation of elk, bison, and other species. Alston Chase (1987) accused park managers of ignoring 50 years of prior research and documentation, suppressing information unfavorable to the natural regulation policy, and using data in a selective manner (Wagner 1994). Kay's research and repeat (before-and-after) photographs led him to conclude that overgrazing by elk and bison on Yellowstone's northern range "denuded the range, [is] destroying plant communities and eliminating critical animal habitat" (Kay 1993, 1997). His photographs indicate dramatic declines in aspens, willows, woody shrubs, and buffaloberry, a plummeting population of beavers over the last 60 years, highly eroded and collapsed streambanks due to the absence of aspens, willows, and cottonwoods, and other negative ecological impacts. Kay concluded that adherence to the natural regulation policy was reducing biodiversity in YNP.

In 1986, Congress appropriated funds for an NPS study of whether Yellowstone's northern range was being overgrazed. Results of the study indicated that "vegetation of the northern range was not overgrazed...the range's ecological processes seemed to be working and...traditional definitions of overgrazing did not seem applicable" (Schullery 1997). The study was criticized on grounds that research plans were absent or deficient, vegetation research focused on herbaceous rather than woody vegetation (grazing impacts the latter), and conclusions were faulty (Wagner 1994). The NPS responded to criticisms of natural regulation by arguing that ecological changes observed during periods of high elk populations could be attributed to other factors, such as climate change, fire suppression, and plant succession (Wagner 1994). Schullery (1997) points out that Yellowstone's history indicates the disappearance of plant species and shifts in species abundance long before natural regulation was implemented, and that the study of natural regulation by Houston (1982) showed that "rather than being an example of an unhealthy landscape, [Yellowstone] was instead an example of a normal wild landscape."

In 2002, the National Research Council (NRC) released a study requested by Congress that evaluated implementation of natural regulation in YNP's northern range (National Research Council 2002). The study gave two reasons why YNP could not be actively managed to sustain a particular environmental state. First, YNP is periodically influenced by large-scale natural and human disturbances. Second, there is insufficient knowledge, resources, and capability to manage the ecosystem to achieve some agreed-upon stable ecosystem state. The approach taken in the NRC study was to determine whether past ungulate grazing caused Yellowstone's northern range to exceed some threshold beyond which aspen and willow communities could not regenerate themselves if grazing were decreased, or equivalently, whether the northern range had changed to a different ecosystem state than existed in the past few thousand years.

The NRC report concluded that "despite claims to the contrary, we found no evidence that the northern range is approaching a threshold after which we would observe irreversible changes, such as loss of local reproductive potential of key plant species (e.g., sagebrush and aspen), that would not have occurred if the park actively controlled ungulates." While the report concluded that ungulate populations did not appear to be causing irreversible damages to the northern range, it admitted that a smaller population of ungulates would allow recovery of some plant communities (willows, aspen, and sagebrush). The report reached several other conclusions. First, due to impacts of development outside the park's borders and human impacts within the park, "it is no longer possible [for the northern range] to be identical to the natural state that existed there before European settlement—that is, containing about the same numbers and distributions of all species of plants and animals." Second, it is not feasible to implement natural regulation in the YNP because a natural state has not existed in the park for more than a century. Third, Yellowstone's natural regulation policy is just as likely to maintain the northern range ecosystem as any other management approach. Fourth, policies that involve smaller interventions, such as using exclosures to evaluate impacts of grazing, are preferable to major interventions like culling of ungulates. This is based on the fact that small interventions are likely to have fewer unintended ecological consequences than large interventions, and would not interfere with the ability to observe ecosystem responses to the introduction of wolves in YNP, which occurred in 1995.

Based on these considerations, NRC offered several recommendations for park management. One recommendation is that "to the degree possible, all management at YNP should be done as adaptive management." In particular, the NRC report pointed out that active adaptive management improves scientific understanding of the consequences of different management actions. In this regard, the report states that the natural regulation policy for ungulates could be continued within the context of adaptive management. This means that populations of elk, bison, and other ungulates would be allowed to vary over time without direct controls within the park. This approach allows a combination of factors, including weather, wolves, range conditions, and factors external to the park, such as killing bison that leave the park, to influence ungulate populations. One hypothesis that could be tested in an adaptive management framework is whether the elk population has reached a dynamic equilibrium since the low populations that occurred during the elk reduction program of the 1950s and 1960s.

Alternatively, the park might consider doing adaptive management experiments to test hypotheses about how reductions in elk population influence recruitment of tree-sized aspen, expansion of willow communities, and sagebrush growth. This type of adaptive management is time-consuming, costly, and generally infeasible to implement because it requires a control (no reduction in elk population) and replicated treatments (various reductions in elk population from the control level). In particular, it would require implementing the control and treatments each with replications for a number of years and monitoring impacts on vegetation. If control and treatments are done concurrently to reduce biases from temporal variability in weather and other natural and human disturbances, then it is necessary to physically separate the subpopulations of elk in the various treatments. In theory, this could be done by confining the elk subpopulations to large fenced enclosures. In practice, this treatment of elk is likely to be unacceptable to park managers and the general public. Additionally, it would bias the results because elk are free roaming and interact with the ecosystem over a broad spatial scale. Doing the experiment in a sequential manner (implementing the control and each treatment in separate time periods) would require physically reducing the size of the elk population. This action not only would raise public ire, but also would require experimenters to move elk into or out of the park from one experimental period to the next, which is not feasible.

Adaptive management experiments that involve reducing elk populations have other drawbacks. Reducing elk population on an experimental basis would impair the ability of park managers to test hypotheses about whether introduction of wolves to Yellowstone reduces the elk population. Another drawback is that reducing elk population could reduce park-based food availability for wolves and cause them to leave the park in search of another food base. Finally, the public is likely to object to experiments that involve culling elk and possibly other ungulates.

In summary, strict application of active adaptive management to ungulates does not appear feasible in YNP and, for that matter, other protected areas. Some sort of second-best adaptive management approach, as discussed in chapter 9, or passive adaptive management would need to be used.

Water Management in Everglades and Kruger National Parks

The two case studies that follow address policies for reducing the impacts of insufficient water supplies on the ecological integrity of Everglades National Park in Florida (United States) and Kruger National Park in South Africa.

Florida Everglades

Everglades National Park (ENP) was established December 6, 1947. It encompasses the southern end of the Florida peninsula and most of Florida Bay and is

North America's only subtropical preserve designated as a World Heritage Site, International Biosphere Reserve, wilderness area, and Wetland of International Importance. ENP contains temperate and tropical plant communities including sawgrass prairies, mangrove and cypress swamps, pinelands, hardwood hammocks, and marine and estuarine areas. The park is renowned for its large wading birds, such as roseate spoonbill, wood stork, great blue heron, and egrets. ENP is the only place in the world that contains both alligators and crocodiles (National Park Service 2003a).

In 1947, a major storm (rainfall in excess of 2,500 mm) caused flooding of millions of hectares of agricultural and urban lands in South Florida. Congress responded to the flooding by creating the Central and Southern Florida Project for Flood Control (the project). The primary objectives of the project were flood protection and water supply. Navigation and recreation were secondary objectives. Project construction occurred from 1950 to the 1970s. When completed it was the world's largest automated water control system (Light and Dineen 1994) consisting of 720 miles of levees, 1,000 miles of canals, 200 gates and water control structures, and 16 massive pumping stations, all designed to regulate water flow from south of Orlando, Florida, to Florida Bay (Kiker et al. 2001).

Hydrological alterations wrought by the project seriously degraded the ecological integrity of ENP. Five major canals constructed to drain the southern Everglades caused water levels to drop precipitously from 0.3 m (1 foot) below ground surface in 1915 to 1.5 m (5 feet) below ground surface in 1939. Canals removed more than 230 percent of the net annual precipitation (total precipitation minus evapotranspiration) formerly available to ENP. Compared to predrainage levels, water flow in ENP dropped 70 percent and timing and duration of water flow was dramatically altered. A levee constructed around Lake Okeechobee terminated natural overflow from the lake to ENP and construction of the Tamiami Trail separated the northern section from the southern section of the park. Other ecological impacts include a reduction in the original area of ENP by 50 percent due to agricultural and urban development (from 10,000 to 5,000 km^2), impoundment of wetland areas, subsidence in surface elevation of the soils (as much as 1–2 m [6–7 feet] in the Lake Okeechobee area), decrease in organic soils, diminution of flora and fauna (68 plants and animals became threatened or endangered and there was a 90 percent reduction in wading bird populations), increased concentration of soil phosphorus (nutrient enrichment), changes in fire frequency and severity, spread of invasive species (cattails displacing sawgrass, and woody species encroaching into marshes and sloughs) (Kiker et al. 2001; Sklar et al. 2001), and a general loss of ecological resilience (Gunderson 2001).

Even before the project was completed, public attitudes toward the resulting ecological degradation of ENP led to passage of the Florida's Water Resources Act of 1972. The act specified that maintaining natural systems would be a top priority for the South Florida Water Management District (SFWMD). In 1988, the federal government sued the state of Florida and SFWMD over the poor quality of water being released to ENP and other federal lands. The lawsuit helped to crystallize state support for viewing the Kissimmee River watershed, Lake Okeechobee, the Big Cyrus Basin, ENP, Florida Bay, and Florida Keys as one large ecosystem, which laid the foundation for ecological restoration in South Florida. Steps taken to improve the South Florida ecosystem include the Federal Settlement Agreement of 1991 between the Department of the Interior and the state of Florida, which recognized the need to reduce phosphorus concentrations in water and restore natural hydroperiods, and the Everglades Forever Act of 1994, which mandated higher freshwater flows and criteria for phosphorus concentrations in runoff through the Everglades (Sklar et al. 2001).

Perhaps the most significant step in ecological restoration of South Florida was passage of the Federal Water Resources Development Acts of 1992 and 1996 and creation of the South Florida Ecosystem Restoration Task Force in 1993. The acts authorized the U.S. Army Corps of Engineers to review the original Central and Southern Florida Project for Flood Control. The task force's review report, called the Central and Southern Florida Comprehensive Review Study, documented the serious ecological degradation caused by the original project. The primary recommendation of the report was to develop a "comprehensive plan for the restoration, protection and preservation of the water resources of central and southern Florida" (USACE and SFWMD 2000).

The federal government and Florida state government responded to this recommendation by passing the Water Resources Development Act of 2000 and the Everglades Restoration and Investment Act, respectively. These laws established the Comprehensive Everglades Restoration Plan (CERP) (Kiker et al. 2001). CERP is the world's "largest and most ambitious ecosystem restoration" (Best 2000) involving the expenditure of $8 billion over 30 years to restore the hydrology of South Florida (Sklar et al. 2001). CERP attempts to reestablish south Florida's historic rainwater sheetflow, remove 115 km (76 miles) of canals and 198 km (131 miles) of levees, improve southward flow of water, restrict urban development, purchase drained land, and preserve buffer areas on the eastern boundary of ENP (National Parks and Conservation Association 1999).

One element of the CERP process is the flexibility planners have to refine and revise the plan "as part of (an) adaptive assessment process" (USACE and SFWMD 2000). One of the five panels established to ensure the best available science is used in restoration efforts is an Adaptive Assessment team. Viewing restoration of ENP as an adaptive learning experience is justified by the fact that ecosystem restoration at the scale and complexity of South Florida and the Everglades is beyond the current knowledge and experience base of the scientific community. While scientists have a reasonably good understanding of pieces of the ecosystem, achieving the restoration goal of reestablishing "a more natural hydrographic function [and flora and fauna] in the lower Everglades" will require adaptive learning that entails "continuous growth in understanding by scientists, managers, policy makers, political representatives and the public" (Kiker et al. 2001).

Kiker et al. (2001) proposed an adaptive ecological learning approach for restoration of the South Florida ecosystem that has merit in managing other ecosystems. The approach involves discussion and exchange and, one hopes, integration of ideas between two groups of scientists. The first group, consisting of biologists, hydrologists, and engineers, interprets data based on traditional, disciplinary-based science. The second group, consisting of landscape ecologists, systems ecologists, and other scientists and engineers, reviews the results of the first group from a holistic (systems) viewpoint and either agrees or disagrees with the first group's interpretation. If the second group disagrees with the first group's interpretation, then the first group is required to reconsider its original interpretation in light of the analysis done by the second group. The first group can either stick to its original interpretation or revise it; its findings are passed along to the second group for consideration. The two groups of scientists continue to interact and communicate in an iterative manner until sufficient agreement is reached on scientific interpretation of ecological processes. If agreement is not reached, then the viewpoints of the two groups are recorded and possibly revisited at a later time as more information becomes available.

Two planning groups are established along the same lines as the two scientific groups. Their role is to use the socioeconomic and planning information provided by monitoring and research programs to reach agreement on restoration plans/policies and inform politicians and the public about those plans/policies. Finally, a merger process group consisting of a subgroup of senior scientists and planners from the science and planning groups attempts to integrate the findings of the science and planning groups into science-based actions that are presented to politicians and the public for their consideration and adoption.

The upper portion of figure 10.1 illustrates the adaptive learning process. Based on research, monitoring, and modeling results for implemented management actions and scientific/planning knowledge, the scientific and planning groups work toward reaching agreement on how ecological and socioeconomic processes operate. Agreed-upon explanations are translated by the merger group into science-based management actions that are communicated to politicians and stakeholders, and eventually implemented. Implemented actions are followed by research, monitoring, and modeling. The cycle is repeated until desirable outcomes are achieved, for example, ecological restoration in the case of the Florida Everglades. In summary, adaptive learning is "an organized, ongoing open procedure for exploring all aspects of the ecological science pursuit and to relate the science to the ongoing policy process" or, equivalently, to achieve "science-based consensus useful in the policy [or management] process." Adaptive learning is not limited to ecological restoration, but is germane to management of a wide range of ecosystem management issues.

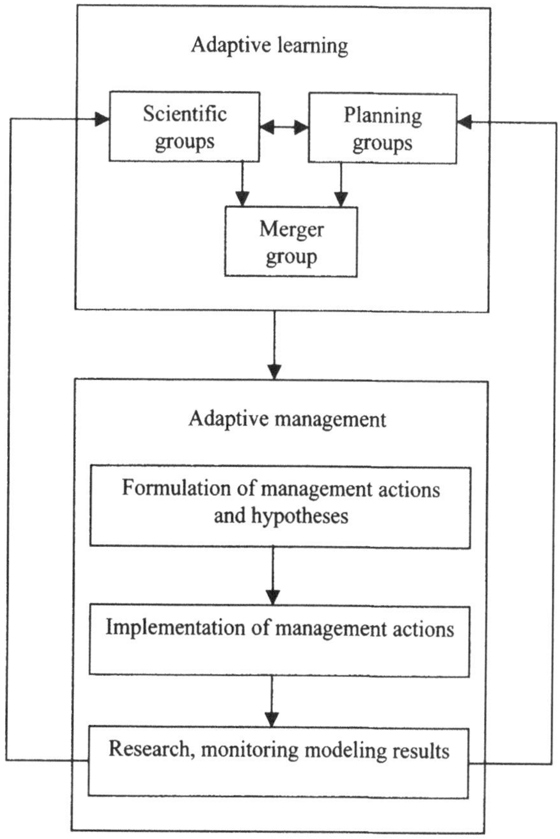

FIG. 10.1. Relationship between adaptive management and adaptive learning.

What is the relationship between adaptive learning and adaptive management? Adaptive learning is the learning component of adaptive management. As shown in figure 10.1, the shared understandings of ecological and socioeconomic processes that emerge from adaptive learning feed into the adaptive management process, which entails formulating management action(s) and, in cases where active adaptive management is being practiced, stating hypotheses about how those actions influence ecological and socioeconomic variables. Management actions are implemented, impacts monitored, and results used to test hypotheses. If passive adaptive management is used, then shared understandings are used to formulate a best management action that is implemented and monitored. Analysis and interpretation of monitoring results are used to determine whether to continue that action or modify it to achieve more desirable outcomes.

Kruger National Park

An example of a national park outside North America that is experiencing major water management problems is Kruger National Park (KNP). KNP is located in the East Transvaal region of the Republic of South Africa. It is one of the oldest and most internationally acclaimed wildlife and wilderness parks and has the world's greatest concentration of species. KNP contains 34 types of amphibians; 505 species of birds; 116 reptilian species; 147 mammal species; including the "Big Five" (buffalo, elephant, leopard, lion, and rhinoceros), and cheetah, impala, and zebra; 49 fish species; and 300 types of trees in 16 ecozones (Kruger National Park 2003). KNP has the highest density of birds of prey on the Earth (South African Tourism Board 1993). The park is 350 km (231 miles) long, 60 km (40 miles) at its widest point, and 1,948,528 hectares (4.8 million acres) in area (see fig. 10.2). KNP is noted for its exceptional wildlife viewing during game walks and night drives in open-air vehicles equipped with powerful spotlights. Night jaunts typically uncover nocturnal birds, civets, nightjar, porcupine, genets, and many predators and scavengers feeding on recent kills. A primary management objective of KNP is "to maintain the ecosystem in its natural state for the enjoyment and enrichment of the visitor" (TOURSAA 2003ab). KNP and many other nature conservation parks in South Africa attract tourists from around the world. Nature-based tourism generates significant socioeconomic benefits at the local, regional, and national levels.

Economic and population growth in the East Transvaal region have increased the demand for water in urban, agricultural (irrigation), silvicultural (aforestation), and industrial sectors. Increased water use in the upper watershed has reduced the supply of water to KNP, which contains five rivers that drain the East Transvaal region. The Limpopo is the most northern river and the Crocodile forms the park's southern boundary. Between the Limpopo and Crocodile Rivers are the Sabie, Letaba, Olifants, and Luvuvhu Rivers. Increased water use and construction of dams and irrigation diversions in upland areas have significantly altered natural flows in KNP's rivers and decreased water quantity, water quality, aquatic biodiversity, and the long-term viability of the park's aquatic and terrestrial ecosystems. In the dry season (April through August), small rivers like the Shisa or Timbavati are usually dry (Kruger National Park 2003).

In 1988, the Departments of Water Affairs and Environmental Affairs, the Foundation for Research Development, the National Parks Board, the Water Research Commission, and other research institutes in South Africa established the KNP Rivers Research Program (KNPRRP). The goal of the KNPRRP is to restore the health and biodiversity of park rivers via a research-based program that informs researchers, natural resource managers, and stakeholders about the water quantity and quality requirements for sustaining the ecological integrity of the rivers in KNP, and developing, testing, and refining models and other methods to predict ecological responses of rivers to changing water quantity and quality conditions. The KNPRRP has four subprograms: (1) information systems development and

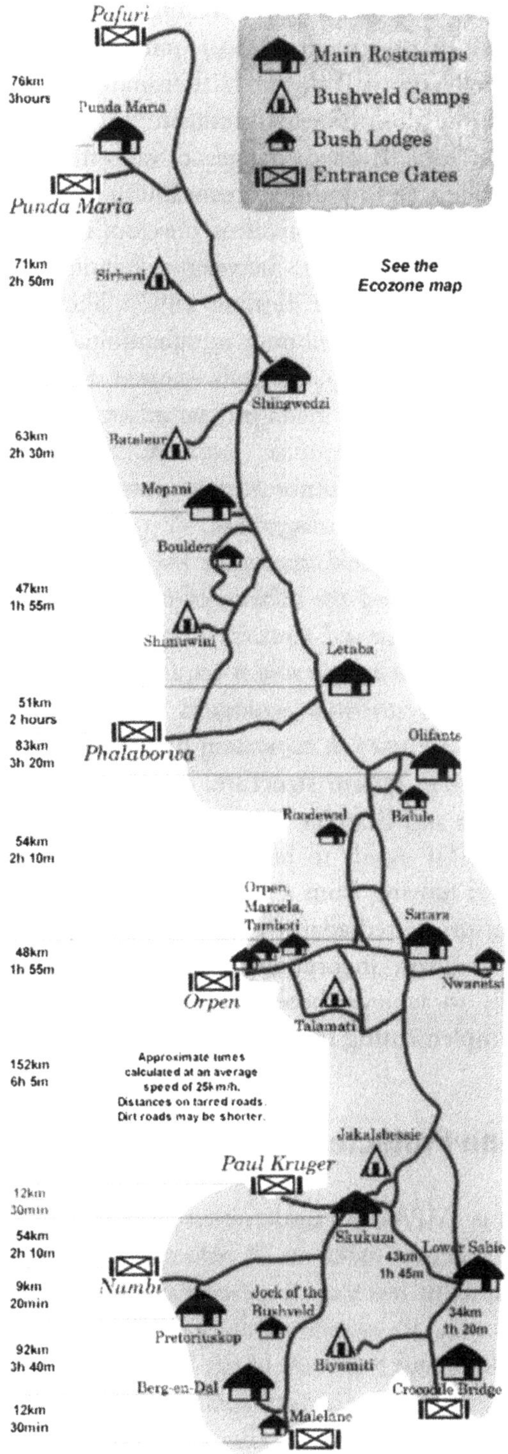

FIG. 10.2. Kruger National Park, South Africa. (From Kruger National Park 2003)

management; (2) decision support system development and management; (3) research development and management; and (4) training, information, and technology transfer (Kruger National Park Rivers Research Program, undated). One project is developing simulation models to evaluate the responses of fish and riparian vegetation in the Sabie River to changes in watershed hydrology, sedimentation, and geomorphology (Jewitt and Görgens 2000).

Implementation of the KNPRRP involves "develop[ing] a consultative management process in which interactions between stakeholders, managers and researchers are facilitated" (Rogers and Bestbier 1997). The management process incorporates three elements: (1) developing an operational framework for setting and evaluating attainable and acceptable goals (desired states of rivers), (2) using a model to predict consequences of management actions and system change, and (3) employing a system response framework to monitor natural disturbance/change and system response to management actions. These same elements are essential for adaptive management.

Adaptive management is a good match for the KNPRRP for other reasons. Many ecologists have abandoned the balance-of-nature paradigm in favor of the flux-of-nature paradigm for natural resource management (Pickett et al. 1992; Rogers 1997). The flux-of-nature paradigm requires shifting the emphasis from managing species for their intrinsic ecological value to adaptively managing species "for their interactive roles in ecosystem functioning, and for their role in promoting heterogeneity in system structure, composition and functioning, in time and space" (Rogers and Bestbier 1997). These same authors maintain that river management in KNP needs to be transparent and accountable to stakeholders, which requires moving from a reactive (wait-and-see) management approach to a more goal-orientated adaptive management approach. The latter entails identifying social goals, incorporating social values into decisions, and managing strategically to achieve those values. In summary, adaptive management is suitable for implementing the results of the KNPRRP.

Preserving Marine Protected Areas

Marine protected areas (MPAs) are underrepresented relative to land-based protected areas. While the oceans make up 70 percent of the Earth's surface, marine protected areas account for less than 1 percent of the Earth's ocean surface. In contrast, land covers only 30 percent of the Earth's surface, but terrestrial protected areas account for about 8 percent of the Earth's total land area. The role of MPAs in conserving marine biodiversity and protecting fisheries is particularly significant in light of dramatic global declines in fisheries populations (Willison 2003). The goal of the World Commission on Protected Areas (WCPA)-Marine Program is "to provide for the protection, restoration, wise use, understanding and enjoyment of the marine heritage of the world in perpetuity through the creation of a global, representative system of marine protected areas and by building the capacity to manage these areas in accordance with the principles of the World Conservation Strategy" (WCPA 2003). One of the themes of the WCPA-Marine Program is to implement "MPAs as exemplary systems of participatory and adaptive management in the context of integrated coastal management."

Wilson's Promontory Marine National Park

Wilson's Promontory National Park (WPNP) in Victoria is the southernmost point of mainland Australia. It forms the northern end of a large granite mass that extends across Bass Strait to northeastern Tasmania. Over time softer rock was worn down by erosion leaving behind large granite mountains. During the last major ice age, subsidence of the oceans exposed the remaining mountains that stretched from present-day Victoria to Tasmania, which is due south of Victoria. A warming trend begun about 15,000 years ago caused the ice to melt and the seas to rise severing the land link between Victoria and Tasmania. The area became an island and was later rejoined to the mainland by the sandy Yanakie isthmus (Parks Victoria 2002a, 2002b, 2003a).

WPNP boasts 700 native plant species; 30 mammal species, including antechinuses, kangaroos, wallabies, wombats, koalas, seals, and bats; about 180 bird species; and numerous insects and invertebrates (Parks Victoria 2002a, 2002b). A significant feature of WPNP is that it is one of the few areas in Victoria that contains rivers and streams unmodified by drainage and engineering structures and free of fish and aquatic weed species (Parks Victoria 2002c). A unique feature about Australia's national parks is that state governments have the authority to create and manage national parks.

Over 90 percent of the plants and animals in the waters off Australia's southern coast are found nowhere else on Earth. In an effort to protect this unique flora and fauna, the Victorian government created 13 Marine National Parks and 11 smaller marine sanctuaries (Parks Victoria 2002a). Currently, 6 percent of Victoria's coastline is designated as a "no-take" zone, which prohibits harvesting of all marine organisms and resource extraction (Meiklejohn and Earle 2003). Wilson's Promontory Marine National Park (WPMNP) is one of Victoria's premier MPAs. It extends beyond 70 km (46 miles) of coastline at the southern tip of WPNP. WPMNP and WPNP cover an area of 15,550 ha (38,320 acres). WPMNP was created in 1986 and is a Biosphere Reserve. It contains smooth-walled granite cliffs, huge granite boulders or pinnacles, sloping headlands, sandy coves, picturesque offshore islands, caves, and ocean waters that are home to a rich assemblage of marine fauna and flora including huge sponges, sea-fans, sea-lace colonies, sea-tulips, sea-whips, kelp and other seaweeds, barber and butterfly perch, moray eels, blue gropers, rays, sharks, giant Maori octopus, several species of whales, Australian fur seals, little penguins, short-tailed shearwaters, fairy prions, silver and Pacific gulls, and many other species of ocean birds. This rich marine wilderness and the presence of numerous shipwrecks provide spectacular underwater scenery and superb diving sites (Parks Victoria 2002a). The management plan for WPMNP prohibits all forms of fishing, netting, spearing, taking, or killing of marine life from the shore or the sea (Parks Victoria 2002c).

Along the western and eastern coastlines of WPNP and north of WPMNP is Wilson's Promontory Marine Park (WPMP). It includes extensive sandy beaches, rocky reefs, offshore islands, and important seagrass meadows. Unlike WPMNP, line fishing and commercial fishing and, with certain restrictions, spearfishing and diving for abalone and rock lobster are permitted in WPMP. Pillar Point near Tidal River and Refuge Cove on the east coast provide outstanding opportunities for diving and snorkeling (Parks Victoria 2003b). In addition, WPMNP and WPMP are popular areas for swimming, scenic sailing and boating, wildlife viewing, surfing, sailboarding, and other waters activities.

The health and integrity of WPNMP, WPMP, and other Australian MPAs, such as Great Barrier Reef Marine Park, are being negatively impacted by various land-based human activities, including water pollution from agricultural, urban, industrial, and shipping activities; high rates of sedimentation; physical disturbance; over-harvesting of fisheries; and introduction of exotic marine organisms. Commercial and recreational fishing, in particular, are having significant impacts on fish and shellfish stocks in offshore areas and overfishing is occurring in the shark and rock lobster fisheries. Illegal take of abalone is another problem. Management of Victoria's MPAs utilizes a cooperative approach in which multiple agencies and stakeholder groups work together to develop management actions for alleviating negative ecological impacts. For example, WPNMP, WPMP, and WPNP, which constitute the Wilson's Promontory biogeographic region, are jointly managed by Parks Victoria (Environmental Conservation Council 2000; Parks Victoria 2002b). While cooperative decision making is not adaptive management, it is an important element of adaptive management. An adaptive management approach is being used in Great Barrier Reef National Park (Day 2003).

Basin Head Lagoon, Prince Edward Island

Basin Head Lagoon (BHL) is a biologically diverse, shallow coastal lagoon on the eastern tip of Canada's Prince Edward Island near the town of Souris (see fig. 10.3). The lagoon is about 5 km (3.3 miles) long and covers 60 ha (148 acres). It used to be a port for fishing vessels. BHL contains a wide array of birds, fish, invertebrates, mammals, and plants. Irish moss (*Chondrus crispus*), which is unique to the BHL, provides transplant stocks for aquaculture. The market value of Irish moss was second only to lobster in the early 1970s, but dropped to eleventh among commercial fisheries by 1999. The area north of BHL is agricultural land (primarily potato fields) and the area south has extensive sand dunes, which are

A

B

FIG. 10.3. Basin Head Lagoon. (A: From Government of Prince Edward Island, Canada 2003; B: From Town of Souris 2003)

protected under the province's Natural Areas Protection Act and Planning Act. BHL is an important tourist and recreational area for the province. In 1998, the museum at BHL had 20,000 visitors and the beach attracted 75,000 people. Other important recreational activities in BHL are swimming, fishing, canoeing, waterfowl hunting, and trapping of muskrat, beaver, and mink (Boyd and Smith 2000; Garrett 2003).

Water quality in BHL has been degraded by fertilizer-, pesticide-, and sediment-laden runoff from potato farms. Sources of sediment are streambank erosion in areas where cattle enter the streams to drink water and highway and road construction sites. To address these and other concerns, governmental agencies and citizen groups worked cooperatively to have BHL designated a protected area. In 1988, the provincial Department of Technology and Environment and the federal Department of Fisheries and Oceans held a public meeting to discuss protected status for BHL. One outcome of the meeting was formation of the Basin Head Lagoon Conservation Committee (BHLCC). This committee includes representatives from agriculture, tourism, fishing, municipal government, the Department of Fisheries and Oceans Canada, and the provincial Department of Fisheries, Aquaculture, and Environment. The BHLCC developed a nominating proposal to designate BHL an Area of Interest in the Marine Protected Areas Program under Canada's Oceans Act and a Marine Conservation Area under provincial legislation. Additionally, BHLCC collected socioeconomic data and worked with federal and provincial agencies to develop a plan for managing BHL. BHLCC's rationale for nominating BHL was to conserve and protect this marine ecosystem, particularly Irish moss, while sustaining the economic benefits of the region's agricultural and tourism industries. The goal of the management plan is to maintain and enhance water quality and the viability of Irish moss in BHL by establishing permitted, limited, and prohibited activities in designated zones in BHL (Boyd and Smith 2000).

To reduce the adverse impacts of agricultural and silvicultural operations on BHL and other MPAs, the provincial and federal governments are restoring streams and improving buffer zones. An important piece of legislation in this regard is the Environmental Protection Act. The act restricts the planting of row crops within 10–20 m (32.5 to 65 feet) of a watercourse depending on slope, the maintenance of cattle under intensive management (17.3 cattle per ha or 7 cattle/acre) within 20–30 m (65 to 97.5 feet) of a watercourse, and cutting of more than 30 percent of the basal area of trees within 20–30 m (65 to 97.5 feet) of a watercourse depending on slope (Boyd and Smith 2000).

The BHL case study has several implications for protected area management. First, the consultative and partnering approach taken by BHLCC, federal, and provincial governmental agencies involved in protecting BHL is a good model for the formation and operation of adaptive management working groups for protected areas. Second, protection of marine ecosystems requires understanding how land use/management activities influence the ecological integrity of MPAs, and if necessary, limiting those activities to reduce adverse ecological impacts. Third, the BHL case study illustrates that it is feasible for federal and state/provincial agencies to collaborate with stakeholders in developing management plans for a protected area.

Prince William Sound–Copper River Watershed and Glacier Bay National Park and Preserve

Going west from Prince Edward Island across the North American continent are two Alaskan ecosystems for which ecosystem-based initiatives were established.

They are the Prince William Sound–Copper River Watershed (PWS-CRW) (see fig. 10.4), which was under the purview of the Prince William Sound–Copper River Natural and Cultural Resources Cooperative and the Glacier Bay National Park and Preserve (GBNPP) (see fig. 10.5), which was the focal area for the Glacier Bay Ecosystem Partnership (Danby 2000). Both cooperative groups included resource managers, scientists, and stakeholders (USGS 2003b). The Prince William Sound/Copper River Natural and Cultural Resources Cooperative consisted of five federal agencies, three Alaskan state agencies, and two native organizations. The Glacier Bay Ecosystem Partnership included seven federal agencies, three Alaskan state agencies, and two native organizations (Danby 2000). In 1994, the National Biological Survey (NBS), now merged with the USGS, established multiagency initiatives in both ecosystems and 10 other ecosystems in the United States. Initiatives were placed in regions having substantive resource and environmental concerns and comanaged by multiple government agencies operating under different mandates (Danby 2000). NBS deemed interagency cooperation and coordination in the 12 regions to be an important prerequisite for successful ecosystem management (Gilbert 1988; Grumbine 1991; Cortner 1994), although implementing ecosystem management was not a goal of the initiatives.

The goal of these initiatives was "to promote and facilitate communication, education, understanding, research, and cooperation to achieve an ecosystem perspective for the use and management of natural, cultural, and other resources [and the] dissemination and use of science in planning and management" (Alaska Science Center 1997a,b; Danby 2000). Leadership for both initiatives was provided by the Alaska Science Center. Initiatives were placed in regions having substantive resource and environmental issues and comanaged by multiple government agencies operating under different mandates (Danby 2000). NBS deemed interagency cooperation and coordination in these regions to be an important prerequisite for successful ecosystem management (Gilbert 1988; Grumbine 1991, Cortner 1994). Implementing ecosystem management was not an objective of the initiatives.

Prince William Sound is located at the northern end of the Gulf of Alaska. Its shoreline contains deep fjords and tidal glaciers. The Copper River watershed, which is 65,000 km^2 (25,096 mi^2) in area, and Copper River delta are east of Prince William Sound (Danby 2000). The western half of the Copper River watershed is in a level-to-rolling lacustrine area containing numerous small lakes and wetlands; the eastern half drains portions of the rugged Chugach, Wrangell, and St. Elias mountains (Gallant et al. 1995). PWS-CRW has high biological diversity and is home for major populations of marine mammals and seabirds, and fish and shellfish species. Significant economic benefits are provided by agriculture, mining, oil transportation (the Port of Valdez is the terminus of the Trans Alaskan Pipeline), forestry, commercial and recreational fishing, outdoor recreation, and tourism. The area is especially

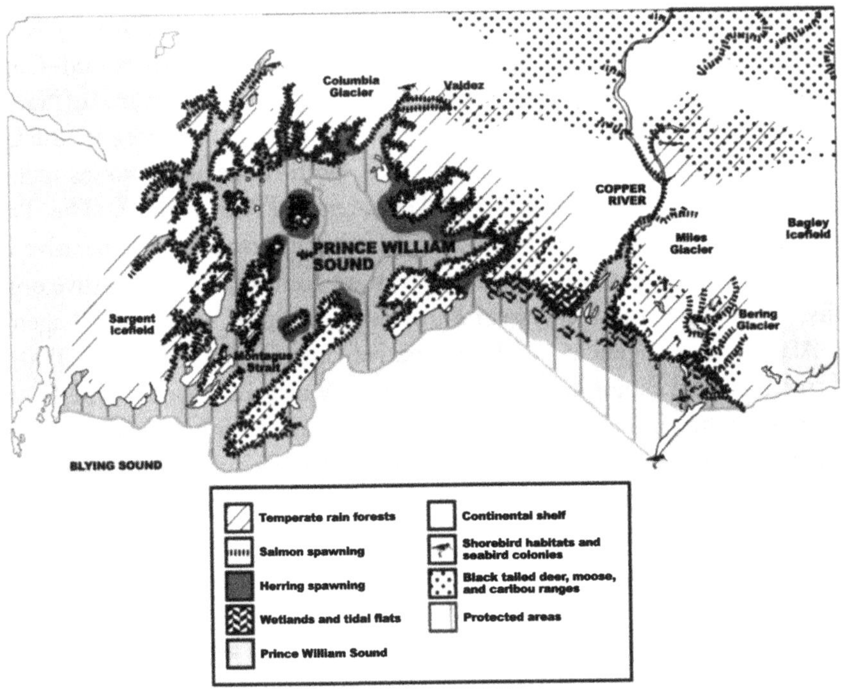

FIG. 10.4. Prince William Sound-Copper River watersheds. (From Inforain 2003b.)

noted for its productive salmon and herring fisheries (see spawning areas in fig. 10.4).

Coastal areas of the PWS-CRW contain temperate rain forests of western hemlock and Sitka spruce. Inland eastern areas contain white spruce at low elevations, willow and alder in subalpine areas, and alpine heath and barrens at high elevations. Inland western areas are dominated by black spruce forests. Most of Prince William Sound is in the Chugach National Forest, and the western half of the Copper River watershed is in the Wrangell-St. Elias National Park and Preserve. Landownership is diverse, including private, native, federal, and state lands. A significant management challenge for the PWS-CRW is to balance the economic benefits against the ecological impacts of natural resource extraction, recreation, and tourism. The most devastating ecological impacts of natural resource extraction occurred on March 24, 1989, when the *Exxon Valdez* oil tanker struck a reef

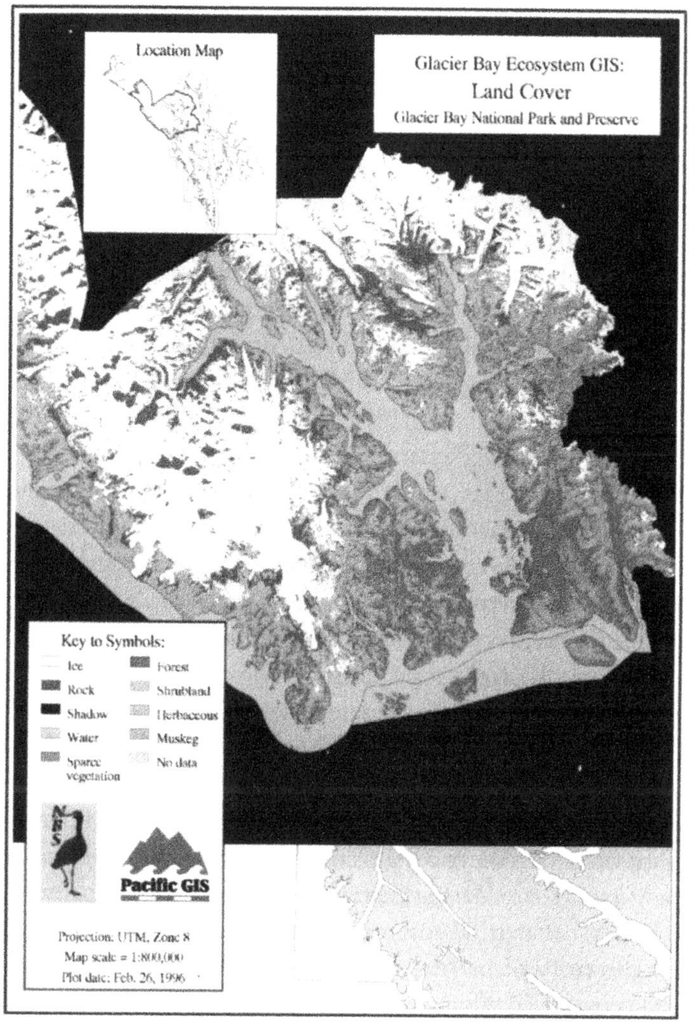

FIG. 10.5. Glacier Bay National Park and Preserve; location map insert is southeast Alaska. (From Inforain 2003a.)

in Prince William Sound and spilled 40 million liters (10.6 million gallons) of crude oil (Danby 2000).

Glacier Bay National Park and Preserve (GBNPP) is a marine park and World Biosphere Reserve located in southeast Alaska at the northwestern end of the inside passage. It is 13,287 km^2 (5,130 mi^2) in area. Glacier Bay is 1,500 km^2 (579 mi^2) in area. GBNPP is interspersed with old-growth forests, productive estuaries and marine waters, and diverse plant and animal populations. Coastal areas provide habitat for many nationally declining or sensitive species, including the Kittlitz's murrelet and endangered marbled murrelet, and critical habitat for fish, harbor seal, sea duck, Steller sea lion, and whale (USGS 2003a). GBNPP was an especially attractive location for an NBS ecosystem-based initiative because it is the only national park in the United States that has a legislative mandate to do scientific research. This mandate stems from the unique opportunities that Glacier Bay provides for doing research on ecological succession on a recently deglaciated area (Danby 2000). GBNPP was deglaciated less than 200 years ago.

Aquatic and terrestrial ecosystems in GBNPP are being adversely impacted by human activities, most notably natural resource extraction and recreation/tourism occurring inland, along the coast line, and on Glacier Bay (Danby 1999; USGS 2003a). Activities for which cumulative impacts are a concern include (1) commercial fishing for halibut, salmon, and crab in Glacier Bay (NPS regulations prohibit commercial fishing in national parks; a plan has been implemented to gradually phase out commercial fishing in Glacier Bay); (2) depletion of fish and shellfish populations in waters outside the park; (3) subsistence hunting and fishing; (4) large-scale clear-cutting of timber; (5) facility and infrastructure development needed to service the hundreds of thousands of people that visit Glacier Bay by air and sea; (6) development of Bartlett Cove, which has a high concentration of land-based visitors; (7) increasing backcountry use; and (8) offshore oil and gas development (Danby 2000; USGS 2003a).

Accomplishments of the two Alaskan initiatives included publication of a quarterly newsletter that informed members of activities and research/monitoring results, a directory of natural resource managers, bibliographies of scientific information, a directory of current research, and development and distribution of an integrated geographic information system (GIS) that synthesized GIS data layers from several organizations and agencies. The land cover map in figure 10.5 is one of the GIS layers developed by the GBNPP initiative.

Congress abolished the NBS in 1996, and former NBS scientists were transferred to the newly created Biological Resources Division of USGS in 1996. Subsequently, the 12 national ecosystem-based initiatives were terminated. Demise of NBS brought loss of political support and lack of funding for the initiatives. Perhaps equally significant is the fact that the two largest federal land management agencies involved in the PWS-CRW and GBNPP initiatives (U.S. Forest Service and NPS) chose not to fill the leadership and funding void for the initiatives created when NBS was abolished. This could be explained several ways. First, it might reflect the agencies' lack of commitment to the collaboration required for ecosystem management (Danby 2000). This seems unlikely given the support expressed by both federal agencies for ecosystem management. Second, each agency might have been committed to the idea of ecosystem management, but constrained by legislative mandates, policies, goals, and clientele, and unwillingness to yield management control to another agency. Third, perhaps the two agencies, and possibly other partners, believed the two ecosystem-based initiatives were not worth continuing because they did not contribute significantly to ecosystem management. Recall that the initiatives were not designed to implement ecosystem management.

Chapter 7 lists seven challenges to ecosystem management that provide additional insights as to why the U.S. Forest Service and NPS decided to suspend the two Alaskan ecosystem-based initiatives. Table 10.1 lists the challenges to ecosystem management (from chapter 7) in the left column and the extent to which those challenges might have limited the success of the two Alaskan initiatives in the right column. Two challenges do not appear to be constraints on the success of the two initiatives. Five challenges appear to have constrained success of the initiatives although not to the same degree.

Table 10.1. Relevance of Ecosystem Management Challenges to Alaskan Initiatives

Challenge to ecosystem management	Extent to which challenge constrained success of initiatives
Group decision making is particularly difficult when stakeholders disagree about management objectives and actions.	Communication, information, and research might have been stymied by ignoring their usefulness in achieving management objectives and evaluating management actions.
Collaborative decision making requires agency personnel to let go of a top-down technocratic approach.	U.S. Forest Service and National Park Service did not relinquish their traditional top-down management style.
Considerable time is required for stakeholders to understand one another and have meaningful interaction.	The program lasted only four years, which may not have been long enough to develop mutual understanding especially in view of the somewhat adversarial relationship between the U.S. Forest Service and National Park Service.
Considerable time, money, and effort are needed to accumulate, organize, and interpret the requisite data and information.	It does not appear that this challenge constrained the success of the initiatives before the elimination of NBS. Lack of funds could have prevented the U.S. Forest Service and National Park Service from continuing the initiatives.
Informational needs require scientists to share ideas with one another and scientists and managers to collaborate.	Annual meeting of executive committee, quarterly newsletter, *Directory of Natural Resource Managers*, bibliographies of scientific information, directory of current research, and integrated geographic information system offered opportunities to share data and ideas.
Current policies governing budgeting and administrative procedures do not support ecosystem management.	This challenge could have constrained the success of the initiatives.
Fear that ecosystem management will infringe on private property rights.	Several participants objected to the term "ecosystem management." Therefore, the memorandum of agreement establishing each imitative did not use this phrase.

The seemingly premature demise of the two Alaskan ecosystem-based initiatives does not mean they were failures. Their termination has as much to do with institutional change, notably termination of NBS, as any other factor. While cooperation and coordination among stakeholders is a necessary condition, it is not a sufficient condition for ecosystem management. In other words, continuation of the Alaskan initiatives beyond 1998 would not have guaranteed implementation of ecosystem management for the PWS-CRW and GBNPP, if for no other reason than that the initiatives were not established for this purpose. While it is not possible to generalize from the fate of the PWS-CRW and GBNPP initiatives, that experience brings into question whether it is appropriate to decouple cooperative activities, even those limited to communication, education, and data synthesis, from the broader process of ecosystem management. In other words, it is possible that the long-term viability and usefulness of the PWS-CRW and GB-NPP initiatives would have been enhanced by embedding them in a full-blown ecosystem management initiative, as difficult and challenging as that would have been.

Indiana Dunes National Lakeshore

The Indiana Dunes National Lakeshore (IDNL) is a prime example of a biologically rich and diverse area that suffers from extreme landscape fragmentation that poses significant management challenges. IDNL is a necklace of protected landscape units strung along the southern tip of Lake Michigan in northern Indiana. It is situated between Chicago, Illinois, and its suburbs that together constitute the third most populous cityscape in the United States, and Michigan City near the Michigan-Indiana border. In 1966, what was left of the wild dunes, wetlands, and oak forests was saved by a passionate citizenry and congressional intervention. Five units, isolated or with connecting road corridors, currently are interspersed with steel manufacturing plants, power generation facilities, rail transportation lines, numerous highways and county roads, and towns. Barges and tanker ships cruise nearby on Lake Michigan on their way to ports in Gary, Indiana, or Chicago. Despite the very urban and fragmented nature of IDNL, it stills retains unusual plant diversity with nearly as many species as parks many times its size. Overall, IDNL is ranked seventh among U.S. national parks in native plant diversity. Over 1,418 vascular plant species have been documented on only 6,071 ha (15,000 acres) because of complex interactions between water, shifting sand, underlying substrates, and plant colonization that result in beaches, bogs, swamps, shallow wetlands, desertlike sand flats, prairie habitats, oak savannas, and various forests. This is where Dr. H. C. Cowles in 1899 first proposed the essential concepts of ecological succession that have since been applied globally and form one of the hallmark organizing principles of ecological thought. Thus, IDNL is of great historical importance in addition to its value as a biodiversity reserve.

Numerous challenges face IDNL managers, which include being completely surrounded by intensely developed areas, being subjected to heavy human visitation and impacts, having significant air and water pollution, and having multiple and overlapping jurisdictions and authorities with surrounding towns, communities, corporations, and utilities as well as state and local law enforcement and natural resource agencies. Nearly all of these issues are aggravated by the size and spatial distribution of this protected area. The perimeter (park boundary) to area (park size) ratio is larger for numerous smaller units than for a few larger units. Nearly all of the protected area is heavily influenced by external stressors in the former case, but not in the latter case. Put another way, there is virtually no core area. Some of the boundary issues are unique to national park units, such as the toxic sludge ponds from the adjoining steel factory that are separated from IDNL woodlands by a chain-link fence. In addition, when IDNL was created there were numerous homes within the area that became part of IDNL and much of the "core" of this protected area had to be restored by systematically buying and removing numerous buildings (approximately four hundred homes).

Park managers have been acutely aware of the tenuous nature of their ability to manage valued natural resources within the constraints imposed by the spatial distribution of protected landscapes. Targeted purchases of private land and agreements with large adjacent landowners (e.g., utility companies) have strengthened the value of existing corridors and created new ones, reduced the perimeter-to-area ratio by increasing the size and making units more compact in shape, and eliminated private inholdings within core areas. This has created problems with property owners and adjoining towns that fear a larger federal footprint in the area and stakeholders that do not value the biodiversity of IDNL.

While the 1966 authorizing legislation included only 3,371 ha (8,330 acres) of land and water, the National Park Service and others continued to seek expansion of the boundaries of preservation. Four subsequent expansion bills for the park (1976, 1980, 1986, and 1992) have increased its size to more than 6,071 ha (15,000 acres). One land acquisition was designed to improve water quality by purchasing riparian areas upstream of IDNL and managing them as buffer areas for agricultural runoff. The value of controlling a riparian area outweighs its disadvantages in creating a linear park unit with a high perimeter-to-area ratio. The era of land acquisition is probably past as population increases and local development have consumed eligible land parcels. The challenge facing IDNL managers now is to encourage "compatible" land management on adjoining private lands to act as buffers for the external threats to park resources. This style of management necessitates intensive engagement with local constituencies, political finesse, and clear articulation of park goals and needs for the benefit of all stakeholders.

Importance of Research in Managing Ozark National Scenic Riverways

The Ozark National Scenic Riverways (ONSR) epitomizes the role and importance of scientific research in protected area management. ONSR is a national park consisting of a 216 km (143 mile)-reach of the Current and Jacks Fork Rivers in the Ozark Highlands area of southeastern Missouri. Both rivers are recharged by an extensive system of springs that represents the largest concentration of first-magnitude springs (water flow in excess of 441 liters [65 million gallons] per day) anywhere in the world (National Park Service 2003d). Seven major springs and 51 other springs provide 60 percent of the flows in ONSR. Big Spring is not only the largest spring in the ONSR system, but also one of the largest springs in the United States with an average daily flow of 1,873 million liters (276 million gallons) (National Park Service 2003d). Due to the spring-based hydrology of the Current and Jacks Fork Rivers, their waters are very clean and clear and have a constant flow making them ideal for johnboating, canoeing, swimming, fishing, and tubing (Chilman et al. 1996; National Park Service 2003d).

Landscapes in the Ozark Highlands (also known as the Ozark Mountains) are characterized by narrow steep-sided hollows, numerous streams, and bluffs. Vegetation includes a mixture of eastern hardwood forests (mostly white oak), Great Plains, and southern pine forests (mostly short leaf pine, which is Missouri's only native pine species), a large variety of wildflowers, and abundant wildlife. Geology in the park is classical karst consisting of soluble limestone and dolomite, sinkholes, caves, losing streams, and springs. Over three hundred caves have been identified in the park. ONSR has the largest number of caves and springs of any unit in the national park system (Chilman et al. 1996; National Park Service 2003c,d).

From 1880 to 1910, extensive unregulated logging and associated deforestation occurred in the Ozark Highlands and eventually led to the region's economic decline. Starting in 1920, state development interests promoted the area for sport fishing, hunting, and tourism. These efforts met with limited success due to the devastating impacts of logging on terrestrial and aquatic ecosystems. Mark Twain National Forest was established in the area in 1933 and with it the concept of scientific timber management. In the 1930s, Congress authorized the Corps of Engineers to construct 50 dams in Missouri. After World War II, the Corps pursued plans to construct a dam on the Current River for hydroelectric power and flood control. This plan was foiled by strong public opposition. In 1964, Congress created ONSR as the nation's first national riverways system. The park was established for the purpose of "conserving and interpreting unique scenic and other natural values and objects of historic interest, including preservation of portions of the Current River and the Jacks Fork River in Missouri as free-flowing streams, preservation of springs and caves, management of wildlife, and provisions for use and enjoyment of the outdoor recreation resources thereof by the people of the United States" (Legal Information Institute 2003).

From 1968 to 1972, boat use in ONSR increased from 40,000 floater days to 142,850 floater days (Marnell et al. 1978). In 1970, an estimated 450 to 500 canoes passed Cave Spring on the Current River (Jackson 1970). The next year up to 700 canoes per day were documented on a 30-km (20-mile) tretch of the Current River (Renken 1971). By 1979, annual canoe use jumped to 295,000 floater days. Total recreation visits in 2001 amounted to 1,497,988 (National Park Service 2003d). High levels of river use caused ecological damages at entry points, picnic and camping areas, and gravel bars, and complaints about congestion, rowdiness, and drunkenness (Chilman et al. 1996).

In 1972, the National Park Service (NPS) initiated a long-term research program in ONSR under the able leadership of Leo Marnell, a fisheries biologist with research experience in Yellowstone and Yosemite National Parks. The initial five-year research program was aimed at the Current and Jacks Fork Rivers and focused on use monitoring, environmental impact assessment, sociological evaluation, and water quality analysis (Marnell et al. 1978). Use monitoring was conducted in 10 river reaches employing time-lapsed photography to count canoes. Environmental impacts focused on canoe landing sites and water quality. Sociological evaluations consisted of surveys to assess visitor perceptions of crowding and its impacts on enjoyment of the rivers. Surface and ground water quality were monitored. Long-term monitoring in the Hurricane Creek Barometer watershed was used to determine the impacts of land management on water quality. The NPS and U.S. Forest Service did several groundwater trace studies in the recharge areas for several springs.

This research developed and refined methods for doing long-distance groundwater traces to springs with large flows, delineated recharge areas for important springs, and evaluated how land-use activities in recharge areas for springs influenced spring water quality. Marnell's original research on canoe counts, visitor perception surveys at major access points, and water quality monitoring continues to this day (Runge 2003).

The ONSR research and monitoring programs yielded several important benefits in terms of NPS management and protection of the riverways. These include:

1. Knowing how particular land-use activities in the recharge areas for springs are most likely to influence spring water quality
2. Improving interpretive programs for the public
3. Resolving court cases that eventually allowed NPS to place limitations on the number of rental canoes owned by each rental company permitted to use the river and a moratorium on additional permits (the judge's ruling indicated

imposing NPS's research on visitor perceptions of crowding provided sufficient justification for imposing use limitations)
4. Developing an alternative in the Draft River Use Management Plan and Environmental Assessment to restrict canoe use in various zones of the rivers and establish zone-specific horsepower limits on outboard motors
5. Demonstrating that U.S. Steel Corporation's lease applications to mine for lead, zinc, and associated metals in the Mark Twain National Forest would adversely affect water quality in Big Spring and ONSR
6. Providing data that were essential in NPS's petition to EPA to have the recharge area for Big Spring designated as a Sole Source Aquifer under the Safe Drinking Water Act of 1974

Research efforts in ONSR have led to scientific discoveries that benefited other protected areas (Chilman et al. 1996).

In summary, research and monitoring programs not only improved scientific understanding of the park's aquatic and terrestrial ecosystems, but also provided invaluable information for managing and protecting those ecosystems for the benefit of future generations. The importance of research in managing national parks was reaffirmed in 1999 through establishment of The Natural Resource Challenge: The National Park Service's Action Plan for Preserving Natural Resources (National Park Service 1999a). One goal of the plan is to substantially increase "the role of science in decision-making, revitalizing and expanding natural resource programs, gathering baseline data on resource conditions, strengthening partnerships with the scientific community, and sharing knowledge with educational institutions and the public."

Summary

The management issues examined in the extended case studies of this chapter are whether natural regulation of ungulates (elk and bison) is harming Yellowstone National Park (YNP) and how to reestablish water flows to conserve biodiversity in Everglades National Park (ENP) and Kruger National Park (KNP). An arid high plateau is the environmental setting for YNP and lowland glades in a humid area for ENP. YNP is the world's first national park and is noted for its abundant wildlife, deep canyons, waterfalls and mountains. ENP is noted for its extensive sawgrass glades and subtropical flora and fauna. Despite the dissimilarity in management issues and environmental settings, adaptive management has been recommended for managing ungulates in YNP and adaptive ecological learning for reestablishing adequate water flows in ENP. Adaptive learning is essentially the learning component of adaptive management, which has the potential for resolving management concerns in both national parks.

Kruger National Park (KNP) faces a challenge similar to ENP in that diversion of water to other uses has reduced water supplies to the point of threatening biodiversity. Unlike South Florida where a large and expensive program is restoring water supplies to the ENP and other ecosystems, conditions in KNP are bleak.

KNP managers have had difficulty convincing the South African government that upland water diversions are threats to the park's ecological integrity. KNP responded to this threat by establishing the Rivers Research Program. The goal of the program is to inform managers and stakeholders about water quantity and quality requirements for sustaining the ecological integrity of the park's rivers, and to develop models and other tools to predict ecological responses of rivers to changing water quantity and quality conditions. A consultative management process has been established by the Rivers Research Program that has three elements in common with adaptive management, namely, defining desired states of rivers, using models to predict consequences of management actions, and monitoring ecosystem responses to natural disturbance and management actions. It has been recommended that river management in KNP be based on a more goal-orientated adaptive management approach that identifies social goals, incorporates social values into decisions, and manages strategically to achieve those values.

Marine protected areas (MPAs) are underrepresented relative to the proportion of the Earth's surface covered by oceans and are increasingly important for conserving marine biodiversity. The WCPA-Marine Program has a theme of implementing MPAs using systems of participatory and adaptive management. Wilson's Promontory Marine National Park (WPMNP), a biologically rich ecosystem located at the southern extreme of the Australian continent, provides many opportunities for water-based recreation and tourism. WPMNP is being threatened by water pollution, high rates of sedimentation, physical disturbance, overharvesting of fisheries, and introduction of exotic marine organisms. Management actions for reducing adverse impacts to WPMNP and other MPAs in Victoria are being identified using a multiple agency/stakeholder cooperative approach. While this cooperative approach is not per se adaptive management, the cooperative elements are a prerequisite for adaptive management.

Basin Head Lagoon (BHL) in Prince Edward Island, Canada, is a biologically rich marine ecosystem that contributes significantly to tourism, recreation, and production of Irish moss. The latter is used in aquaculture. Fertilizers and pesticides in runoff from potato farms and sediment from livestock use and road construction in upland areas of the watershed have degraded water quality in BHL. In response to these threats, federal and state governmental agencies and stakeholder groups formed the Basin Head Lagoon Conservation Committee. Accomplishments of the committee include developing a nomination proposal to designate BHL an Area of Interest in the national Marine Protected Areas Program, collecting socioeconomic data, and helping to prepare a management plan for BHL that establishes permitted, limited, and prohibited activities in designated zones. While the Basin Head Lagoon Conservation Committee was not established as an adaptive management working group, it is an excellent template for forming and operating such a group.

The former National Biological Survey provided the leadership in establishing multiagency initiatives in 12 areas of the United States to promote ecosystem-based management. A primary goal of the initiatives was to foster the use of science in planning and management for protected areas. Two initiatives were established in Prince William Sound–Copper River Watershed (PWS-CRW) and Glacier Bay National Park and Preserve (GBNNP) in Alaska. PWS-CRW supports agriculture, mining, oil transportation, forestry, fishing (especially salmon and herring), outdoor recreation, and tourism. In 1989, the PWS-CRW was severely impacted by the *Exxon Valdez* oil spill. GBNNP is a biologically rich marine park that is the only national park in the United States with a legislative mandate for scientific research. The park's aquatic and terrestrial ecosystems are being heavily impacted by natural resource extraction, recreation, and tourism.

Although the two initiatives had significant accomplishments, both were terminated shortly after Congress disbanded the National Biological Survey. The largest federal land management agencies involved in the two initiatives (U.S. Forest Service and National Park Service) did not provide leadership and funding to continue the initiatives after NBS was disbanded. This case study illustrates the importance of institutional stability in maintaining ecosystem-based initiatives and raises the question of whether decoupling cooperative activities associated with improving communication, education, and data synthesis from the broader process of ecosystem management is appropriate.

Indiana Dunes National Lakeshore (IDNL) is unique both for its biodiversity and its position in a semiurban area. The original land area set aside in 1966 created small isolated units that pose natural resource management problems. Subsequent expansion of IDNL has nearly doubled the area under protection and has reduced the perimeter-to-area ratio. This means that core areas with lessened external influences now exist and have increased IDNL's ability to manage its threatened plant species. The spatial arrangement of the protected areas still leaves IDNL vulnerable to negative impacts from the surrounding landscape and to future changes in adjoining land uses. This vulnerability is compounded by heavy visitor use. IDNL managers will need to find creative ways to expand the effective protected area by encouraging compatible use of surrounding lands.

The Ozark National Scenic Riverways (ONSR) contains segments of the relatively pristine Current and Jacks Fork Rivers in Missouri. ONSR contains the largest concentration of first-magnitude springs anywhere in the world and one of the largest springs in the United States. The park contains other karst features such as sinkholes, caves (over 300 in the park), and losing streams. Due to the excellent opportunities ONSR provides for johnboating, canoeing, swimming, fishing, and tubing, use of the riverways park grew very rapidly after its establishment. Fortunately, park scientists had the foresight to initiate a long-term research and monitoring program to study visitor perceptions of crowding on the river and in campgrounds and its impacts on public enjoyment of the park, track changes in surface and ground water quality, and determine impacts of land use in recharge areas for springs on water quality in springs. The ONSR research and monitoring programs not only improved scientific understanding of the park's ecosystems, but also provided information that was invaluable for management and protection of ONSR.

References

Alaska Science Center. 1997a. *Prince William Sound-Copper River Ecosystem News: First quarter, 1997.* Biological Resources Division, U.S. Geological Survey, Anchorage, AK.

———. 1997b. *Glacier Bay Ecosystem Partnership News: First Quarter, 1997.* Biological Resources Division, U.S. Geological Survey, Anchorage, AK.

Bath, A. J. Public attitudes about wolf restoration in Yellowstone National Park. 1997. In *The Greater Yellowstone Ecosystem: Redefining America's wilderness heritage,* ed. R. B. Keiter and M. S. Boyce, 367–376. New Haven, CT: Yale University Press.

Best, G. R. (Ed.). 2000. Greater Everglades ecosystem restoration (GEER) science conference: Defining success. Proceedings, South Florida Ecosystem Restoration Task Force and Working Group and U.S. Geological Survey, Miami, FL.

Boyce, M. S. 1991. Natural regulation or the control of nature? In *The Greater Yellowstone Ecosystem: Redefining America's wilderness heritage,* R. B. Keiter and M. S. Boyce, 183–208. New Haven, CT: Yale University Press.

Boyd, P., and B. Smith. 2000. Working together to conserve and protect a unique ecosystem: Basin Head, Prince Edward Island. In *Managing protected areas in a changing world.* Proceedings of the Fourth International Conference on Science and Management of Protected Areas, Science and Management of Protected Areas Association, Nova Scotia, Canada, ed. S. Bondrup-Nielsen et al., 1367–1386.

Caughley, G., and J. H. Lawton. 1981. Plant-herbivore systems. In *Theoretical ecology: Principles and applications,* ed. M. S. Boyce and L. D. Hayden-Wing, 132–166. Oxford, England: Blackwell Science.

Chase, A. 1987. *Playing God in Yellowstone: The destruction of America's first national park.* Boston, MA: Atlantic Monthly Press.

Chilman, K., D. Foster, and T. Aley. 1996. River management at Ozark National Scenic Riverways. In *Science and ecosystem management in national parks,* ed. W. L. Halvorson and G. E. Davis, 295–317. Tucson: The University of Arizona Press.

Cortner, H. J. 1994. Intergovernmental coordination in ecosystem management. In *Ecosystem management: Status and potential.* Proceedings from a Congressional Research Service Symposium, Washington, DC, pp. 229–242.

Craighead, J. J. 1991. Yellowstone in transition. In *The Greater Yellowstone Ecosystem: Redefining America's wilderness heritage,* ed. R. B. Keiter and M. S. Boyce, 27–39. New Haven, CT: Yale University Press.

Danby, R. K. 1999. Regional ecology of the St. Elias mountain parks: A synthesis with management implications. Masters Thesis, Department of Geography and Environmental Studies, Wilfrid Laurier University, Waterloo, ON.

———. 2000. Fostering an ecosystem perspective through intergovernmental cooperation: A look at two Alaskan examples. In *Managing protected areas in a changing world.* Proceedings of the Fourth International Conference on Science and Management of Protected Areas, Science and Management of Protected Areas Association, Nova Scotia, Canada, ed. S. Bondrup-Nielsen et al., 722–735.

Day, J. C. 2003. Marine park management and monitoring-lessons for adaptive management from the Great Barrier Reef. In *Managing protected areas in a changing world.* Proceedings of the Fourth International Conference on Science and Management of Protected Areas, Science and Management of Protected Areas Association, Nova Scotia, Canada, ed. S. Bondrup-Nielsen et al., 1258–1282.

Despain, D., D. Houston, M. Meagher, and P. Schullery. 1986. *Wildlife in transition: Man and nature on Yellowstone's northern range.* Boulder, CO: Roberts Renehart.

Environmental Conservation Council. 2000. Marine coastal and estuarine investigation, final report. East Melbourne, Victoria (Australia). http://www.veac.vic.gov.au/ecc/marine/report2000.htm (accessed February 27, 2003).

Gallant, A. L., E. F. Binnian, J. M. Omernik, and M. B. Shasby. 1995. Ecoregions of Alaska. Professional Paper 1567. U.S. Department of the Interior, Geological Survey, Washington, DC.

Garrett, A. H. 2003. *History of Basin Head.* http://www.edu.pe.ca/easternkings/fishing/basinhead.htm (accessed February 26, 2003).

Gilbert, V. 1988. Cooperation in ecosystem management. In *Ecosystem management for parks and wilderness,* ed. J. K. Agee and D. R. Johnson, 180–192. Seattle, WA: University of Washington Press.

Good, J. 2000. Reminiscence from the firing line. *Yellowstone Science* 8: 3–6.

Government of Prince Edward Island, Canada. 2003. Basin Head, Prince Edward Island. http://www.gov.pe.ca/placefinder/index.php3?cgndb=BAAGY&city=Basin+Head (accessed February 26, 2003).

Grumbine, R. E. 1991. Cooperation or conflict? Interagency relationships and the future of biodiversity for U.S. parks and forests. *Environmental Management* 15: 27–37.

Gunderson, L. H. Managing surprising ecosystems in southern Florida. 2001. *Ecological Economics* 37: 371–378.

Houston, D. B. 1982. Research in ungulates in northern Yellowstone National Park. *Research in the parks: Transactions of the National Park Centennial Symposium,* 11–25. National Park Service, Washington, DC.

Inforain. 2003a. *Glacier Bay ecosystem GIS: Sample maps. USGS Interim Land Classification, Glacier Bay National Park.* http://www.inforain.org/alaska/glabaycd/CATALOG/htm/plots.htm (accessed February 26, 2003).

———. 2003b. *Prince William Sound-Copper River ecosystem: Biological and ecological resources.* http://www.inforain.org/maparchive/copperriver_biology.htm (accessed February 26, 2003).

Jackson, J. 1970. People problems on the riverways. *National Parks and Conservation Association Magazine* 44: 24–27.

Jewitt, G. P. W., and A. H. M. Görgens. 2000. Facilitation of interdisciplinary collaboration in research: Lessons from a Kruger National Park Rivers Research Program project. *South African Journal of Science* 96: 410–420.

Kay, C. E. 1993. Browsing by native ungulates: Effects on shrub and seed production in the Greater Yellowstone ecosystem. In *Proceedings: Wildland shrub and arid land restoration symposium.* Las Vegas, NV, October 10–21, Intermountain Research Station, Forest Service, Ogden, UT, comp. B. A. Roundy, E. Durant McArthur, J. S. Haley, and D. K. Mann, 310–320.

———. 1997. Yellowstone: Ecological malpractice. *PERC (Political Economy Research Center) Reports* 15:5–39.

Kiker, C. F., J. W. Milon, and A. W. Hodges. 2001. Adaptive learning for science-based policy: The Everglades restoration. *Ecological Economics* 37: 403–416.

Kruger National Park. 2003. *General information.* http://www.ecoafrica.com/krugerpark/general.htm (accessed February 20, 2003).

Kruger National Park Rivers Research Program. Undated brochure.

Legal Information Institute. 2003. Title 16, Chapter 1, subchapter LXX, section 460m. http://www4.law.cornell.edu/uscode/16/460m.html (accessed March 1, 2003).

Light, S. S., and J. W. Dineen. 1994. Water control in the Everglades: A historical perspective. In *Everglades: The ecosystem and its restoration,* ed. S. M. Davis and J. C. Ogden, 47–84. Delray Beach, FL: St. Lucie Press.

Marnell, L., D. Foster, and K. Chilman. 1978. River recreation research conducted at Ozark National Scenic Riverways, 1970–1977: A summary of research projects and findings. National Park Service, Van Buren, MO.

Mech, L. D. 1997. Returning the wolf to Yellowstone. In *The Greater Yellowstone Ecosystem: Redefining America's wilderness heritage,* ed. R. B. Keiter and M. S. Boyce, 309–322. New Haven, CT: Yale University Press.

Merrill, E. H., and M. S. Boyce. 1991. Summer range and elk population dynamics in Yellowstone National Park. In *The Greater Yellowstone Ecosystem: Redefining America's wilderness heritage,* ed. R. B. Keiter and M. S. Boyce, 263–273.

Meiklejohn, B., and S. Earle. 2003. Sylvia Earle: Our oceans and ourselves. *Wild Earth* 12: 19–23.

National Parks and Conservation Association. 1999. River of grass ready to heal. *National Parks* 7(1–2; January/February).

National Park Service. 1999a. National resource challenge. *Protecting America's natural heritage: National Park Service announces new plan to strengthen and revitalize natural resource programs in national parks.* http://www1.nature.nps.gov/challenge/NRC.htm (accessed March 6, 2003).

———. 1999b. *Wolf restoration to Yellowstone.* http://www.nps.gov/yell/nature/animals/wolf/wolfrest.html (accessed April 24, 2003).

———. 2003a. *Everglades National Park.* http://www.nps.gov/ever/ (accessed February 18, 2003).

———. 2003b. *Indiana Dunes National Lakeshore map.* http://www.nps.gov/indu/maps/ (accessed December 15, 2003)

———. 2003c. *Ozark National Scenic Riverways: Nature and science.* http://data2.itc.nps.gov/nature/index.cfm?alphacode=ozar (accessed March 1, 2003).

———. 2003d. *Ozark National Scenic Riverways.* http://www.nps.gov/ozar/ (accessed March 1, 2003).

———. 2003e. *Yellowstone wildlife: Update 2002.* http://www.nps.gov/yell/nature/animals/wolf/wolfup.html (accessed April 24, 2003).

National Research Council. 2002. *Ecological dynamics on Yellowstone's northern range: The report of the National Academy of Sciences.* Chapter 5: Conclusions and recommendations. *Yellowstone Science* 10:3–11.

Parks Victoria. 2002a. *Marine notes: Wilson's Promontory Marine National Park.* East Melbourne, Victoria.

———. 2002b. *Victoria's system of marine national parks and marine sanctuaries: Draft management strategy for public comment.* Melbourne, Australia. http://www.parkweb.vic.gov.au/resources/08_0703.pdf (accessed February 27, 2003).

———. 2002c. *Wilson's Promontory National Park: Management plan.* East Melbourne, Victoria.

———. 2003a. *Wilson's Promontory Marine National Park.* http://www.parkweb.vic.gov.au/1park_display.cfm (accessed February 25, 2003).

———. 2003b. Wilson's Promontory National Park-Victoria's blue wilderness. http://www.parkweb.vic.gov.au/resources/05_0787.pdf (accessed February 1, 2004).

Phillips, M. K., and D. W. Smith. 1997. *Yellowstone wolf project biennial report: 1995 and 1996.* YCR-NR-97-4, National Park Service, Yellowstone Center for Resources, Yellowstone National Park, WY.

Pickett, S. T. A., V. T. Parker, and P. L. Fiedler. 1992. The new paradigm in ecology: Implications for conservation biology above the species level. In *Conservation biology: The theory and practice of nature conservation, preservation and management,* ed. P. L. Fiedler and S. K. Jain, 65–88. New York: Chapman and Hall.

Renken, T. 1971. Canoes jamming, damming Current River. *St. Louis Post-Dispatch,* St. Louis, MO, September 10.

Rogers, K., and R. Bestbier. 1997. *Development of a protocol for the definition of the desired state of riverine systems in South Africa.* Centre for Water in the Environment, University of the Witwatersrand, South Africa. http://www.wits.ac.za/cwe/desired.pdf (accessed February 25, 2003).

Rogers, K. H. 1997. Operationalizing ecology under a new paradigm. In *Enhancing the ecological basis of conservation: Heterogeneity, ecosystem function and biodiversity,* ed. S. T. A. Pickett, R. S. Ostfeld, M. Shachak, and G. E. Likens, 60–77. New York: Chapman and Hall.

Runge, R. W. 2003. Chief of Resource Management and Education, Ozark National Scenic Riverways. Personal communication. March 7.

Schaffer, W. M. 1988. Predicting order in the chaos of nature. In *Evolution of life histories of animals,* ed. M. S. Boyce, 313–350. New Haven, CT: Yale University Press.

Schullery, P. 1997. *Searching for Yellowstone: Ecology and wonder in the last wilderness.* New York: Houghton Mifflin Co..

Sklar, F. H., H. C. Fitz, Y. Wu, R. Van Zee, and C. McVoy. 2001. The design of ecological landscape models for Everglades restoration. *Ecological Economics* 37: 379–401.

South African Tourism Board. 1993. *South Africa: Wildlife and safari county.* Pretoria, South Africa.

Town of Souris, Prince Edward Island. 2003. http://www.peisland.com/souris/photos.htm#boyd (accessed February 26, 2003).

TOURSAA (SA Tours and Safaris with African Alternatives). 2003a. *Destination Kruger Park-Kruger Park mobile camping safari.* http://www.ecoafrica.com/krugerparksafari/index.htm (accessed February 20, 2003).

———. 2003b. *Kruger National Park.* http://www.ecoafrica.com/saparks/idxkrug.htm (accessed February 20, 2003).

U.S. Army Corps of Engineers and South Florida Water Management District (USACE and SFWMD). 2000. *Master program management plan, comprehensive Everglades restoration plan.* South Florida Water Management District, West Palm Beach, FL.

U.S. Geological Survey (USGS). 2003a. *Alaska Science, Biological Science Office, Glacier Bay National Parks Project, Glacier Bay Ecosystem GIS* http://www.absc.usgs.gov/ecosys/glba/glba_ecosys.htm (accessed February 26, 2003).

———. 2003b. *Developing frameworks for scientific research and resource management in Glacier Bay, Denali, Prince William Sound, and Wrangell-St. Elias.* http://www.absc.usgs.gov/research/sis_summaries/terrestrial_sis/Developing_frameworks.htm (accessed February 26, 2003).

U.S. News & World Report. 1996. Yellowstone's unraveling. 121: 80–82.

Wagner, F. H. 1994. Scientist says Yellowstone is being destroyed. *High Country News,* May 30, 1994, pp. 14–15.

Willison, M. 2003. Fishing for a wild ocean. *Wild Earth* 12: 10–12.

World Commission on Protected Areas (WCPA). 2003. *Key issues.* http://iucn.org/themes/wcpa/biome/marine/issues.htm (accessed February 25, 2003).

CHAPTER

11

Future Planning and Management

It is, of course, impossible for us to view our every action in terms of its relevance to the environment, but it is possible to understand the importance of our actions in relation to our environment. The importance of our learning cannot be overstated; if we do not learn, we have little hope.

—John Stadler

Previous chapters addressed social, economic, and ecological values of protected areas and methodological approaches for establishing, preserving, and managing them. Case studies in chapters 9 and 10 considered particular management problems and issues, potential actions for redressing those problems and issues, and whether or not management actions have acceptable outcomes. Understanding the motivation for and consequences of past management actions is important for several reasons, not the least of which is that it affords opportunities to learn about how ecosystems respond to human interventions. In order for protected area managers, policy actors, scientists, environmental interest groups, gateway communities, and other stakeholders to balance the future values of protected areas, they need to agree on desirable or sustainable future ecosystem states and, when conditions indicate an undesirable or unsustainable state, select management actions and policies to achieve a desirable state.

The most concentrated effort aimed at future planning and management of protected areas occurs during preparation of general management plans (GMPs). These plans lay out an overall management strategy for a protected area for the next 10 to 15 years. A GMP

> focus[es] on why the park was established, and what management prescriptions (i.e., resource conditions, visitor experiences, and appropriate types of management actions) should be achieved and maintained over time. The plan will consider the park in its full ecological, scenic, and cultural contexts as a unit of the national park system and as part of a surrounding region...[and] will also establish a common management direction for all park divisions and districts. (National Park Service 2004)

GMPs determine how to manage a protected area in a manner consistent with the area's organic act and the mission of the National Park Service (National Park Service 2003b).

Each IUCN category of protected areas is managed according to a specific set of goals (see table 2.3). Even within the same IUCN protected area category, the relative importance of the management goals and constraints varies from country to country and from one location to another in the same country. Hence, future planning and management for protected area ecosystems require careful assessment of site-specific attributes of and management goals for ecosystems as is done in GMPs. While site-specific planning and management are very important, they are not the aim of this book.

This chapter describes a general framework for future planning and management of protected area ecosystems. A protected area ecosystem is the entire ecosystem in which a protected area is located, not just the area within the boundaries of the protected area. The framework keys on desirable ecosystem states for the future as defined by long-term social, economic, and ecological goals, and the need for protected area managers to take action to achieve those goals. The goals are:

1. Reducing external threats
2. Improving managerial responses to changing demographic, cultural, ecological and economic conditions (referred to collectively as ecosystem conditions)
3. Balancing resource conservation and visitor satisfaction
4. Enhancing planning and management capacity

The four goals are by no means inclusive. Other goals can be incorporated in the framework. The remainder of this chapter discusses each of the four goals and a method for integrating them to determine whether the future state of a protected area ecosystem is desirable.

Reducing External Threats

Many natural resource-based parks and protected areas in North America and elsewhere face external threats and impairments that will persist in the future (Dilsaver 1994; EarthVision Environmental News 2003). External threats originate outside the boundaries of protected areas and have impacts within and outside those boundaries (see chapters 9 and 10 for more details). They include:

- Fragmented, degraded, and destroyed habitat for wildlife species
- Severe limitations on the movement of wide-ranging species

- Reduced populations of native fish that are unable to compete with invasive nonnative species
- Water diversions upstream of protected areas and water pollution
- Loss of native vegetation
- Air pollution
- Alteration of hydrologic regimes
- Ecological impacts from natural resource extraction, recreation, and tourism

While reducing external threats is essential to the goal of conserving the natural and cultural resources of protected areas, federal land management agencies typically believe they have limited authority to alleviate external threats (Keiter 1996). This attitude prevails due to limited statutory authority, agency culture, ingrained political sensitivities, and deference to adjacent landowners. For example, the National Park Service (NPS) has this attitude despite the fact that the Organic Act establishing the National Park System requires the agency to conserve scenic, natural, wildlife, and historic resources. Conservation cannot be achieved without alleviating external threats. NPS's Vail Agenda recognizes this point in stating, "the mandate of resource protection means that prevention of external and transboundary impairment of park resources and their attendant values should be a central objective of park system policy" (National Park Service 1992). In addition, the Endangered Species Act, Clean Water Act, and Clean Air Act provide statutory authority to alleviate negative environmental impacts on protected areas. The U.S. Congress has not legitimized ecosystem management or natural regulation, which makes it more difficult for protected area managers to address external threats (Keiter 1996).

How can managers define desirable future ecosystem states with respect to external threats? One way is to define ecosystem-wide conditions that conserve the ecological integrity of protected area ecosystems and monitor future management actions for achievement of those conditions. This strategy is supported by the Vail Agenda, which states that "the mission of preserving and protecting the natural treasures that belong in the national park system can only be met if the [National Park] Service can confront the threats to park resources and has the means of dealing with those threats" (National Park Service 1992).

The importance of natural resource protection in national park units was confirmed by the Natural Resource Challenge. One of the primary goals of the challenge is to preserve national parks so that current and future generations can enjoy, benefit, and learn from them (National Park Service 1999). In addressing external threats, the challenge poses the following questions: "What is the condition of resources outside of park boundaries (air, water, nonnative and migratory species)? What impact do these resources have on park resources?" These questions clearly indicate that the NPS understands the potential adverse consequences of external threats.

Recognizing external threats have adverse ecological consequences for protected areas is a necessary but not a sufficient condition for alleviating them. The latter requires protected area managers to implement and monitor management actions designed to reduce external threats and their impacts. Implementation and monitoring elements are now examined. The Natural Resource Challenge makes two recommendations pertinent to the implementation of management actions to prevent external and transboundary impairment of natural resources and associated values. The first recommendation is that NPS provide public and private parties, such as those in gateway communities, with technical and planning assistance designed to significantly reduce external and transboundary threats to natural and cultural resources in national park units. This involves offering NPS expertise to planning departments, developers, and other local and state organizations that make decisions influencing external threats.

The second recommendation is that NPS devote sufficient human and financial resources to develop technical expertise and cooperative relationships that ensure compliance with laws and policies governing protected area management. Most protected area managers have developed cooperative relationships with other federal and state natural resource management agencies, private landowners, and planning/zoning departments in gateway communities. Collaboration and cooperative management could be accomplished through an adaptive management working group whose goal is to conserve the natural and cultural resources of a particular protected area while sustaining social and economic values in gateway communities. In an adaptive management approach, alternative forms of cooperative management are tried on an experimental basis (perhaps different forms are implemented in different protected areas) and the results of those experiments are used to improve cooperative efforts (see the following sidebar for an example of cooperative management). Motivation for cooperation stems from the symbiotic relationship between protected areas and gateway communities. Specifically, gateway communities have a vested interest in preserving the social, cultural, and ecological values of protected areas, and protected areas have a vested interest in sustaining the economic well-being of gateway communities.

Obstacles to cooperation include differences in the statutory authority under which federal agencies operate (multiple use management for the Forest Service and Bureau of Land Management versus resource conservation and public use for the NPS) and conflict between the profit motive for private lands and the resource conservation mandate for certain protected areas. In commenting on obstacles to cooperation, Keiter (1996) notes that despite the absence of laws mandating ecosystem management, national park managers have sufficient legal authority to address transboundary management issues and protect park resources from degradation. An example of interagency cooperation is the Greater Yellowstone Coordinating Committee, a voluntary group of NPS and Forest Service managers that coordinates

Cooperative Management in the Valles Caldera National Preserve

There is growing recognition that protected areas can no longer be treated as islands but must be seen in the context of overall land use, and that successful managers are adopting more inclusive, collaborative approaches in working closely with local communities.

—Jessica Brown

Several of the case studies in this chapter involve the establishment of an adaptive management working group. The group can either advise a protected area manager on how to adaptively manage an area or share management responsibility with the manager. Both arrangements, particularly the latter one, raise the issue of whether collaborative management of public lands is feasible.

An experimental collaborative management approach is being used in Valles Caldera National Preserve (VCNA) west of Los Alamos, New Mexico. VCNA is an 89,000-acre unit of the National Forest System that was formerly part of the Baca Ranch. In July 2000, Congress purchased the acreage from a Texas oil family for $101 million. The preserve includes forest, high country meadows, the headwaters of the Jemez River, cultural and religious sites, and a dormant volcano.

Normally a national preserve would be managed by a federal agency such as the National Park Service, the Bureau of Land Management, or the U.S. Department of Agriculture. To appease Senator Pete Domenici (R-NM), who opposed purchase of the ranch for a national preserve, the VCNA is being managed by the Valles Caldera Trust, a wholly owned government corporation. The trust is governed by a nine-member board of trustees made up of the forest supervisor of the Santa Fe National Forest, the superintendent of Bandelier National Monument, and seven individuals. Trustee members are appointed by the president of the United States in consultation with the New Mexico Congressional Delegation.

The board of trustees has almost complete control over management of VCNA. The Valles Caldera Preservation Act requires the board to "protect and preserve the scenic, geological, watershed, fish, wildlife, historic, cultural and recreational values of the preserve, and to provide for multiple use and sustained yield of renewable resources within the preserve." Current trustees have expertise in livestock management, game and nongame wildlife and fish populations, sustainable forestry, nonprofit conservation, financial management, cultural and natural history of the region, and New Mexico state or local governance.

The board has developed a general management plan and funded studies to assemble baseline ecological information on the diversity and distribution of species in the preserve. Activities permitted under the management plan include elk hunting, limited grazing, small-scale timber thinning in high fire hazard areas, and very limited public bus tours of the preserve. Hunting is allowed to help manage herd size and quality, protect the preserve's outstanding natural resources, and raise funds to finance management of the preserve. The trust provides management and administrative services for the preserve, establishes and implements management policies that best achieve the purposes and requirements of the Valles Caldera Preservation Act, receives and collects funds from private and public sources, disburses funds in support of management and administration of the preserve, cooperates with federal, state, and local governmental units, and Indian tribes and Pueblos, and furthers the purposes for which the preserve was established.

The Political Economy Research Center in Bozeman, Montana, views the Valles Caldera Trust as a way of funding environmental protection in the face of stringent budgets for federal and state land management agencies. They maintain that if the trust proves successful, it can be implemented in other protected areas such as Grand Staircase-Escalante National Monument or the Artic National Wildlife Refuge. Environmental groups are concerned that income-generating extractive resource activities in Valles Caldera might compromise ecological preservation. This concern is fueled in part by the fact that the organic act for the preserve requires it to be financially self-sufficient by 2017 and the possibility that future trustees may be less sympathetic to ecological preservation than current trustees.

It is too early to judge the success of the Valles Caldera Trust experiment. If it proves successful, then federal land management agencies might use it to manage other protected areas. On the other hand, if it fails, then collaborative management of protected areas is likely to be viewed with skepticism. Of course, traditional public land management arrangements have their share of drawbacks. Whether it proves to be successful or not, the Valles Caldera Trust signifies that collaborative management of public lands is possible and worthy of consideration.

The National Park Service (NPS) collaboratively plans and manages units that are part of the National Park System, National Heritage Areas, National Wild and Scenic Rivers System, National Trails System, and affiliated areas. The latter includes congressionally designated properties with high historic or scenic value that are not federally owned or managed by the NPS. Examples of collaboration include (1) providing technical and financial assistance to locally initiated projects in the Chesapeake Bay Gateways Network;

(2) partnership management in the nonprofit Cuyahoga Valley Scenic Railroad in Ohio; (3) cooperative management of the Maurice National Wild and Scenic River in New Jersey; (4) the Chimney Rock National Historic Site, an NPS affiliated site, in western Nebraska that is one of the first landmarks along the Oregon Trail; (5) the Appalachian National Scenic Trail, a unit of the National Park System jointly managed with the U.S. Forest Service, state agencies, the Appalachian Trail Conference, and other hiking clubs; and (6) the Natchez Trace National Scenic Trail between the Tennessee and Mississippi Rivers and the Pacific Crest National Scenic Trail that goes from Mexico, through California, Oregon, and Washington, to Canada. NPS involvement in these areas is not without challenges. Yet they provide opportunities for shared investment and management among public and private organizations.

Sources: Conservation Study Institute (2001); *High Country News* (2001); Valles Caldera Trust (2003); Cuyahoga Valley Scenic Railway (2004).

federal land management policies in the Greater Yellowstone Ecosystem. Another example is the Crown Manager's Partnership, an informal voluntary affiliation of Glacier National Park in Montana, Waterton Lakes National Park in Alberta, Canada, and the Forest Service, whose purpose is "to build awareness of common interests and issues in the Crown of the Continent Ecosystem, build relationships and opportunities for collaboration across mandates and borders, and identify collaborative work already under way and opportunities for further cooperation" (Miistakis Institute for the Rockies 2003). Greater reliance on interagency cooperation would go a long way toward alleviating future external threats to protected area ecosystems.

Alleviating future external threats is unlikely unless protected area managers fully understand the potential adverse consequences of external threats and receive adequate funds for that purpose. Nongovernmental organizations have helped the general public to understand the nature and extent of external threats. For example, the National Parks Conservation Association's State of the Parks Program issued reports that examined the resource conditions and threats to selected park units (National Parks Conservation Association 2003).

Another way to alleviate external threats is to monitor whether future management actions are reducing external threats. This is best accomplished by comparing monitored attributes for a protected area ecosystem to performance standards for those attributes. For example, species-specific performance standards can be established for a suite of landscape metrics like the ones discussed in chapter 8. Future monitoring of attributes determines whether performance standards are violated. Violation of a standard means the ecosystem state is not desirable, which triggers corrective action.

The capacity of protected area managers to reduce external threats can be enhanced in other ways. Congress could enact a law that requires protected area managers to implement ecosystem management and authorize sufficient funds to address external threats. Gateway communities for protected areas can address external threats by implementing smart growth policies that allow economic development to occur in a manner that does not jeopardize the ecological integrity of protected areas. Federal departments, like the U.S. Department of the Interior and the U.S. Department of Agriculture, can cooperate to reduce external threats to protected areas from management of multiple-use lands. For example, the U.S. Forest Service and NPS can work together to ensure that timber sales and subsequent harvesting and thinning operations in national forests do not threaten national park values. Similarly, the Bureau of Land Management (BLM) and NPS can collaborate to ensure that oil and gas development on BLM land does not degrade the ecological integrity of parks and protected areas.

Improving Managerial Responses to Changing Ecosystem Conditions

Success in achieving desirable future ecosystem states necessitates effective managerial responses to changing demographic, cultural, ecological, and economic conditions, referred to collectively as ecosystem conditions. Managers can make more effective responses by developing expectations about how ecosystem states are likely to change in the future, implementing management actions designed to achieve desirable ecosystem states, and monitoring whether implemented management actions achieve desirable ecosystem states. This is easier said than done.

A classic study of management gone awry was done by Gunderson et al. (1995). They identified three common management responses and conditions based on 23 case studies of natural resource systems. First, managers took action to reduce perceived adverse socioeconomic impacts (particularly income and employment) of natural variability in an ecological variable, such as insect damage, fire hazard, or the decline in salmon populations. For example, an aggressive pesticide application program was initiated in New Brunswick, Canada, to control spruce budworm infestation. Second, as time progressed, resource managers shifted their focus from reducing natural variability to improving operational efficiency. Third, political pressure was exerted on managers to sustain the high flow of natural resources (e.g., volume of timber harvested or tons of fish landed) made possible by reducing natural variability. This trilogy of management responses and conditions had three negative consequences. Ecosystems became less resilient and more vulnerable, management agencies became more rigid and unresponsive, and societies became more dependent on the flow of natural resource commodities. In short, the ecosystem state becomes less desirable. Gunderson et al. claim that managerial pathologies in natural resource systems can be avoided by using an adaptive policy framework (essentially adaptive management) that employs a scientific perspective, embraces complex nonlinear systems, and incorporates an improved understanding of human behavior.

Other management responses influence whether desirable future ecosystem states are achieved. For example, consider managerial responses to increased visitation in a protected area. Responding effectively to higher visitation requires managers to anticipate growth in future visitation, which calls for an understanding of the determinants of visitation. Major determinants of visitation are the level and age distribution of the population, particularly that segment of the population most inclined to visit protected areas, and visitor preferences for protected areas. Future population is highly dependent on the replacement rate (average number of children per woman). According to Wilson (2002), the worldwide average replacement rate dropped from 4.3 in 1960 to 2.6 in 2000. A replacement rate of 2.1 is needed to achieve zero population growth. When the replacement rate exceeds 2.1, population grows at an exponential rate. Population declines at an exponential rate if the replacement rate falls below 2.1. Replacement rates in all western European countries, Thailand, and the United States (excluding immigrants) fell below 2.1 in 2000. If the replacement rate had dropped to 2.0 beginning in 2000, then world population would level off at approximately 7.3 billion in 2050. A more probable population estimate for the late twenty-first century is 9 to 10 billion people, which is 43 percent to 59 percent higher than the current (2003) world population of 6.3 billion.

World population growth over the next 50 years will significantly increase the number of visitors to protected areas. For example, suppose the visitation rate for U.S. national park units is the same in 2050 as it was in 2000, namely 70.75 visitors per 1,000 of world population. Based on this visitation rate and a world population of 9 billion in 2050, there could be 636.7 million visitors to U.S. national parks in 2050 compared to 429.8 million in 2000, a 48 percent increase (U.S. Census Bureau 2000; National Park Service 2001a). Visitation increases of this magnitude are likely to increase congestion, reduce public enjoyment, and degrade the natural and cultural resources of national parks. Not planning for future increases in visitation runs a high risk of degrading ecosystem conditions and reducing public enjoyment, which imply undesirable ecosystem states for national parks. Stated differently, "without proper management, public access can degrade park resources such that the very values that public access is intended to provide cannot be perpetuated" (National Park Service 1992).

To prevent future increases in visitation from having unacceptable impacts on public enjoyment and natural/cultural resources, protected area managers need to respond in an effective manner. Since the law requires national park units to be managed for the enjoyment of current and future generations, it is not socially acceptable to turn away visitors in order to protect natural and cultural resources. On the other hand, if visitor use and activities are not managed, then natural and cultural resources are likely to deteriorate resulting in an undesirable ecosystem state. Therefore, it may be necessary to limit the number of visitors and/or visitor activities in popular protected areas. The social and ecological carrying capacity method (Multiple Attribute Scoring Test for Capacity) described in chapter 7 can be used to assess the need for visitor restrictions and rank future management actions for complying with established social and ecological carrying capacities.

Balancing Resource Conservation and Visitor Satisfaction

Many protected area ecosystems, such as national parks in the United States, are mandated by law to balance resource conservation and visitor satisfaction (use and enjoyment). Undesirable ecosystem conditions signify imbalance between these two objectives. Achieving such balance would be especially difficult if visitation increased by 48 percent over the next 50 years as discussed in the previous section. Balancing resource conservation and visitor satisfaction is similar to the goal of improving managerial responses to changing ecosystem conditions because visitor use influences ecosystem conditions and vice versa. There is a difference, however. The need to balance resource conservation and visitor satisfaction would still be an issue even if the number of people visiting protected areas remained constant in the future. This is so because other factors influence visitor impacts on natural and cultural resources, namely, mode of travel, season of use, group size, visitor behavior, and type of recreational equipment.

Mode of travel influences visitor impacts on natural and cultural resources. Along with projected increases in visitation comes vehicle congestion, potentially lower visitor enjoyment, and negative resource impacts unless alternative transportation modes are developed. Possible alternative transportation modes include conventional bus, articulated bus, monorail, self-propelled rail transit, cog railway, and others. These modes are not equally suited to all protected areas. Yellowstone National Park, for example, evaluated alternative transportation modes in terms of accommodating expected visitation or ridership levels, reducing private vehicle traffic, lowering vehicle-related emissions, decreasing energy consumed by transportation, employing proven, available transportation systems, complying with Americans with Disabilities Act, allowing unrestricted movement of wildlife across and along transportation systems, controlling visitor encroachment into sensitive natural areas, and others (U.S. Department of the Interior 1994).

The temporal (seasonal) distribution of visitor use has changed dramatically in some national parks. For example, there has been tremendous growth in snowmobile use in the Greater Yellowstone Ecosystem and associated impacts on human health, wildlife, and other natural resources (see chapter 9 for more details). Greater winter use in protected areas is likely to harm natural resources and place greater demands on park staff. Resource impacts of group size can be controlled by restricting the number and size of campgrounds in frontcountry and backcountry areas and the number of people per campsite, restricting backcountry camping to designated campsites through a campsite reservation system, and limiting road access to ecologically sensitive areas of the park. Visitor behavior influences visitor safety and enjoyment as well as natural resources. In the early days following establishment, visitor behavior in Yellowstone National Park was so destructive (e.g., stuffing mattresses down Old Faithful and walking on geothermal features) that management of the park was turned over to the cavalry for a period of time. Other behavior like poaching, exceeding bag limits on fish, defacing public property, rock climbing in sacred places (like Devils Tower National Monument), starting forest fires, marring and cutting trees, and illegal camping need to be controlled for the future benefit of people and natural resources. Controls should not be too restrictive or else visitor satisfaction will decrease.

Recreational equipment influences visitor safety and impacts of visitors on natural and cultural resources. For example, before the advent of large recreational vehicles, it was not necessary to restrict the length and width of vehicles on Going-to-the-Sun Road in Glacier National Park. As recreational vehicles grew in size and number, safety concerns posed by the road's narrow width, sharp curves, and steep drop-offs led park managers to restrict vehicle length to less than 6.5 m (21 feet) and width to less than 2.5 m (8 feet) on the steepest sections of the road. Natural resource impacts of snowmobiling in Denali National Park and Preserve in Alaska have increased over time due to improved engineering designs that allow snowmobiles to be driven up steeper, more fragile slopes than in the past. The issue of whether or not mountain bikes should be allowed in wilderness areas was not an issue prior to 1981 (when they were first mass produced) (Hasenauer 2002). Improvements in clothing and outdoor equipment now allow visitors to penetrate regions of protected areas once considered too remote and unsafe for human use. Carefully crafted management actions will be required to reduce technology-driven impacts of human activity on natural and cultural resources.

Interactions between visitor use and management policies can have unintended consequences in the form of imbalances between resource conservation and visitor satisfaction. For example, in 1996 a Recreational Fee Demonstration Program was initiated in one hundred federally managed units in the United States (National Park Service 2001b). The program allows participating units to retain

80 percent of the revenue generated by admission fees. The added revenue is used to pay operating and maintenance costs. For instance, Olympic National Park retained $2 million in admission fees that were used to fund 25 park projects (National Park Service 2003a). Before the fee program, all admission fees were returned to the management agency's headquarters. The fee program potentially creates an incentive for participating units to maintain or even increase visitation (more visitors mean higher revenues), which can thwart resource conservation. This does not mean the program is unsound, but rather that it establishes a link between visitor use and revenue generation that could work against balancing resource conservation and visitor satisfaction.

Interactions between visitor use and natural disturbance (such as fire and disease) can reduce the capacity to balance resource conservation and visitor satisfaction. Consider wildfire on public lands. The natural regulation policy adopted by NPS in 1967 allowed wildfires to burn in backcountry areas as long as they do not threaten life and property. Negative public reaction to the 1988 fires in Yellowstone National Park, notably the perception that the fires were destroying the park, eventually led to a huge fire-fighting campaign to extinguish the fires. While wildfires are essential for proper functioning of ecosystems, they typically engender negative public reaction. Suppressing future wildfires to appease the public can have long-term negative consequences for ecological integrity and public welfare. An example of the latter is the long-term policy of suppressing fires in national forests. This policy contributed to dramatic increases in the frequency and extent of large, destructive fires and higher risk to people and properties in the wildland-urban interface. Based on the historical fire record and current directions in fire-management policy (e.g., the Healthy Forests Initiative), there is a high likelihood that future fire-management policy will call for more prescribed burning and aggressive thinning of trees and underbrush to reduce fire hazards to people and property. Both actions influence ecological integrity.

Method for Evaluating Trade-offs

The remainder of this section describes a method for evaluating whether future management actions result in ecosystem states that provide efficient, sustainable, and preferred combinations of resource conservation (C) and visitor satisfaction (S). The method is based on concepts discussed in chapters 4 and 7 (see figs. 4.8 through 4.11, and figs. 7.6 and 7.7). A hypothetical example is used to illustrate the method. It assumes managers and stakeholders have identified four future management actions for a protected area ecosystem and have estimated the expected levels of C and S with the four actions. Figure 11.1 illustrates the four hypothetical combinations of C and S provided by those actions. Specifically, future management actions a_1, a_2, a_3, and a_4 are expected to result in ecosystem states that provide combinations of C and S given by b_1, b_2, b_3, and b_4, respectively.

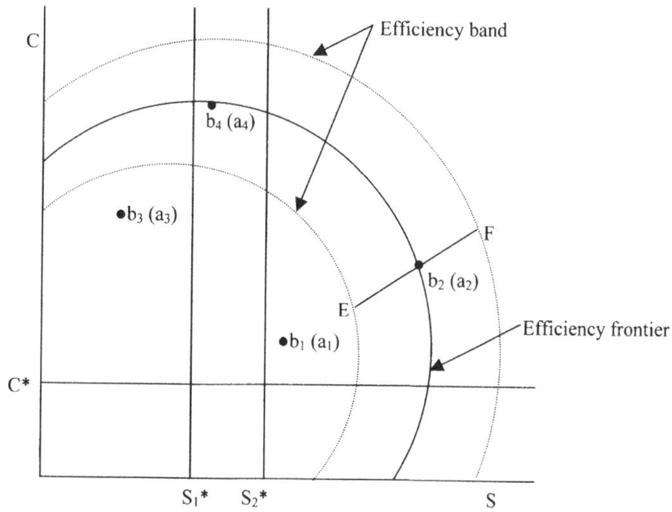

FIG. 11.1. Combinations of resource conservation (C) and visitor satisfaction (S) indicated by b_1, b_2, b_3, and b_4, achieved by four future management actions indicated by a_1, a_2, a_3, and a_4, efficiency frontier and efficiency band.

The first task for the protected area manager is to weed out those management actions that provide inefficient combinations of C and S. Inefficient management actions provide a combination of C and S that is dominated by at least one other management action. In figure 11.1, b_1 is dominated by b_2 because b_2 provides more of both C and S than b_1. Similarly, b_3 is dominated by b_4 because b_4 provides more of both C and S than b_3. This implies a_2 and a_4 are efficient and a_1 and a_3 are inefficient management actions. If preferences for C and S are conventional (indifference curves have the conventional shape illustrated in fig. 4.10), then b_2 and b_4 lie on higher indifference curves than b_1 and b_3. Therefore, management actions a_2 and a_4 and corresponding C-S combinations b_2 and b_4 are preferred to management actions a_1 and a_3 and corresponding C-S combinations b_1 and b_3. In summary, only management actions a_2 and a_4 are acceptable based on efficiency criteria and preferences.

Since b_4 involves more C and less S than b_2, it is not possible to say b_4 is preferred to b_2 or b_2 is preferred to b_4 without additional information about stakeholders' preferences for C and S. The inability to establish a preference ordering between b_2 and b_4 occurs because the two combinations involve trade-offs between C and S. A trade-off exists between two management actions when one provides more C but less S than the other. When there is a large number of efficient management actions, the combinations of C and S provided by those actions define a continuous efficiency frontier like the one illustrated in figure 11.1. Management actions that provide combinations of C and S below the efficiency frontier are inefficient and management actions that provide combinations of C and S above the efficiency frontier are not feasible.

The next step for the protected area manager is to determine whether management actions on the efficiency frontier are sustainable. This can be done by establishing sustainability thresholds like the ones described in chapter 7 and illustrated in figure 7.6. Suppose the protected area manager and stakeholders determine that levels of resource conservation below C* do not maintain ecological functions and processes (values of C less than C* are not ecologically conserving). Similarly, suppose levels of visitor satisfaction below S_1* are judged to be socially and/or economically unacceptable. Therefore, only management actions that provide combinations of C and S along the efficiency frontier between C* and S_1* are sustainable. Based on these sustainability thresholds, management actions a_2 and a_4 are sustainable and management action a_3 is unsustainable. Of course, a_3 has already been ruled out for being inefficient. Higher values of C* and/or S* can alter the set of sustainable management actions. For example, if the sustainability threshold for visitor satisfaction is increased from S_1* to S_2*, then C-S combination b_4 and associated management action a_4 are unsustainable.

Ecosystem states resulting from future management actions also depend on the nature and extent of natural disturbances, such as fire, disease, and climate change. Since a protected area manager does not know when and where future natural disturbances will occur and their resource impacts, it is more realistic for the plausible efficient and sustainable values of C and S for a particular management action to lie in some range. The range can be determined using a Delphi method in which experts reach agreement about the possible values of C and S for each future management action being considered. Suppose a Delphi method indicates that management action a_2 is likely to generate a range of values for C and S that lie along line EF in figure 11.1. When efficient management actions provide ranges of values for C and S, the efficient combinations of C and S are defined by an efficiency band like the one shown in figure 11.1. Whereas the efficiency frontier implies that each management action results in a unique combination of C and S, the efficiency band implies that values of C and S resulting from a given management action fall in a range; the latter is more realistic. Management actions that provide C-S combinations within the efficiency band are efficient relative to management actions that provide C-S combinations below the band.

Once the efficient and sustainable future management actions and associated C-S combinations have been identified, it is possible to identify the most preferred desirable ecosystem state for the future. Based on the discussion in chapter 4 (see fig. 4.11), differences in stakeholders' preferences for C and S are likely to result in different preferred management actions. This is illustrated in figure 11.2, which shows hypothetical indifference curves for C and S for two stakeholders, I

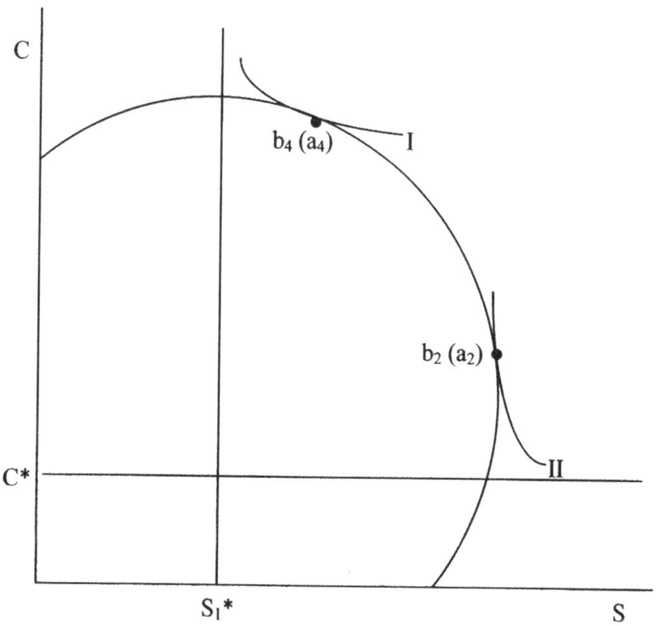

FIG. 11.2. Preferred combinations of resource conservation (C) and visitor satisfaction (S) (indicated by b_2 and b_4) and associated management actions (indicated by a_2 and a_4) for two hypothetical stakeholders, I and II.

and II. The indifference curves represent stakeholders' future preferences for C and S. A stakeholder is indifferent between all combinations of C and S on the same indifference curve and prefers combinations of C and S on a higher indifference curve to combinations on a lower indifference curve. The highest indifference curve stakeholder I can attain is labeled I; it indicates a stronger preference for C than S. In contrast, the highest indifference curve stakeholder II can attain is labeled II; it indicates a stronger preference for S than C. Therefore, a_4 and b_4 are the most preferred management action and C-S combination, respectively, for stakeholder I and a_2 and b_2 are the most preferred management action and C-S combination, respectively, for stakeholder II.

The following three-step procedure can be used to select the most preferred future management action. First, the protected area manager and stakeholders form a negotiation group to rank future management actions based on the values of C and S. Stakeholders' preferences for C and S are summarized by the weights they place on C and S. Second, group weights for C and S are derived by taking a weighted average of all stakeholders' weights for C and S where weights indicate the relative importance of stakeholders to the manager (Pavlikakis and Tsihrintzis

2003). For example, if stakeholder I's weights are 0.8 for C and 0.2 for S, stakeholder II's weights are 0.2 for C and 0.8 for S, and the manager attaches a weight of 0.6 to stakeholder I and 0.4 to stakeholder II, then the group weight for C is (0.6)(0.8) + (0.4)(0.2) = 0.56 and the group weight for S is (0.6)(0.2) + (0.4)(0.8) = 0.44. Steps 1 and 2 make up the negotiation group process.

Third, efficient and sustainable management actions for balancing C and S are scored based on the following utility function:

$$U(a) = w_C(C')^{1/2} + w_S(S')^{1/2}$$

This function states the satisfaction provided by future management action "a" is the weighted sum of the square roots of the values of C and S. The w_C and w_S are the group weights for C and S, respectively, where $0 \leq w_C \leq 1$, $0 \leq w_S \leq 1$, and $w_C + w_S = 1$. C' and S' are normalized values of C and S, respectively. Normalization avoids biases in the scores for management actions that can arise when C and S are measured in different units. Taking the square root of the normalized values of C and S implies decreasing marginal utility for C and S, which means that additional units of both C and S provide decreasing increments in utility. An additive utility function assumes that C and S are independent in the sense that the added (marginal) utility provided by another unit of C is independent of the amount of S, and the added (marginal) utility provided by another unit of S is independent of the amount of C. It also assumes stakeholders are risk neutral (see Prato 1999 for a discussion of risk preferences).

The most preferred future management action for the negotiation group in terms of balancing C and S is the one with the highest score. Other procedures beside negotiation groups can be used to resolve stakeholder differences in preferences for C and S. They include applying consensus-based decision making to the stakeholders' rankings of management actions, and allowing stakeholders to vote directly for their most preferred management action (Hafkamp and Nijkamp 1993; Bentham and Greer 1998).

Enhancing Planning and Management Capacity

Continued pressure from external threats, the likelihood of ineffective managerial responses to changing ecosystem conditions, conflicts between resource conservation and visitor satisfaction, inadequate funding of protected areas, and other challenges make it difficult to do future planning and management for protected areas. Fortunately, future innovations in scientific knowledge, management techniques, coverage and accuracy of spatial data, spatial information technologies, and computer hardware and software have the potential to enhance the capacity to plan for and manage protected areas.

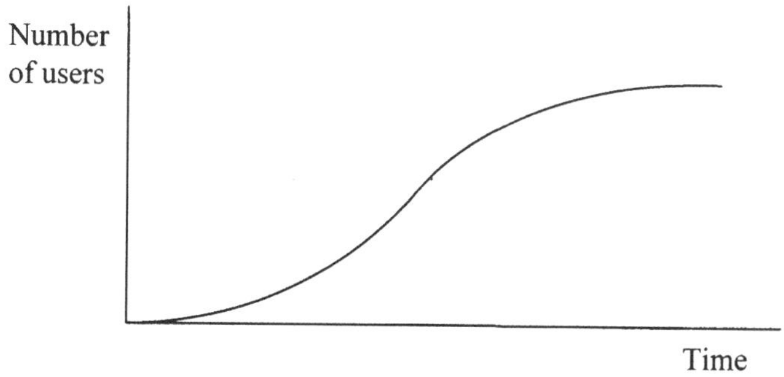

FIG. 11.3. Adoption-diffusion curve.

An innovation is a new method, idea, or device. New is relative. An innovation that has been around for many years is still considered new to managers who neither know about it nor have assessed its potential benefits in protected area management. Adoption and diffusion theory indicates that an innovation is first implemented by so-called early adopters. In the early adoption phase, there is exponential growth in adoption as knowledge of the innovation and its benefits spread to other potential users. Eventually the rate of adoption begins to decline and use reaches a saturation level. Figure 11.3 illustrates a classical adoption-diffusion curve. It shows that use of an innovation increases at an increasing rate (exponential growth) and then increases at a decreasing rate until a saturation level is reached after which use remains relatively constant. The saturation phase can be followed by a contraction phase (not shown in fig. 11.3) in which individuals abandon the original innovation (use decreases) in favor of another innovation.

Key determinants of the rate of adoption of an innovation are (1) awareness of the innovation, (2) knowledge of its benefits and costs, (3) willingness to adopt the innovation, and (4) nature of previous experiences with innovations. It is not possible for a protected area manager to adopt an innovation he or she is not aware of. Awareness of certain innovations, such as ecosystem management, natural regulation, and geographic information systems, is much greater than for less well known innovations, such as using multiple attribute decision making to evaluate and rank management actions or Bayes' rule to evaluate whether a protected area ecosystem is sustainable. Willingness to adopt an innovation depends on a comparison of expected benefits and expected costs, budget constraints, and the manager's attitudes toward new concepts and ideas. Even if expected benefits exceed expected costs, restrictive budgets, unfavorable attitudes toward new methods, ideas, and devices, and negative experiences with previous innovations work against adoption. If an innovation is adopted and falls short of the adopter's expectations, then it is likely to be abandoned. These determinants of adoption assume a manager has the freedom to adopt innovations apart from budget constraints. Adoption is often constrained by institutional culture. An example of the latter is when upper management dictates agencywide use of a new personnel and financial accounting system or brand of geographic information system.

The remainder of this section evaluates the potential benefits and costs of adopting innovations (ideas and methods) relevant to protected area planning and management. Examples of innovations in the form of ideas are ecosystem management (particularly ecological functioning, ecological resilience, and ecosystem dynamics), ecological economics, adaptive management, natural regulation, carrying capacity, and regional scale conservation (covered in chapter 7). Examples of innovations in the form of methods are benefit-cost analysis, cost effectiveness analysis, economic impact analysis, landscape change detection and analysis, multiple attribute decision making, Bayesian approaches to handling of uncertainty, optimization and simulation, integrated ecological economic modeling, and spatial analytical techniques, such as geographic information systems, remote sensing, global positioning systems, interactive mapping, and computerized landscape visualization (covered in chapter 8).

Innovative ideas and methods have a wide range of potential benefits in terms of enhancing planning and management capacity for protected areas. Examples of innovative ideas and methods are given in table 11.1. Consider how to determine expected benefits and costs of an innovation that uses ecological economic models to eliminate inefficient and unsustainable management actions for balancing resource conservation (C) and visitor satisfaction (S) as discussed in the previous section. If the C and S attributes of future management actions are not quantified, then it is possible for a manager to choose an inefficient and unsustainable management action. This possibility is illustrated with reference to figure 11.1. Suppose attributes C and S have not been quantified and this lack of information makes the manager assign equal probabilities to management actions. Since there are four management actions (a_1, a_2, a_3, and a_4), the probability of each management action being selected is $^1/_4$. This implies a probability of 0.5 that an inefficient management action (a_1 or a_3) is chosen and a probability of 0.5 that an efficient management action (a_2 or a_4) is chosen. Expected loss in utility from not using ecological economic models to quantify the C and S attributes of management actions and not eliminating inefficient and unsustainable management actions based on those attributes is $^1/_4$ [$EL(a_1) + EL(a_3)$]. $EL(a_1)$ is the expected loss in utility from choosing a_1 and $EL(a_3)$ is the expected loss in utility from choosing a_3. There is no loss in utility from choosing a_2 and a_4 because they are efficient. Values of $EL(a_1)$ and $EL(a_3)$ depend on whether a_2 or a_4 is optimal, which cannot be determined without further information about the manager's preferences for C and S. If the sustainability threshold for visitor satisfaction is S_2* instead of S^*_1, then the expected loss in utility is even higher be-

Table 11.1. Examples of Innovations Having Potential Benefits for Protected Areas

Innovation	Benefit
Ideas	
Ecological economic concepts are used to eliminate inefficient and unsustainable management actions.	More efficient and sustainable ecosystems
Fire and other natural disturbances are managed based on natural regulation.	Improved ecological functioning
Adaptive management approach is used to evaluate and select management actions.	Uncertainty regarding impacts of management actions on a protected area ecosystem taken into account
Landscape ecology is used to understand how landscape patterns influence ecological connectivity.	Continental scale conservation of biodiversity
Methods	
A GIS is used to display and evaluate data from inventorying and monitoring programs.	Reduced cost and/or improved quality of planning and management activities
A GPS collar is used to monitor movements of endangered species.	Greater success in recovery of endangered species
Multiple criteria evaluation is used to rank proposed management actions for different stakeholder groups.	Increased ability to account for community-based values in decision making
Bayesian statistics is used to determine compliance of an ecosystem with human and ecological carrying capacities.	Better assessments of carrying capacity

cause the probability the manager would choose an inefficient and unsustainable management action increases from $^1/_2$ to $^3/_4$. The expected benefit of using ecological economic models to eliminate inefficient and unsustainable management actions equals avoidance of the expected losses indicated above.

The expected cost of using ecological economic models to eliminate inefficient and unsustainable management actions equals the cost of (1) quantifying the attributes of alternative management actions; (2) determining the efficiency frontier, which is low once the C-S values of management actions have been determined; and (3) reaching consensus for the sustainability thresholds (C^* and S^*), which is likely to be high. In general, expected benefits increase relative to expected costs as the percentage of management actions that are inefficient and unsustainable increases.

Integrated Assessment of Ecosystem States

The preceding sections discussed four goals for evaluating desirable future states for protected area ecosystems. Goals can be complementary or competitive over their entire range of values, or complementary over one range and competitive over another range. Two goals are complementary when increasing the value of one goal causes the value of the other goal to increase. For example, improving managerial responses to changing ecosystem conditions can reduce external threats. Taking advantage of new innovations in planning and management can help to balance resource conservation and visitor satisfaction. Two goals are competitive when increasing the value of one goal causes the value of the other goal to decrease. Goals are likely to become competitive as the operating budget for a protected area becomes more limiting. For example, a protected area manager may delay efforts to alleviate external threats in order to allocate more operating budget to balancing resource conservation and visitor satisfaction. As a general rule, good planning and management become more important as budgets tighten.

Regardless of the relationships between the four goals (complementary and/or competitive), evaluating the desirability of future ecosystem states based on these goals requires their integration. Managers can integrate the goals using a three-step procedure. First, expected values of the four goals corresponding to alternative future management actions are determined for a protected area ecosystem. Second, future ecosystem states are classified as either desirable or undesirable based on the extent to which they achieve the goals. Third, the goals achieved by implemented management actions are monitored to determine whether they result in a desirable or undesirable ecosystem state. The remainder of this section describes implementation of the three steps.

Expected values of the four goals for alternative future management actions are determined based on the impacts of past management actions and biophysical and socioeconomic simulation models. For example, if the gateway communities for a protected area are expected to experience significant economic development in the future, then reducing external threats to that protected area requires that growth be directed into areas that are not critical habitats for vulnerable plant and animal species. Possible future impacts of alternative development policies on ecological integrity and economic development can be simulated using an integrated ecological economic model as discussed in chapter 8.

The second step classifies ecosystem states as desirable or undesirable based on the expected attainment of the four goals. At one extreme is a protected area ecosystem for which external threats are severe, managerial responses to changing ecosystem conditions are ineffective, there is imbalance between resource conservation and visitor satisfaction, and the capacity for planning and management is limited. An ecosystem state with these goal values is highly undesirable. At the other extreme is a protected area ecosystem for which external threats are in check, managerial responses to changing ecosystem conditions are effective, balance is achieved between resource conservation and visitor satisfaction, and there is significant capacity for planning and management. An ecosystem state with these goal values is highly desirable. Since most ecosystem states are likely to generate goal values between these two extremes, it will be necessary to decide which ecosystem states are desirable or undesirable based on the goals. This type of decision is ideally suited to application of Bayes' rule that was described in chapter 8. Bayes' rule requires calculating posterior probabilities for ecosystem states and using them to determine the most likely state of the protected area ecosystem.

The third step monitors implemented management actions to determine the extent to which those actions produce desirable ecosystem states. If future goal values indicate a desirable ecosystem state, then there is no need to change the management action. Conversely, if future goal values imply an undesirable ecosystem state, then there are grounds to change the management action to achieve a desirable ecosystem state.

Summary

Sustaining the flow of ecological economic goods and services provided by protected areas requires managers and stakeholders to plan and manage for the future. Most future planning and management for protected areas occurs in developing general management plans. The future planning and management framework proposed here defines desirable or sustainable states for protected area ecosystems in terms of four broad goals: (1) reducing external threats, (2) improving managerial responses to changing ecosystem conditions, (3) balancing resource conservation and visitor satisfaction, and (4) enhancing planning and management capacity. The framework can be used with other goals as well.

External threats can adversely influence the ecological integrity of protected area ecosystems. The importance of external threats was recognized by the National Park Service's Vail Agenda and Natural Resource Challenge. In order for managers to respond effectively to future external threats, they must understand the nature of those threats, design management actions to alleviate them, and monitor the effectiveness of management actions in reducing threats. Developing cooperative relationships with stakeholders is helpful in reducing future external threats.

Achieving desirable future ecosystem states requires managers to respond effectively to changing ecosystem conditions. This entails developing expectations about how ecosystem states are likely to change in the future, designing and implementing management actions to achieve desirable ecosystem states, and monitoring whether implemented management actions achieve desirable ecosystem states. Research indicates that it is common for natural resource managers to respond in ways that cause ecosystems to become less resilient and more vulnerable, management agencies to become more rigid and unresponsive, and societies to become more dependent on extraction of natural resources.

Accommodating future growth in protected area visitation without jeopardizing visitor satisfaction and the quality of natural and cultural resources is especially difficult. Factors influencing the extent to which increased visitation damages natural and cultural resources are the mode of travel, season of use, group size, visitor behavior, and type of recreational equipment. Interactions between visitor use and management policies, as exemplified by the Recreational Fee Demonstration Program, and between visitor use and natural disturbance, as evidenced by fire-management policy, can reduce the capacity of protected area managers to balance resource conservation and visitor satisfaction. Future management actions that provide efficient combinations of resource conservation and visitor satisfaction can be determined from the efficiency frontier. Management actions that provide unsustainable combinations of resource conservation and visitor satisfaction are identified by using sustainability thresholds. Identifying the most preferred management action requires information about the manager's and stakeholders' preferences for resource conservation and visitor satisfaction.

External threats, a high likelihood of ineffective managerial responses to changing ecosystem conditions, conflicts between resource conservation and visitor satisfaction, chronically low budgets for protected areas, and other challenges limit achievement of the four goals. Fortunately, innovations in scientific knowledge, management techniques, coverage and accuracy of spatial data, spatial information technologies, and computer hardware and software should enhance the capacity to plan for and manage protected areas.

The four goals for future planning and management of protected areas can be integrated using a three-step procedure that requires protected area managers and stakeholders to (1) determine expected values of the four goals corresponding to alternative future management actions, (2) classify future ecosystem states as either desirable or undesirable based on the goals, and (3) monitor future management actions to determine whether they result in a desirable or undesirable ecosystem state.

References

Bentham, M. J., and J. E. Greer. 1998. Conflict resolution among heterogeneous reusable agents in the PARA DSS. In *Multiple objective decision making for land, water and environmental management,* ed. S. A. El-Swaify and D. S. Yakowitz, 299–311. Delray Beach, FL: St. Lucie Press.

Conservation Study Institute. 2001. *Collaboration and conservation: Lessons learned in areas managed through national park service partnerships. A report on a workshop.* Marsh-Billings-Rockefeller National Historic Park, Woodstock, VT.

Dilsaver, L. M. (Ed.). 1994. *America's national park system: The critical documents.* Lanham, MD: Rowman & Littlefield Publishers.

EarthVision Environmental News. 2003. *State parks threatened says report.* http://www.earthvision.net/ColdFusion/News_Page1.cfm?NewsID=12052 (accessed May 1, 2003).

Gunderson, L. H., C. S. Holling, and S. S. Light (Eds). 1995. *Barriers and bridges to the renewal of ecosystems and institutions.* New York: Columbia University Press.

Hafkamp, W., and P. Nijkamp. 1993. Conflict analysis and compromise strategies in integrated spatial systems. *Regional Science and Urban Economics* 13: 115–140.

Hasenauer, J. 2002. Mountain biking in wilderness: A niche for bicycles. *Wild Earth* 13: 21–22.

High Country News. 2001. Western roundup, cooperating on the Valles Caldera. 33: 5.

Keiter, R. B. 1996. Ecosystem management: Exploring the legal-political framework (chap. 5). In *National parks and protected areas: Their role in environmental protection,* ed. R. G. Wright, 63–88. Cambridge, MA: Blackwell Science.

Miistakis Institute for the Rockies. 2003. *Crown of the continent managers partnership.* http://www.rockies.ca/programs/program1a.htm (accessed October 31, 2003).

National Parks Conservation Association. 2003. *Across the nation: State of the parks program.* http://www.npca.org/across_the_nation/park_pulse/default.asp (accessed May 1, 2003).

National Park Service. 1992. *National parks for the 21st century, the Vail Agenda.* Post Mills, VT: Chelsea Green Publishing Co.

———. 1999. *Natural resource challenge: The Park Service's action plan for preserving natural resources.* http://www.nature.nps.gov/challengedoc/index.htm (accessed May 6, 2003).

———. 2001a. *Public Use Statistics Office.* (http://www2.nature.nps.gov/stats/) (accessed April 2, 2001).

———. 2001b. *Recreational Fee Demonstration Program fact sheet.* http://www.nps.gov/feedemo/facts.htm (accessed December 3, 2001).

———. 2003a. *Olympic News. Fee demonstration program.* http://www.nps.gov/olym/feedemo.htm (accessed May 8, 2003).

———. 2003b. *Record of decision: General management plan, abbreviated final environmental impact statement, Mojave National Preserve, California.* http://www.nps.gov/moja/rod&fp.htm (accessed June 10, 2003).

———. 2004. *2001 NPS management policies,* Chapter 2: Park System planning. Office of Planning. http://www.nps.gov/policy/mp/chapter2.htm (accessed April 11, 2004).

Pavlikakis, G. E., and V. A. Tsihrintzis. 2003. Integrating humans in ecosystem management using multiple-criteria decision making. *Journal of the American Water Resources Association* 39: 277–288.

Prato, T. 1999. Risk-based multiattribute decision making in property and watershed management. *Natural Resource Modeling* 12: 307–334.

U.S. Census Bureau. 2000. *Population Projections Program, Population Division, Washington, DC.* http://www.census.gov/population/www/projections/natsum-T1.htmln (accessed May 6, 2003).

U.S. Department of the Interior. 1994. Alternative transportation modes feasibility study, vol. III: Yellowstone National Park, Wyoming. National Park Service, Denver Service Center.

Valles Caldera Trust. 2003. *Valles Caldera National Preserve.* http://www.vallescaldera.gov/about.php (accessed April 8, 2003).

Wilson, E. O. 2002. *The future of life.* New York: Alfred A. Knopf

CHAPTER

12

Synthesis

Integration of diverse perspectives and approaches...is one of the most difficult challenges confronting the social, natural, and physical sciences and is at the forefront of the scientific endeavor.

—R. A. Haeuber and V. H. Dale

The preceding eleven chapters describe and explain the status, importance, planning, and management of national parks and protected areas with particular reference to North America. In this final chapter, major contributions of the book are synthesized into ten major themes. Each theme is briefly discussed. Because themes are highly interdependent, the order in which they are presented is no indication of their relative importance.

Theme 1

Protected areas provide important social, cultural, economic, and ecological values that are worth preserving.

Henry David Thoreau said that "in wilderness is the preservation of the world." John Muir echoed Thoreau's sentiment in writing, "Thousands of tired, nerve-shaken people are beginning to find out that going to the mountains is going home; that wilderness is a necessity." Whether or not national parks and protected areas contain wilderness, they provide outstanding social values in the form of solitude, tranquillity, inspiration, reflection, recreation, spiritual and physical renewal, adventure, and education. The social values of protected areas are likely to increase in importance and significance as society becomes more urbanized, materialistic, competitive, and estranged from nature.

Human culture and protected area ecosystems are coevolutionary and interdependent (see chapter 3). Culture influences the meaning and significance of protected areas, and the way protected areas are established and managed influences the long-term well-being of local culture. The coevolutionary and interdependent relationship between human culture and protected areas exists because many protected areas contain sacred places and natural resources that are vital to the subsistence of local cultures. Establishing and managing protected areas in a manner that preserves deep-seated cultural and spiritual values of native peoples contribute to social cohesion, vitality, and cultural diversity. Protected area managers need to consider the implications of their actions on cultural resources by involving culturally impacted people in planning and management, and encouraging the use of protected areas by people from a broad spectrum of cultures and ethnic groups.

Total economic value of protected areas, which is the standard measure of economic worth, is the sum of direct use, indirect use, option, quasi-option, existence, and bequest values (see chapter 4). Direct-use value of protected areas is derived from direct consumption of ecological goods (timber, petroleum, and water), or direct use of ecological services (hiking, camping, and nature viewing). Indirect-use value of protected areas comes from ecological services that are not directly consumed, such as conservation of biodiversity, purification of air and water, and watershed protection. Ecological goods have market prices, whereas most ecological services do not have market prices because they are not traded in the marketplace. Nonmarketed, nonuse values of ecological goods and services are typically estimated by what people are willing to pay for them. Demand and supply of ecological goods and services are influenced by consumer income, tastes, and preferences, market prices, population, and technology, all of which vary over time and space. Option, quasi-option, bequest, and existence values are nonuse values. While the total economic value of national parks and protected areas has not been estimated, it is likely to be high based on studies of the value of nonmarketed ecological goods and services. Protected areas also provide intrinsic, spiritual, and ethical values.

Social value equals total economic value minus total social cost. The latter equals the private cost plus external cost of producing a good or service. External cost is positive (a disbenefit) when there are negative externalities and negative (a benefit) when there are positive externalities. While maximizing social value results in an efficient allocation of resources, goods, and services, the resulting allocation is not necessarily fair (equitable) or ecologically sustainable. Protected area managers should attempt to make decisions in a manner that balances social efficiency, distributional fairness, and ecological sustainability.

Theme 2

Management actions and policies need to be tailored to site-specific conditions in protected areas.

A protected area is an "area of land and/or sea especially dedicated to the protection and maintenance of biological diversity, and of natural and associated cultural resources, and managed through legal or other effective means" (IUCN 1994).

Uniqueness of protected areas and other landscapes is buttressed by landscape ecology, which maintains that "each site or region [within a landscape] has a unique set of organisms and abiotic conditions influencing and constraining ecological processes" (Turner et al. 2001). Protected areas are located in a variety of places and serve a variety of purposes. There are 30,350 protected areas covering 13,236 km^2 (34,281 mi^2) as of 1997, or 8.8 percent of the Earth's land area. National parks account for 30 percent of the total land area in protected status. A protected area falls in a particular IUCN category and is managed based on the multiple objectives for that category and the statute (organic act) or executive order that created the area.

IUCN has established seven categories and nine management objectives for protected areas. The seven protected area categories are strict nature reserve (Ia), wilderness area (Ib), national park (II), natural monument (III), habitat/species management area (IV), protected landscape/seascape (V), and managed resource protected area (VI) (see chapter 2 and fig. 2.1). IUCN categories are useful because they provide international consistency in the definition of protected areas. The nine IUCN management objectives are scientific research, wilderness protection, preservation of species and genetic diversity, maintenance of environmental services, protection of specific natural and cultural features, tourism and recreation, education, sustainable use of resources from natural ecosystems, and maintenance of cultural and traditional attributes. Protected areas in each category are managed for between four and six objectives that are designated as primary, secondary, potentially applicable, or not applicable (see table 2.3). Another important determinant of the long-term management of protected areas is the General Management Plan. These plans "ensure that each park has a clearly defined direction for resource preservation and visitor use" (National Park Service 1998).

Management of protected areas varies from one location to another and over time even for protected areas in the same IUCN category and country. This variability results from numerous factors including national, regional, and local differences in the relative importance of management objectives to protected area managers and stakeholders; biophysical, sociocultural, and economic characteristics of the protected area and ecosystem in which it is located; and the protected area manager's commitment and capacity to manage for particular objectives (see theme 3). Due to the unique social, cultural, economic, and ecological setting for each protected area, management actions and policies that work in one protected area may not work in another protected area. In other words, management actions and policies are unique to each protected area.

Theme 3

Several institutions influence the ability of planners and managers to achieve the purposes of protected areas, and the performance of protected area agencies and organizations.

Formal and informal institutions determine the "rules of the game" that shape how people and organizations behave, interact, and perform (see chapter 5). Examples of formal institutions are the organic acts establishing federal land management agencies, enabling legislation for individual protected areas, environmental laws, regulations, executive orders, and court decrees. Norms, social conventions, mores, beliefs, etiquette, traditions, and taboos are informal institutions that govern expectations regarding individual and social behaviors. Institutions influence how people and organizations behave by creating incentives or disincentives that affect the reward structure.

There are numerous protected area institutions. Consider four examples. First, the IUCN has established protected categories and associated management objectives that guide the designation and management of protected areas worldwide (see theme 2). Second, the World Commission on Protected Areas contributes to the IUCN mission by promoting the establishment and effective management of a worldwide representative network of terrestrial and marine protected areas. Third, the National Park Service Organic Act requires the National Park Service to "conserve the scenery and the natural and historic objects and the wildlife therein and to provide for the enjoyment of the same in such manner and by such means as will leave them unimpaired for the enjoyment of future generations." Fourth, several U.S. laws, particularly the National Environmental Policy Act, Endangered Species Act, Clean Air Act, and Clean Water Act, strongly influence planning and management for protected areas.

Institutions influence the capacity of protected area managers to balance the multiple values of protected areas. Consider the consequences of different land uses along the boundary between Targhee National Forest and Yellowstone National Park. The forest side of the boundary is managed according to the Multiple Use-Sustained Yield Act of 1960, which requires national forests to be managed for multiple objectives including "outdoor recreation, range, timber, watershed and wildlife and fish and a sustained yield of products and services." The Targhee National Forest side is essentially a clear-cut. Yellowstone National Park was created by the Yellowstone Park Act, which states that the park be "reserved and withdrawn from settlement, occupancy, or sale under the laws of the U.S., and dedicated and set apart as a public park or pleasuring-ground for the benefit and enjoyment of the people." It also instructs the secretary of the interior "to make and publish such rules and regulations as he may deem necessary or proper for the care and management of the [park, and that]...[s]uch regulations shall provide for the preservation, from injury or spoilation, of all timber, mineral deposits, natural curiosities, or wonders within said park, and their retention in their natural condition" (Haines 1974). The park side of the boundary is an uncut forest. Having a clear-cut adjacent to Yellowstone National Park degrades wildlife habitat and makes it more difficult to maintain the park's ecological integrity.

The theory of institutions states that decision making and organizational performance are strongly influenced by the incentives and disincentives facing individuals and organizations (see chapter 5). A national park that is in the Recreational Fee Demonstration Program is allowed to retain 80 percent of the gate receipts. This program creates an incentive to maintain and increase visitation even though it might conflict with resource protection goals. Institutional change occurs when new leaders take control and/or organizational performance is unacceptable. The high incidence of poaching in a wildlife reserve (poor performance) can be reduced by passing and enforcing a law (creating a formal institution) that makes it illegal to export wildlife and their parts. The Convention on International Trade in Endangered Species of Wild Fauna and Flora is a prime example of an institution that controls international trade in endangered species. Institutional change and improved organizational performance are dampened when individuals and organizations resist change (institutional rigidity).

New institutional mechanisms have been proposed to improve the capacity to manage public lands in a manner that supports the public trust doctrine and captures the economic efficiency benefits of market-based incentives. New resource economics (NRE) argues that mismanagement of public land can be decreased by establishing well-defined property rights, institutions, and private incentives. The latter could improve individual and organizational performance in public land management by linking the self-interest of managers with the desires of stakeholders.

Theme 4

Protected areas should be managed according to the principles of ecosystem management.

Since protected landscapes are embedded in greater ecosystems, they should be managed using an ecosystem management approach (see chapter 7). Ecosystem management entails long timeframes, large geographic areas, and integration of multiple social, cultural, economic, and ecological values. Its popularity stems from the fact that "many resource managers are shifting their management from an approach that focuses on specific resources, such as fish, wildlife, and water, to one focused on the integrity of entire systems" (Turner et al. 2001). Other features of ecosystem management are accounting for competition between multiple values of ecosystems, considering stakeholders' preferences for ecosystem values, and handling uncertainty regarding the outcomes of management actions. Each element of ecosystem management is briefly discussed.

Managing protected areas over long temporal scales requires assessing the long-term (intergenerational) consequences of management actions. Managing over large spatial scales necessitates monitoring and evaluation of conditions within the protected area and the greater ecosystem in which it is located. Multiple attribute decision making is a useful tool for ecosystem management (see chapter 8). While ecosystem management is more difficult to implement than, say, identifying native habitat at risk from rural residential development, it is essential for evaluating the impacts of management actions on diverse stakeholders.

Balancing the multiple and often competing values of protected areas involves assessing the impacts of management actions on a wide range of social, cultural, economic, and ecological values. Managing for multiple values is more difficult than managing for single values because it requires integrating values that are measured in noncommensurate units. For example, a new park visitor center might increase visitation, as measured by the number of visitor days, increase park revenues, as measured in monetary terms, and decrease biodiversity, as measured by species richness. Multiple attribute evaluation facilitates integration of protected area values measured in noncommensurate units (see chapter 8). Management values often compete with one another. Competition implies that an increase in one value comes at the expense of a decrease in another value. For example, the new visitor center could increase social, cultural, and economic values but decrease ecological integrity.

Matters are further complicated when, as commonly occurs, managers and stakeholders disagree about the mix of values that should be provided by a protected area. Building contractors and trades are likely to be concerned with the positive economic benefits (higher income and employment) and environmental groups with the potential negative ecological impacts of the visitor center. When the weights stakeholders place on the multiple values of a protected area are sufficiently different to cause the rank ordering of management actions to vary across stakeholders, then preferences for management actions will vary across stakeholders. This can give rise to conflicts among stakeholders that need to be resolved using conflict resolution methods (see chapter 8).

Theme 5

Adaptive management allows protected area managers to embrace uncertainty regarding the outcomes of management actions.

Traditional approaches to natural resource management implicitly assume that managers understand how ecosystems respond to management actions prior to their implementation. In particular, managers are assumed to know with certainty the social, economic, and ecological consequences of management actions. This is not the case because variability in environmental, demographic, and other factors makes ecosystem responses to management actions uncertain, sampling and measurement errors make it difficult to precisely measure ecosystem responses to management actions, and incomplete understanding and the complexity of ecosystems prevent accurate prediction of ecosystem responses to management actions.

Adaptive management allows protected area managers to embrace uncertainty regarding the outcomes of management actions (see chapter 8). There are two types of adaptive management, passive and active. Passive adaptive management formulates predictive models of ecosystem response to management actions, makes management decisions based on those models, and revises the models based on monitoring data on ecosystem responses. While passive adaptive management is easier and less expensive to apply than active adaptive management, it is a nonexperimental approach. As such, it lacks statistical validity and does not provide reliable knowledge about ecosystem responses to management actions.

Active adaptive management uses statistically designed experiments to test assumptions or hypotheses about ecosystem responses to management actions. Experimental results are used to determine whether a particular management action achieves a desired outcome, for example, recovery of an endangered species. Active adaptive management experiments incorporate replication and randomization of management actions (treatments), which allows managers to test various hypotheses about how ecosystems respond to management actions. Active adaptive management is best carried out by a collaborative working group that identifies desirable outcomes for a protected area ecosystem; determines feasible management actions for achieving those outcomes; compares the social, economic, and ecological consequences of management actions based on experimental results and other information; and selects preferred management actions. Membership in the group is based on the principles of inclusivity, self-selection, and diversity of representation. Potential working group members include protected area managers, scientists, private landowners, First Nations, businesspersons, developers, community planners, and environmental groups. One approach is to apply passive adaptive management to management actions or elements of management actions that have relatively certain outcomes and, where feasible, active adaptive management to management actions or elements of management actions that have uncertain outcomes.

Adaptive management is being used for implementing Washington State's Timber, Fish, and Wildlife Agreement, restoring Everglades National Park, assessing management policies for improving British Columbia's salmon fishery, implementing a human use management strategy for Banff National Park, and managing snowmobile use in Yellowstone and Grand Teton National Parks. It has been recommended for managing ungulate grazing in Yellowstone National Park.

Theme 6

Protected area managers should select efficient and sustainable management actions.

One of the major contributions of economics to the management of protected areas is the principle of efficient resource use. This principle states that human, reproducible, and natural capital available to protected areas should be used to efficiently produce ecological goods and services. Efficient resource use occurs when it is not possible to increase the amount of one ecological good or service without reducing the amount of another ecological good or service. In economic terms, efficient resource use requires the protected area manager to select amounts of ecological goods or services on the efficiency frontier (see chapters 4 and 11). Since there is uncertainty regarding the amounts of ecological goods and services produced from available resources, it is more appropriate to represent the efficient amounts of ecological goods and services using an efficiency band rather than an efficiency frontier.

Efficient production of ecological goods and services is not necessarily sustainable. Sustainability is ensured by setting minimum acceptable amounts for all ecological goods or services produced by a particular protected area. Efficient and sustainable amounts of ecological goods or services are on the efficiency frontier or efficiency band and exceed the minimum acceptable amounts. Protected area managers should exclude from further consideration management actions that provide inefficient and unsustainable amounts of ecological goods and services.

The most preferred management action is determined based on information about stakeholders' preferences for ecological goods and services. Preferences for ecological goods and services are represented by a utility function or indifference map. The latter consists of a set indifference curves. An indifference curve contains all combinations of ecological goods and services that provide a stakeholder with a constant level of satisfaction or utility (see fig. 4.11 and fig. 11.2). Superimposing the indifference maps for stakeholders on the efficiency frontier yields the most preferred amounts of ecological goods and services for each stakeholder. Since stakeholders usually have different preferences (indifference maps), the most preferred amounts of ecological goods and services will vary across stakeholders. This gives rise to conflict among stakeholders, which can be resolved using conflict resolution methods (see chapter 8).

Theme 7

Internal and external threats to protected areas impair the capacity to sustain visitor enjoyment and protect ecological integrity.

Protected areas face numerous internal and external threats (see chapters 9 and 10). Internal threats include adverse effects of overcrowding, inadequate facilities and budgets, aging infrastructure and personnel shortages on visitor enjoyment and environmental quality, and adverse ecological impacts of increased visitation and invasive species. External threats include fragmented, degraded, and destroyed habitat for wildlife species, severe limitations on the movement of wide-ranging species, water diversions, and water pollution upstream of protected areas, loss of native vegetation, air pollution, alteration of hydrologic regimes, ecological impacts of natural resource extraction, and recreation and tourism on adjacent multiple use lands.

External threats are more difficult to alleviate than internal threats. Factors limiting the capacity of protected area managers to alleviate external threats include statutory limitations, agency culture, ingrained political sensitivities, and deference to adjacent landowners. Since threats are expected to increase in the future, protected area managers need to understand how various external threats are likely to influence protected area ecosystems, implement and monitor planning and management strategies for reducing those threats, and develop cooperative relationships with other federal and state natural resource management agencies, private landowners, planners groups, and other stakeholders. Obstacles to cooperation include differences in the statutory authority under which federal agencies operate and conflicts between the profit motive for private landowners and the resource conservation motive for protected areas, such as national parks.

The capacity of protected area managers to reduce external threats would be enhanced if Congress enacted a law requiring protected area managers to practice ecosystem management and authorized sufficient funds to alleviate external threats, and gateway communities implemented smart growth policies that allow economic development to occur in a manner that does not compromise the ecological integrity of protected areas.

Theme 8

Selection and delineation of protected areas and development of a national system of protected areas are important decisions having long-term consequences.

Establishing and expanding protected areas involves making decisions about the numbers, types, locations, sizes, shapes, and boundaries of protected areas (see chapter 6). Selection and delineation decisions are made within a social, cultural, economic, environmental, and political context that varies by location and changes in response to established priorities and management objectives for protected areas. Decisions regarding selection and delineation of protected areas are especially difficult to make when stakeholders have conflicting preferences for the ecological goods and services provided by protected areas, and disagree over the socioeconomic and environmental consequences of those decisions. Selection and delineation for protected areas are influenced by local (community), state, and national interests; social, cultural, economic, and ecological conditions; relative importance of protected area categories; national system planning; and other factors.

A country might find it advantageous to design and execute a national system plan for protected areas (see chapter 6). Such plans stipulate coordinated management of protected areas included in the national system as well as coordination with other land-use planning activities in the country. Development of a national system of protected areas is accomplished by having stakeholders clearly articulate the purpose or reason for the system; doing a complete inventory and assessment of candidate areas; making a preliminary allocation of candidate areas to protected area categories based on inventory/assessment, constraint analysis, and the protected area categories established by the country; determining in consultation with stakeholders the relative importance of management objectives for the protected area system; assigning candidate areas to each category and ranking them relative to one another based on their capacity to achieve the management objectives for that category; and making a final selection of protected areas subject to limitations on human and financial resources, regional equity, and political acceptability.

Global planning for protected areas attempts to balance management objectives on a global scale. Although there is considerable resistance to global planning for protected areas, international designations of Biosphere Reserves are widespread. Global strategies for protecting threatened biomes can be made more palatable to the countries by offering them financial and technical assistance and other incentives.

Theme 9

It is important to plan and manage protected areas for the future.

Sustaining the flow of ecological goods and services provided by protected areas requires managers and stakeholders to plan and manage for the future. Developing a general management plan for a protected area involves taking a future perspective. The future planning and management framework proposed in this book (see chapter 11) centers on achieving desirable future states for protected area ecosystems. Desirability of a protected area ecosystem is defined in terms of four broad goals, namely, reducing external threats, improving managerial responses to changing ecosystem conditions, balancing resource conservation and visitor satisfaction, and enhancing planning and management capacity. Other goals can be pursued as well. Integration of these goals is accomplished using a three-step procedure in which protected area managers and stakeholders determine expected values of goals for alternative future management actions, classify future ecosystem states as either desirable or undesirable based on those goals, and monitor future management actions to determine whether they result in a desirable or undesirable ecosystem state. If future values of the goals are compatible with a desirable ecosystem state, then there is no need to change management. Conversely, if future values of the goals indicate an undesirable ecosystem state, then it becomes necessary to design, implement, and monitor new management actions.

Theme 10

Integrated planning and management for protected areas would be improved by implementing the concepts, principles, and approaches discussed in this book.

Several of the previous themes incorporate concepts, principles, and approaches described in this book. These include ecosystem management, ecological economics, adaptive management, natural regulation, carrying capacity, regional scale conservation, benefit-cost analysis, cost effectiveness analysis, economic impact analysis, landscape change detection and analysis, multiple attribute decision making, Bayesian approaches to decision making under uncertainty, optimization and simulation, integrated ecological economic modeling, geographic information systems, remote sensing, global positioning systems, interactive mapping, and computerized landscape visualization (see chapters 7 and 8). Application of these concepts, principles, and approaches should substantially improve planning and management for protected areas. A protected area manager is highly unlikely to apply all of these concepts, principles, and approaches. However, it is feasible for managers to become more aware of them, for better funded protected areas to adopt the more promising ones, and for less well funded protected areas to adopt one or two of the most useful ones as needed and appropriate.

Conclusions

Planning and management for national parks and protected areas are complex, challenging, and contentious. Several factors impede the ability of protected area managers to balance the social, cultural, economic, and ecological values of protected areas. These factors include competition among the multiple values of protected areas, conflicts among stakeholders' preferences for values, biological and social complexity of protected areas, uncertainty about how management actions influence multiple values, internal and external threats, insufficient number of personnel, inadequate operating budgets, lack of familiarity with or access to scientific and technical knowledge, and others. Balancing the multiple values of national parks and protected areas is likely to become more challenging in the future due to increases in visitation; greater pressure from internal and external threats; high priorities for homeland security, national defense, and social welfare; and decreases in biodiversity within and outside of protected areas.

On the positive side, application of scientific and technical knowledge, as reflected in the concepts, principles, and approaches discussed in this book, have the potential to enhance the capacity of managers to balance the multiple values of protected areas. Many protected area managers are poorly versed in and/or do not have ready access to these concepts, principles, and approaches. Since the ecological integrity of national parks and protected areas and the economic vitality of gateway communities rest on good planning and management, protected area managers and stakeholders should become familiar with scientific and technical knowledge relevant to protected area planning and management. Two programs contributing to this goal are the U.S. National Park Service's Natural Resource Challenge (National Park Service 1999) and the national network of Cooperative Ecosystem Studies Units (2003).

In summary, protected area managers should consider how the themes presented in this book can be used to balance the social, economic, and ecological values of national parks and protected areas.

References

Cooperative Ecosystem Studies Units. 2003. http://www.cesu.org/cesu/ (accessed July 7, 2003).

Haines, A. L. 1974. *Yellowstone National Park: Its exploration and establishment.* Washington, DC: U.S. Department of the Interior, National Park Service.

IUCN. 1994. *Guidelines for protected areas management categories.* Cambridge, UK, and Gland, Switzerland.

National Park Service. 1998. *Director's Order #2: Park planning.* http://www.nps.gov/refdesk/DOrders/DOrder2.html (accessed June 20, 2003).

———. 1999. *Natural resource challenge: The Park Service's action plan for preserving natural resources.* http://www.nature.nps.gov/challengedoc/index.htm (accessed May 6, 2003).

Turner, M. G., R. H. Gardner, and R. V. O'Neill. 2001. *Landscape ecology in theory and practice: Pattern and process.* New York: Springer-Verlag.

INDEX

Printed and bound by CPI Group (UK) Ltd, Croydon, CR0 4YY
18/09/2026
14965006-0001